W9-CFI-576

Encyclopedia of Animal Rights and Animal Welfare

Encyclopedia of Animal Rights and Animal Welfare

Second Edition
VOLUME 1: A–H

Edited by Marc Bekoff

Foreword by Jane Goodall

GREENWOOD PRESS
An Imprint of ABC-CLIO, LLC

A B C ☰ C L I O

Santa Barbara, California • Denver, Colorado • Oxford, England

Library of Congress Cataloging-in-Publication Data

Encyclopedia of animal rights and animal welfare / edited by Marc Bekoff ; foreword by
Jane Goodall.—2nd ed.
  v. cm.
  Includes bibliographical references and index.
  ISBN 978-0-313-35255-3 (set : alk. paper) — ISBN 978-0-313-35257-7 (vol. 1 : alk.
paper) — ISBN 978-0-313-35259-1 (vol. 2 : alk. paper) — ISBN 978-0-313-35256-0
(ebook) — ISBN 978-0-313-35258-4 (vol. 1 : ebook) — ISBN 978-0-313-35260-7
(vol. 2 : ebook)
1. Animal rights—Encyclopedias.   2. Animal welfare—Encyclopedias.   I. Bekoff, Marc.
  HV4708.E53   2009
   179′.3—dc22   2009022275

14   13   12   11   10   1   2   3   4   5

This book is also available on the World Wide Web as an eBook.
Visit www.abc-clio.com for details.

ABC-CLIO, LLC
130 Cremona Drive, P.O. Box 1911
Santa Barbara, California 93116-1911

This book is printed on acid-free paper (∞)
Manufactured in the United States of America

Contents

Volume 1

Alphabetical List of Entries   *vii*

Guide to Related Topics   *xi*

Foreword by Jane Goodall   *xxi*

Preface   *xxv*

Introduction: Why Animal Rights and Animal Welfare Matter   *xxix*

Entries A–H   **1**

Volume 2

Alphabetical List of Entries   *vii*

Guide to Related Topics   *xi*

Entries I–Z

Chronology of Historical Events in Animal Rights and Animal Welfare   *635*

Resources on Animal Rights and Animal Welfare   *643*

About the Editor and Contributors   *651*

Index   *665*

Alphabetical List of Entries

Volume 1

Abolitionist Approach to
  Animal Rights
Affective Ethology
Alternatives to Animal Experiments in
  the Life Sciences
Alternatives to Animal Experiments:
  Reduction, Refinement, and
  Replacement
The American Society for the
  Prevention of Cruelty to Animals
  (ASPCA)
Amphibians
Animal Body, Alteration of
Animal Liberation Ethics
Animal Models and Animal Welfare
Animal Protection: The Future of
  Activism
Animal Reproduction, Human Control
Animal Rights
Animal Rights Movement, New
  Welfarism
Animal Studies
Animal Subjectivity
Animal Welfare
Animal Welfare and Animal Rights,
  A Comparison
Animal Welfare: Assessment
Animal Welfare: Coping
Animal Welfare: Freedom
Animal Welfare: Risk Assessment
Animal-Assisted Therapy

Animals in Space
Anthropocentrism
Anthropomorphism
Anthropomorphism: Critical
  Anthropomorphism
Antivivisectionism
Art, Animals, and Ethics
Association of Veterinarians for Animal
  Rights (AVAR)
Autonomy of Animals
Bestiality
Bestiality: History of Attitudes
Blessing of the Animals Rituals
Blood Sports
Bullfighting
Captive Breeding Ethics
Cats
Chickens
Chimpanzees in Captivity
China: Animal Rights and Animal
  Welfare
China: Moon Bears and the Bear Bile
  Industry
Cockfighting
Companion Animals
Companion Animals, Welfare, and the
  Human-Animal Bond
Consciousness, Animal
Conservation Ethics, Elephants
Cosmic Justice
Cruelty to Animals and Human
  Violence

Cruelty to Animals: Enforcement of Anti-Cruelty Laws

Cruelty to Animals: Prosecuting Anti-Cruelty Laws

Deep Ethology

Deviance and Animals

Disasters and Animals

Disasters and Animals: Legal Treatment in the United States

Disneyfication

Dissection in Science and Health Education

Distress in Animals

Docking

Dogfighting

Dogs

Domestication

Dominionism

Donkeys

Ecofeminism and Animal Rights

Ecological Inclusion: Unity among Animals

Embryo Research

Empathy with Animals

Endangered Species Act

Endangered Species and Ethical Perspectives

Enrichment and Well-Being for Zoo Animals

Entertainment and Amusement: Animals in the Performing Arts

Entertainment and Amusement: Circuses, Rodeos, and Zoos

Environmental Ethics

Equal Consideration

Euthanasia

Evolutionary Continuity

Exotic Species

Experimentation and Research with Animals

Extinction and Ethical Perspectives

Factory Farms

Factory Farms and Emerging Infectious Diseases

Field Studies and Ethics

Field Studies: Animal Immobilization

Field Studies: Ethics of Communication Research with Wild Animals

Field Studies: Noninvasive Wildlife Research

Fish

Fishing as Sport

Food Animals: Ethics and Methods of Raising Animals

The Gender Gap and Policies toward Animals

Genetic Engineering

Genetic Engineering and Farmed Animal Cloning

Genetic Engineering: Genethics

Global Warming and Animals

The Great Ape Project

Great Apes and Language Research

Horse Slaughter

Human Effects on Animal Behavior

Humane Education

Humane Education, Animal Welfare, and Conservation

Humane Education Movement

Humane Education Movement in Schools

Humane Education: The Humane University

Hunting, History of Ideas

Volume 2

India: Animal Experimentation

Institutional Animal Care and Use Committees (IACUCs)

Institutional Animal Care and Use Committees (IACUCs): Nonaffiliated Members

Institutional Animal Care and Use Committees (IACUCs): Regulatory Requirements

Israel: Animal Protection

Kenya: Conservation and Ethics

Krogh Principle

Laboratory Animal Use: Sacrifice

Laboratory Animal Welfare

Law and Animals

Law and Animals: Australia

Law and Animals: European Union

Law and Animals: United States

Marginal Cases

Medical Research with Animals

Mice

Misothery

Moral Standing of Animals

Museums and Representation of Animals

Native Americans and Early Uses of Animals in Medicine and Research

Native Americans' Relationships with Animals: All Our Relations

Objectification of Animals

Pain, Invertebrates

Pain, Suffering, and Behavior

Painism

People for the Ethical Treatment of Animals (PETA)

Pet Renting

Pigs

Pleasure and Animal Welfare

Poetry and Representation of Animals

The Political Subjectivity of Animals

Polyism

Practical Ethics and Human-Animal Relationships

Predator Control and Ethics

Puppy Mills

Quality of Life for Animals

Rabbits

Rats

Religion and Animals

Religion and Animals: Animal Theology

Religion and Animals: Buddhism

Religion and Animals: Christianity

Religion and Animals: Daoism

Religion and Animals: Disensoulment

Religion and Animals: Hinduism

Religion and Animals: Islam

Religion and Animals: Jainism

Religion and Animals: Judaism

Religion and Animals: Judaism and Animal Sacrifice

Religion and Animals: Pantheism and Panentheism

Religion and Animals: Reverence for Life

Religion and Animals: Saints

Religion and Animals: Theodicy

Religion and Animals: Theos Rights

Religion and Animals: Veganism and the Bible

Religion, History, and the Animal Protection Movement

Reptiles

Rescue Groups

Royal Society for the Prevention of Cruelty to Animals (RSPCA), History

Royal Society for the Prevention of Cruelty to Animals (RSPCA) Reform Group

Sanctuaries

Sanctuaries, Ethics of Keeping Chimpanzees in

Scholarship and Advocacy

Sentience and Animal Protection

Sentientism

Shelters, No-Kill

Signals and Rituals of Humans and Animals

The Silver Spring Monkeys

Sizeism

Sociology of the Animal Rights Movement

Species Essentialism

Speciesism

Speciesism: Biological Classification

Speciesism: Ethics, Law, and Policy

Sports and Animals

Stereotypies in Animals

Stress and Laboratory Routines

Stress Assessment, Reduction, and Science

Student Attitudes toward Animals

Student Objections to Dissection

Student Rights and the First Amendment

Teleology and Telos

Toxicity Testing and Animals

Trapping, Behavior, and Welfare

Urban Wildlife

Utilitarianism

Utilitarianism and Assessment of Animal Experimentation

Veganism

Vegetarianism

Veterinary Medicine and Ethics

Virtue Ethics

War and Animals

War: Using Animals in Transport

Whales and Dolphins: Culture and Human Interactions

Whales and Dolphins: Sentience and Suffering

Whales and Dolphins: Solitary Dolphin Welfare

Wild Animals and Ethical Perspectives

Wildlife Abuse

Wildlife Contraception

Wildlife Services

Wolves and Ethical Perspectives

Xenograft

Zoos: History

Zoos: Roles

Zoos: Welfare Concerns

Guide to Related Topics

Below are the headwords for all entries in *The Encyclopedia of Animal Rights and Animal Welfare,* arranged under broad topics.

Activism (*See also* Global Efforts for Animal Protection)

Abolitionist Approach to Animal Rights

American Society for the Prevention of Cruelty to Animals (ASPCA), The

Animal Protection: The Future of Activism

Antivivisectionism

People for the Ethical Treatment of Animals (PETA)

Rescue Groups

Royal Society for the Prevention of Cruelty to Animals (RSPCA) Reform Group

Scholarship and Advocacy

Silver Spring Monkeys, The

Student Objections to Dissection

Student Rights and the First Amendment

Alternatives to Animal Use (*See also* Experimentation and Models)

Alternatives to Animal Experiments in the Life Sciences

Alternatives to Animal Experiments: Reduction, Refinement, and Replacement

The Animal Body

Animal Body, Alteration of

Docking

Domestication

Animal Reproduction, Human Control of

Animal Reproduction: Human Control

Wildlife Contraception

Animal Welfare

Animal Welfare

Animal Welfare: Assessment

Animal Welfare: Coping
Animal Welfare: Freedom
Animal Welfare: Risk Assessment
Animal Welfare and Animal Rights, A Comparison

Animal-Assisted Therapy
Animal-Assisted Therapy

Animals in Space
Animals in Space

Anthropocentrism
Anthropocentrism

Anthropomorphism
Anthropomorphism
Anthropomorphism: Critical Anthropomorphism

Anthrozoology: Human-Animal Interactions
Animal Studies
Human Effects on Animal Behavior
Practical Ethics and Human-Animal Relationships
Signals and Rituals of Humans and Animals
Whales and Dolphins: Culture and Human Interactions

Attitudes
Empathy with Animals
Gender Gap and Policies toward Animals, The
Objectification of Animals
Sociology of the Animal Rights Movement
Student Attitudes toward Animals

Bestiality
Bestiality
Bestiality: History of Attitudes

Climate Change
Global Warming and Animals

Cognition and Sentience
Affective Ethology
Animal Subjectivity
Consciousness, Animal
Deep Ethology
Great Apes and Language Research
Pleasure and Animal Welfare
Sentience and Animal Protection
Sentientism
Whales and Dolphins: Sentience and Suffering in

Companion Animals
Companion Animals
Companion Animals, Welfare, and the Human-Animal Bond
Pet Renting
Puppy Mills
Shelters, No-Kill

Conservation Ethics. *See* **Wildlife Ethics**

Cruelty (*See also*** Law)**
Cruelty to Animals and Human Violence
Cruelty to Animals: Enforcement of Anti-Cruelty Laws
Cruelty to Animals: Prosecuting Anti-Cruelty Laws
Deviance and Animals

Disasters
Disasters and Animals
Disasters and Animals: Legal Treatment in the United States

Dissection
Dissection in Science and Health Education
Student Objections to Dissection

Education
Humane Education
Humane Education, Animal Welfare, and Conservation

Humane Education: The Humane University
Humane Education Movement
Humane Education Movement in Schools

Endangered Species
Endangered Species Act
Endangered Species and Ethical Perspectives

Enrichment
Enrichment and Well-Being for Zoo Animals

Entertainment and Animals
Disneyfication
Entertainment and Amusement: Animals in the Performing Arts
Entertainment and Amusement: Circuses, Rodeos, and Zoos

Euthanasia and Sacrifice
Euthanasia
Laboratory Animal Use: Sacrifice

Exotic Species
Exotic Species

Experimentation and Models
Animal Models and Animal Welfare
Embryo Research
Experimentation and Research with Animals
Laboratory Animal Welfare
Medical Research with Animals
Toxicity Testing and Animals
Xenograft

Extinction
Extinction and Ethical Perspectives

Evolutionary Continuity
Evolutionary Continuity

Food Animals
Factory Farms
Factory Farms and Emerging Infectious Diseases

Food Animals: Ethics and Methods of Raising Animals

Genetic Engineering and Farmed Animal Cloning

Horse Slaughter

Veganism

Vegetarianism

Gender and Animal Issues

Ecofeminism and Animal Rights

Gender Gap and Policies toward Animals, The

Genetic Engineering

Genetic Engineering

Genetic Engineering and Farmed Animal Cloning

Genetic Engineering: Genethics

Global Efforts for Animal Protection

China: Animal Rights and Animal Welfare

China: Moon Bears and the Bear Bile Industry

India: Animal Experimentation

Israel: Animal Protection

Kenya: Conservation and Ethics

Law and Animals: Australia

Law and Animals: European Union

Royal Society for the Prevention of Cruelty to Animals (RSPCA) Reform Group

History

Royal Society for the Prevention of Cruelty to Animals (RSPCA)

Horse Slaughter

Horse Slaughter

Institutional Animal Care and Use Committees

Institutional Animal Care and Use Committees (IACUCs)

Institutional Animal Care and Use Committees (IACUCs): Nonaffiliated Members

Institutional Animal Care and Use Committees (IACUCs): Regulatory Requirements

Law and Animals

Cruelty to Animals: Enforcement of Anti-Cruelty Laws

Cruelty to Animals, Prosecuting Anti-Cruelty Laws

Law and Animals
Law and Animals: Australia
Law and Animals: European Union
Law and Animals: United States

Native American Relationships with Animals
Native Americans and Early Uses of Animals in Medicine and Research
Native Americans' Relationships with Animals: All Our Relations

Pain, Stress, and Suffering
Distress in Animals
Pain, Invertebrates
Painism
Pain, Suffering, and Behavior
Stress and Laboratory Routines
Stress Assessment, Reduction, and Science

Philosophical Principles
Animal Liberation Ethics
Animal Rights
Animal Rights Movement, New Welfarism
Animal Welfare and Animal Rights, A Comparison
Autonomy of Animals
Cosmic Justice
Dominionism
Ecofeminism and Animal Rights
Ecological Inclusion: Unity among Animals
Environmental Ethics
Equal Consideration
Krogh Principle
Marginal Cases
Misothery
Moral Standing of Animals
Polyism
Quality of Life for Animals
Sizeism
Species Essentialism

Speciesism

Teleology and Telos

Utilitarianism

Utilitarianism and Assessment of Animal Experimentation

Virtue Ethics

Political Rights of Animals

Political Subjectivity of Animals, The

Religion

Religion and Animals

Religion and Animals: Animal Theology

Religion and Animals: Buddhism

Religion and Animals: Christianity

Religion and Animals: Daoism

Religion and Animals: Disensoulment

Religion and Animals: Hinduism

Religion and Animals: Islam

Religion and Animals: Jainism

Religion and Animals: Judaism

Religion and Animals: Judaism and Animal Sacrifice

Religion and Animals: Pantheism and Panentheism

Religion and Animals: Reverence for Life

Religion and Animals: Saints

Religion and Animals: Theodicy

Religion and Animals: Theos Rights

Religion and Animals: Veganism and the Bible

Religion, History, and the Animal Protection Movement

Representation of Animals

Animal Body, Alteration of

Art, Animals, and Ethics

Disneyfication

Docking

Entertainment and Amusement: Animals in the Performing Arts

Museums and Representation of Animals

Objectification of Animals

Poetry and the Representation of Animals

Sanctuaries
Sanctuaries
Sanctuaries, Ethics of Keeping Chimpanzees in

Sentience. *See* **Cognition and Sentience**

Species
Amphibians
Cats
Chickens
Dogs
Donkeys
Fish
Great Ape Project, The
Mice
Pigs
Rabbits
Rats
Reptiles
Sanctuaries
Sanctuaries, Ethics of Keeping Chimpanzees in
Whales and Dolphins: Solitary Dolphin Welfare
Wolves and Ethical Perspectives

Speciesism
Sizeism
Speciesism
Speciesism: Biological Classification
Speciesism: Ethics, Law, and Policy

Sports and Animals
Blood Sports
Bullfighting
Cockfighting
Dogfighting
Fishing as Sport
Hunting, History of Ideas
Sports and Animals

Stereotypies
Stereotypies in Animals

Veterinary Medicine
Association of Veterinarians for Animal
Rights (AVAR)
Veterinary Medicine and Ethics

War and Animals
War and Animals
War: Using Animals for Transport

Wildlife Ethics
Captive Breeding Ethics
Conservation Ethics, Elephants
Field Studies and Ethics
Field Studies: Animal Immobilization
Field Studies: Ethics of Communication Research with Wild Animals
Field Studies: Noninvasive Wildlife Research
Predator Control and Ethics
Trapping, Behavior, and Welfare
Urban Wildlife
Wild Animals and Ethical Perspectives
Wildlife Abuse
Wildlife Contraception
Wildlife Services

Zoos
Enrichment and Well-Being for Zoo Animals
Zoos: History
Zoos: Roles of
Zoos: Welfare Concerns

—

Foreword

It is an honor for me to have the preface that I wrote for the first edition of the *Encyclopedia of Animal Rights and Animal Welfare* be included in the updated and expanded revision of this unique collection of essays. An incredible amount has happened in the field of animal protection in the eleven years since the first edition appeared, and these two volumes highlight much of what we have learned and accomplished during that period.

As I wrote in my preface to the first edition, never before had an attempt been made to gather together comprehensive information about the use and abuse of nonhuman animals by our own human species, along with the complex issues that must be understood by those who are concerned with animal welfare and animal rights, and some of the ways in which different groups are tackling these issues. Because human beings are animals, this book could have been expanded to include the horrible abuse and torture to which we subject other humans; theoretically, there could be a whole section on human rights. But that is not the purpose of the editor. This encyclopedia is concerned with the essential dignity of the wondrous nonhuman beings with whom we share this planet, and our human responsibilities toward them. These are the beings known in common balance as animals, which is how I shall refer to them here.

Of course, we humans are much more like other animals than was once thought, much more so then many people like to, or are prepared to, believe. I have been privileged to spend 50 years learning about and from the chimpanzees, our closest living relatives. A detailed understanding of chimpanzee nature has helped, perhaps more than anything else, to blur the line, once thought to be so clear and sharp, dividing humans from the rest of the animal kingdom. Once we are prepared to accept that it is not only humans who have personalities, not only humans who are capable of rational thought and simple problem solving, and above all, not only humans who can experience emotions such as joy, sorrow, fear, despair, and mental as well as physical suffering, then we are surely compelled to have new respect not only for chimpanzees, but also for so many other amazing animal species. (In fact, I received my first lessons about the amazing capabilities of nonhumans from my dog, Rusty, before I was 10 years old.)

The only thing that we humans do, that no other animals do in the same way, is to communicate by means of a sophisticated spoken and written language, and this I believe lays on us certain responsibilities toward the rest of the animal kingdom. It might be mentioned here that in the Book of Psalms in the Old Testament, one word was mistranslated. "Dominion" is not the best translation of the original Hebrew word *Tam Shilayhu*. Rather the word implies a "respectful and caring attitude toward creation," suggesting a sense of responsibility. This, of course, gives the text a completely different meaning. I have been fortunate. I have been able to spend many years observing chimpanzees and other animals in their own natural environments, thereby gaining unique insights into their true nature. For this reason I believe it is my particular responsibility to share my knowledge with as large an audience as possible for the benefit of the animals themselves. Chimpanzees have given me so much, and I am haunted at the thought of those who are imprisoned in the name of entertainment or science. As I have written elsewhere, the least I can do is to speak out for the hundreds of chimpanzees who, right now, sit hunched, miserable, and without hope, staring out with dead eyes from their metal prisons. They cannot speak for themselves.

This is why I am so very glad that this encyclopedia has been put together, for it speaks out for animals, for all kinds of animals. It broadcasts a simple message, a plea, that needs desperately to be heard as we move ahead in the 21st century: Give animals the respect that, as sentient beings, is their due. And this simple message is delivered here by a multitude of voices from many different disciplines, from biology, including ethology, the study of behavior, to ecology, anthropology, psychology, philosophy, sociology, education, law, ethnology, history, politics, theology, veterinary science, and public administration. This multidisciplinary collection of contributors means that the essays discuss the central theme from different perspectives; collectively they provide an astonishingly rich overview of the extent of animal suffering in our modern society, and the various steps that have been taken by those fighting for animal welfare and animal rights. And, importantly, the material is presented in a straightforward way intended to appeal to the general public as well as scientists. Once this encyclopedia reaches the shelves of libraries in schools and universities, many young people, as well as their teachers, will have access to this valuable information.

This reference work provides the reader with an opportunity to acquire in-depth understanding of complex issues. And because different contributors voice differing opinions, the reader will also be able to develop his or her own carefully reasoned arguments to use when discussing controversial issues with people who hold different views. This is important. The more passionate one feels about animal abuse, the more important it becomes to try to understand what is behind it. However distasteful it may seem, it really is necessary to become fully informed about a given issue. Dogmatism, a refusal to listen to any point of view differing from one's own, results in moral and intellectual arrogance. This is far from helpful, and is most unlikely to lead to any kind of progress. The us versus them attitude brings useful dialogue to an end. In fact, most issues are quite complex and can seldom be described in simple terms of black and white, and until we become fully cognizant of all that is involved, we had better not start arguing, let alone throwing bricks at anyone.

Let me give an example. During a semi-official visit to South Korea, my host organization set up a press conference. The subject of cruelty came up. I said that I would like to discuss their habit of eating dogs. My interpreter blanched. Quite clearly she felt that this was politically insensitive and would embarrass my hosts! I explained that in England, the country where I grew up, people typically ate cows and pigs and chickens, and that pigs are at least as intelligent as dogs and, in fact, make wonderful pets. Yet only too often they are kept in horrendous conditions. I suggested that the most important issue, if one was going to eat an animal at all, which I did not, was not so much the species as how it was treated in life. At this point one of the journalists assured me that the dogs they ate were bred for eating. This led to discussions about whether or not this made any difference, the ways in which dogs and pigs were kept, and a variety of other issues. The point was that an almost taboo subject was aired in public, and this led, for a number of people, to new ways of thinking about animals in general.

Perhaps the bitterest pill that we who care about animals have to swallow is that, only too often, it is through a series of compromises that progress is actually made, and this seems agonizingly slow. There are, of course, situations when the cruelty inflicted is so great that no compromise is possible. Then it is vitally important to know as much as possible about the situation. This encyclopedia may provide the animal activist with information about how similar situations have been successfully tackled.

The essays in this volume are necessarily brief, summarizing information which, in some cases, is extensive. Each essay can serve to stimulate the reader to pursue a particular issue in greater depth, guided by the extensive lists of references and key organizations that have been compiled for the encyclopedia. These lists will be a goldmine for all those who care about animal issues. All in all, these two volumes are a unique contribution to the field of animal protection.

Albert Schweitzer once said, "We need a boundless ethic that includes animals, too." At present our ethic concerning animals is limited and confused. For me, cruelty, in any shape or form, whether it be directed toward humans or sentient nonhumans, is the very worst of human sins. To fight cruelty brings us into direct conflict with that unfortunate streak of inhumanity that lurks in all of us. For all who like me, are committed to joining this particular battle, this encyclopedia will prove invaluable. A great deal of the behavior that we deem cruel is not deliberate, but due to a lack of understanding. It is that lack of understanding that we must overcome. And every time cruelty is overcome by compassion, we are moving toward that new and boundless ethic that will respect all living beings. Then indeed we shall stand at the threshold of a new era in human evolution—the realization of our most unique quality: Humanity.

Jane Goodall, PhD, DBE
Founder—The Jane Goodall Institute
U.N. Messenger of Peace
www.janegoodall.org
U.K. January 2009

Preface

Currently, there is growing interest in the nature of human-nonhuman animal (hereafter animal) interactions as we head into the 21st century, for it is clear that there are many important associated issues that demand immediate and careful attention.

That is how I began my introduction to the first edition of this encyclopedia, the first of its kind, more than a decade ago. These statements are just as true today, in 2009, when there is even greater and growing interest in human-animal interactions in general and in the field of animal protection more specifically. This expanded and updated revision of the *Encyclopedia of Animal Rights and Animal Welfare* will address the needs of students, researchers, and the general public.

There is unprecedented and growing global interest in the well-being of animals. Many people come to these issues from very different walks of life, both academic and nonacademic, and from many points of interest, for example, social, political, educational, philosophical, psychological, legal, zoological, ethological, ecological, theological, anthropological, sociological, historical, biographical, medical and veterinary sciences, ethnological, and public health, which are represented in this volume. We thought it important, therefore, to collect as much information as possible in one easy-to-read reference book.

The issues with which humans need to deal to develop informed views about human-animal interactions require that people from many different disciplines get involved in the discussions. And, of course, these exchanges of ideas must be open, and people must be sensitive to all different views if we are to make progress. We hope that we have been successful in presenting different viewpoints, because us versus them interactions are not very helpful and tend to alienate, rather than unite, individuals who share common concerns and goals. It is important for all people to listen to one another, and for all of us to listen to the animals with whom we are privileged to share the planet and interact. Respect for the dignity of all animals' lives needs to underlie consideration of how humans interact with other animals. Thus, we hope that we and our authors have covered the issues from varied approaches, including theoretical matters and practical applications, using information gathered from animals living in highly controlled laboratory environments and those living in the wild. All types of data are important, and much useful information about the complexity, diversity, and richness of animals' lives has come from the study of free-living animals.

It also is important to stress that there is a long, rich, diverse, and sometimes painful history of events that center on how animals have been used by humans in various sorts of activities. We had to make some difficult choices of which topics to include and which to exclude. Because of space considerations, we decided not to include entries on individuals, even though they may have made profound contributions to the history of animal welfare and animal rights. Although we have only some historical information in the *Encyclopedia,* we call readers' attention to the historical account of the people who contributed to the anti-vivisection movement, published by the American Anti-Vivisection Society in Fall 2008. This account is extremely useful: http://www.aavs.org/images/spring2008.pdf.

We were thrilled that many extremely busy and over-committed people, a veritable who's who of people working on topics related to animal protection, thought that this revision was important enough for them to free the time to write new essays that reflect the growing interest and the accumulation of scientific information in hot fields such as conservation ethics, the use of noninvasive field techniques to study wildlife, exotic species, wildlife contraception, the importance of animal sanctuaries, the emotional lives of animals and animal sentience, puppy mills, no-kill shelters, dogfighting, cockfighting, bullfighting, animals in the performing arts, stress and well-being, the gender gap in the animal protection movement, factory farming and disease, climate change and its effect on animals, pet renting, the welfare of fish, the legacy of captive chimpanzees, animals in disasters, the Endangered Species Act, animal law from an international perspective, the nature and importance of human-animal interactions in general (anthrozoology), and the welfare of whales and dolphins.

In addition to many new essays, we have pieces written by founders and leaders of major animal protection groups, and people directly involved in humane education in the United States and abroad, including China, India, Kenya, Israel, Australia, and the European Union. There are a number of essays in this edition on various cultural and religious views of animals, which bear on issues of animal protection. Having these kinds of firsthand contributions from people who are actually doing the work is invaluable.

This revised and expanded encyclopedia offers a discussion of just about all of the major issues that need to be considered in discussions of animal protection. Essays vary in length; some are short, covering topics succinctly, with others more wide-ranging and detailed. All in all, the information in these two volumes is both broad in scope and unprecedented. While the vast majority of essays are presented in a neutral manner, a few are more personal, because it is very difficult to be impartial when writing about our animal kin. All humans have unique responsibilities to animals that need to be taken seriously. We and the animals whom we use should be viewed as partners in a joint venture. We can teach one another respect and trust, and animals can facilitate contact among us and help us learn about our place in this complex, challenging, and awe-inspiring world.

It is my hope that the information in these volumes will be useful to all people who are interested in animal rights and welfare, and will help us increase what I call our compassion footprint as we head into the future.

HOW TO USE THIS ENCYCLOPEDIA

The 207 entries are arranged in alphabetical order. All of the essays in these volumes, and the list of further readings that follows nearly every one, contain information about what has been done and what remains to be done in specific areas in animal rights and welfare. The Chronology of Historical Events in Animal Protection at the end of the second volume of this encyclopedia, and the Resources on Animal Rights and Animal Welfare section, also at the end of the second volume, provide more information for further study of animal rights and welfare. Finally, included with the affiliation of some contributors are Web sites that are outstanding interdisciplinary and international resources, containing details about the authors and various educational programs, projects, and organizations concerned with animal rights, animal welfare, and human-animal interactions.

While each of the entries generally presents an extensive summary of the issues at hand, successes that are being made in animal protection, and information about where more work is needed, entries should not be read as comprehensive treatments, nor should the list of further readings at the end be thought of as exhaustive coverage. Rather, each entry and the summary of resources should be viewed as points of departure for further investigations, like kindling that can be used to ignite larger fires. I hope that you enjoy this reference book and that the essays stimulate you to learn more about the animals with whom we share our planet.

GIVING THANKS

Suffice it to say, I could never have completed this project on my own. When Kevin Downing and Anne Thompson asked me if I'd consider revising the first edition of this encyclopedia, I jumped at the chance. How exciting it would be to update all that has happened in the eleven years since the first edition appeared! Contacting former and new authors, preventing and putting out fires, and editing and editing and editing, were extremely time-consuming. In and of itself, there is an interesting sociological story that can be told at another time. As usual, Anne Thompson was always there, as she had been for two of my previous encyclopedias, the *Encyclopedia of Animal Behavior* (Greenwood, 2004) and the *Encyclopedia of Human-Animal Relationships* (Greenwood, 2007). Thank you so very much, Anne. The people at Apex CoVantage also helped to bring this encyclopedia to life. And, of course, many thanks to all the contributors who took time out of their busy days to write new essays or revise their excellent entries from the first edition. The many and different perspectives on animal protection that are presented here show just how rich and complicated our relationships with other animals can be.

Introduction

WHY ANIMAL RIGHTS AND ANIMAL WELFARE MATTER

The growing general concern around the world about how humans interact with other animals, as well as the field of animal protection more specifically, includes academics, activists, and animal lovers, many of whom wear all three hats. No longer is someone who is interested in animal rights or animal welfare automatically dismissed as a radical. The animal protection movement is not a fringe cause made up of extremists. In the past five years, I have had the good fortune to visit numerous countries in many different parts of the world, and seen firsthand how people yearn not only to learn more about the lives of animals, but also what they can do to grant animals more protection. There are essays and news articles in the popular media daily about the use and abuse of animals across all cultures. That is how much interest there really is, and this is why I am revising, updating, and expanding the first edition of the *Encyclopedia of Animal Rights and Animal Welfare.* One audience for which this encyclopedia holds special interest is young people, especially teenagers, who have a rapidly growing concern about animal protection.

Animals are in. It's the century of the animal. Every day I receive numerous stories from around the world about the amazing intellectual skills of animals and what they're feeling; it's impossible to keep up with them all. Sometimes when I log on to my e-mail at 5:00 A.M., I'm inundated, but I am also pleased to read both down-home anecdotes and hard-data scientific papers that bear on the emotional lives of animals, human arrogance and, most welcomed, stories about all the wonderful things that people around the world are doing for animals. Popular media regularly feature articles about animals, and it's clear that animals are on the agenda of millions of people around the world. The *New York Times* published obituaries for two famous animals whose language abilities startled the word, Washoe, a "chimpanzee of many words" (http://www.nytimes.com/2007/11/01/science/01chimp.html?_r=1&scp=1&sq=washoe&st=cse&oref=slogin) and Alex, an African gray parrot, who mastered English and could count and recognize different shapes and colors (http://www.nytimes.com/2007/09/11/science/11parrot.html?scp=2&sq=ALEX+PARROT&st=nyt). In the May 7, 2007 issue of *Newsweek,* there was an essay about the emotional lives of elephants, and how they deserve far more respect then we're giving them ("Deserving of Respect: Is it acceptable to kill the elephants of South Africa even when it is necessary to save other species? The answer is no longer an automatic 'yes'": http://www.newsweek.com/id/35114).

We now know that elephants suffer from posttraumatic stress disorder (PTSD). During recent years there have also been many other surprises. We know that mice are empathic rodents—they feel the pain of other mice—and that whales possess spindle cells, which are important in processing emotions. Before this discovery, it was thought that only humans and other great apes possessed spindle cells. We've also learned that fish have distinct personalities, ranging from shy and timid to bold, and are very intelligent and possess long-term memory, that turtles mourn the loss of friends, and that birds plan for the future, and are more sophisticated in making and using tools than chimpanzees. These cognitive and emotional capacities factor in to how we should treat other animals.

I met traumatized elephants at the David Sheldrick Wildlife Trust (http://www.sheldrickwildlifetrust.org/index.asp) outside of Nairobi, Kenya, and saw the marvelous work that was being done there to rehabilitate individuals so that they could be returned to the wild. Based on the fact that zoos can't satisfy the social, emotional, or physical needs of elephants, the Bronx Zoo and zoos in Detroit, Chicago, San Francisco, and Philadelphia are phasing out their elephant exhibits, despite the fact that they are moneymakers. A landmark review of survivorship in zoo elephants, written by six eminent biologists and published in the prestigious journal *Science* in December 2008, concluded that, "Overall, bringing elephants into zoos profoundly impairs their viability. The effects of early experience, interzoo transfer, and possibly maternal loss, plus the health and reproductive problems recorded in zoo elephants . . . suggest stress and/or obesity as likely causes." Critics of zoos often ask for hard data to support claims that animals don't do well in zoos, and this incredibly detailed study, headed by Ros Clubb, shows just that.

While we often see the ways in which the lives of animals are compromised, much abuse goes unnoticed. For example, worldwide as many as 300,000 cetaceans (whales, dolphins, porpoises) slowly meet death over the course of many minutes when they get entangled as accidental by-catch in fishing nets. When their bodies are recovered, it is obvious that they had desperately struggled to escape from their entrapment, and that they sustained horrific injures while doing so. Trapped individuals sustain deep cuts and abrasions to the skin from the rope and the netting, and fins and tail flukes can be partially or completely amputated. They also have broken teeth, beaks, or jaws, torn muscles, hemorrhaging, and serious internal injuries ("Shrouded by the Sea," http://www.wdcs.org/submissions_bin/wdcs_bycatchreport_2008.pdf). The suffering of these sentient beings often goes unnoticed because it is hidden in the sea, a part of the world where human beings are less prevalent than other animals, but it's safe to say that this kind of treatment would not be tolerated if it happened on land in situations such as commercial meat production. What is simply unacceptable is that there isn't any legislation that is concerned with this problem.

THE NATURE OF HUMAN-ANIMAL RELATIONSHIPS

Our relationships with nonhuman animals are complicated, frustrating, ambiguous, paradoxical, and range all over the place. The growing field of anthrozoology (http://www.isaz.net/; http://www.anthrozoology.org/) is concerned with reaching a more complete understanding of how and why we interact with animals in the many different ways that we do. When people tell me that they love animals, and then harm or kill them, I tell them I'm glad they don't love me. We observe animals, gawk at them in wonder, experiment on them, eat them, wear them, write about them, draw and paint them, move them from here to there as we redecorate nature, make decisions for them without their consent, and represent them in many varied ways, yet we often dispassionately ignore who they are and what they want and need. Surely we can do better in our relationships with animals.

A very good example of how difficult our relationships with animals can be centers on the keeping of exotic animals as our household companions or pets. In February 2009, a chimpanzee named Travis, who had lived in a home for years, attacked and maimed a close friend of his female human companion. As a result, Travis' long-time friend had to stab him to stop the attack, and ultimately Travis was killed by a police officer.

In the past, Travis had been allowed to drink wine and brush his teeth with his human companion (http://www.stamfordadvocate.com/ci_11717191?source=most_emailed). Numerous people were saddened by this tragedy, and outraged that Travis had been kept as if he were a dog or a cat. I pointed out that this terrible situation could easily have been avoided if Travis had been living at a sanctuary, and not in a private home being treated as if he were a human. Chimpanzees do not typically drink wine or brush their teeth with a WaterPik, and while it may seem cute, asking a chimpanzee to do these things is an insult to who they are. Furthermore, a story published by the Associated Press called Travis a domesticated chimpanzee, but this is a complete misrepresentation of who he was. Domestication is an evolutionary process that results in animals such as our companion dogs and cats undergoing substantial behavioral, anatomical, physiological, and genetic changes during the process. Travis was a socialized chimpanzee, an exotic pet, who usually got along with humans, but he was not a domesticated being. He still had his wild genes, as do wolves, cougars, and bears, who sometimes live with humans, causing tragedies to occur, because these are wild animals, despite being treated as if they're human.

Many people were surprised by what seemed to be an unprovoked attack. But to say there was no known provocation for the attack is to ignore the basic fact that Travis was still genetically a wild chimpanzee. Wild animals do not belong in human homes; they can be highly unpredictable (consider other attacks by famous animals on their handlers), and they should be allowed to live at sanctuaries that are dedicated to respecting their lives while minimizing human contact.

In an editorial, the local paper, The *New Haven Advocate,* called for a ban on the keeping of wild animals as pets (http://www.stamfordadvocate.com/opinion/ci_11733105). I hope that this tragic situation serves to stimulate people to send the wild friends who share their homes to places that are safe for everyone (http://www.stamfordadvocate.com/letters/ci_11724995).

Bucknell University philosopher Gary Steiner argues in his book *Animals and the Moral Community* that there is profound historical prejudice against animals, although more and more people are currently working on behalf of animals. While this is so, and there is a growing animal ethic globally, our attitudes and practices remain full of contradictions and ambivalence, as shown in the case of Travis. It's as if we suffer from moral schizophrenia because it's so difficult to live with a consistent morality toward animals. Travis was tolerated as long as he behaved like a human, but killed when he behaved as a wild chimpanzee might when something happened that he found unacceptable. Animal advocate and lawyer Gary Francione notes that while we claim to accept the principle that we should not inflict suffering or death on nonhumans unless it is necessary to do so, we do so in situations in which 99.99 percent of the suffering and death cannot be justified under any plausible notion of coherence. On the one hand, animals are used and abused in a vast array of human-centered activities. On the other hand, animals are revered, worshipped, and form an indispensable part of the tapestry of our own well-being; they make us whole, they shape us, and they make us feel good. Yet animals are sometimes confused and desperate because of the widespread and wanton abuse that they suffer at the hands of humans. Animal advocate Samantha Wilson says animals feel asphyxiated when they try to tell us how much pain we bring to them and we ignore their pleas, and what's really interesting is that the animals aren't the cause of the treatment that they receive, but it is that, rather, there are just too many of us marauding human animals, dominant human beings, who think we can do anything we want because we're superior. People don't like to talk about our own tendency to overpopulate, but at the core that's the major problem.

While many people try to treat animals with respect and dignity, many also agree that good welfare is not good enough. Existing laws and regulations don't adequately protect animals. We're only fooling ourselves when we claim that they prevent pain and suffering. Good welfare, and research performed within existing regulations, allow mice to be shocked and otherwise tortured, rats to be starved or forced-fed, pigs to be castrated without anesthetics, cats to be blinded, dogs to be shot, and primates to have their brains invaded with electrodes.

Only about one percent of animals used in research in the United States are protected by current legislation. For instance, here is a quote from the U.S. *Federal Register*, volume 69, number 108, Friday June 4, 2004:

> We are amending the Animal Welfare Act (AWA) regulations to reflect an amendment to the Act's definition of the term *animal*. The Farm Security and Rural Investment Act of 2002 amended the definition of *animal* to specifically exclude birds, rats of the genus *Rattus,* and mice of the genus *Mus,* bred for use in research.

It may surprise you to hear that birds, rats, and mice are no longer considered animals, but that's the sort of logic that characterizes federal legislators. Since researchers are not allowed to abuse animals, the definition of animal is simply revised until it only refers to creatures that researchers don't need. We now know that mice are empathic beings who feel the pain and suffering of other mice (www.the-scientist.com/news/display/23764/#23829), yet this scientific fact hasn't entered into discussions about the well-being of mice and other animals. (For more information, see the "Mice" and "Rats" entries in this encyclopedia.) It is now known that even hermit crabs suffer and remember situations that caused them pain (http://news.bbc.co.uk/2/hi/uk_news/northern_ireland/7966807.stm).

Concerning the use and well-being of birds, Karen Davis, president of United Poultry Concerns notes that

> Millions of birds suffer miserably each year in government, university, and private corporation laboratories, especially considering the huge numbers of chickens, turkeys, ducks, quails, and pigeons being used in agricultural research throughout the world, in addition to the increasing experimental use of adult chickens and chicken embryos to replace mammalian species in basic and biomedical research.
>
> Slaughter experiments are also routinely performed on live chickens, turkeys, ducks, ostriches, and emus, in which these birds are subjected to varying levels of electric shock in order to test the effect of various voltages on their muscle tissue for the meat industry. (See http://www.upc-online.org/genetic/experimental.htm for specific references.)

For example, the Spring 2002 issue of the *Journal of Applied Poultry Research* featured an article in which USDA researchers describe shocking 250 hens in a laboratory simulation of commercial slaughter conditions to show that "subjecting mature chickens to electrical stimulation will allow breast muscle deboning after two hours in the chiller with little or no additional holding time."

Concern for animals has moved beyond primarily captive situations such as laboratories, zoos and aquaria, rodeos, circuses, slaughterhouses, and fur farms into the field. Many of the new essays in this encyclopedia reflect this growing interest and concern. The lives of individual animals are also now much more centrally located in the conservation or green movement, and animals' points of view, including what they like and what they want, and their fate, is more and more factored into conservation decisions, such as relocation and reintroduction projects. This has been evident in popular reaction to urban animals who become pests.

For example, in July 2008, a mother bear was shot when she returned to Boulder, Colorado, my hometown, to look for her cub, which had been electrocuted by an uninsulated electric wire. The citizens were incensed and made their feelings known. The vast majority of people thought it unnecessary to kill the mother bear, and she should have been relocated so that she could live without bothering people. She had done nothing wrong, and was merely trying to live where bears had previously lived before being displaced because of human development.

In another story, when a bear whose head was stuck in a jar left as trash by humans was killed in Minnesota (http://news.bbc.co.uk/2/hi/americas/7534325.stm), people were outraged by this action as well. They wanted to know why the bear couldn't have been tranquilized instead of killed.

In a very real sense, animals are part of the green movement, and coexistence is the guiding philosophy that drives many decisions about how to treat them without trumping their interests with our own. Fewer and fewer decisions to trade off animals for humans go without discussion and concern by a growing portion of the general population. Much interest is driven by interactions with the companion animals who share the homes of people around the world and by children, who are inherently interested in the lives of animals regardless of where they live.

While most people agree that animals are important to humans and that we must pay attention to their well-being, there is also a good deal of disagreement about the types of obligations, if any, that humans have toward other animals. Despite growing interest in and concern over the use of animals, over the past five years violations of the federal Animal Welfare Act (AWA) in the United States have increased more than 90 percent (http://www.all-creatures.org/saen/). In 2006 alone there were more than 2,100 violations of the AWA, with the highest level of violations occurring in the areas of Institutional Animal Care and Use Committees (IACUCs) (58%), and veterinary care (25%). It has been estimated that about 75 percent of all laboratories violate the AWA at one time or another.

On the other hand, progress is being made. In June 2008, the Spanish government extended legal rights to great apes that include the right to life, protection of individual liberty, and prohibition of torture (www.reuters.com/article/scienceNews/idUSL2565 86320080625?feedType=RSS&; see also http://www.greatapeproject.org/). Kentucky Fried Chicken (KFC) outlets in Canada agreed to require more humane treatment of chickens, including improved slaughtering methods, and to serve vegan chicken items made of soy (http://edmontonsun.com/News/Canada/2008/06/01/5739946.html). In July 2008, Governor Arnold Schwarzenegger signed a law that strengthened the protection of downed cows (http://www.hsus.org/press_and_publications/press releases/schwarzenegger_signs_law_protecting_california_downed_cows_072208.html) who aren't strong enough to survive their trip to a slaughterhouse. In March 2009, the U.S. government banned the use of downer cows for food (http://news.yahoo.com/s/ap/20090314/ap_on_go_pr_wh/mad_cow/print). Farm Sanctuary (www.farmsanctuary.org) achieved a precedent-setting victory after a ten-year battle with the New Jersey Department of Agriculture. Farm Sanctuary's press release notes that, "In a monumental legal decision, the New Jersey Supreme Court unanimously declared that factory farming practices cannot be considered 'humane' simply because they are 'routine husbandry practices'" (http://www.farmsanctuary.org/mediacenter/2008/pr_nj_decision08.html). In addition, in July 2008, great apes that had been used to make movies were moved to the Great Ape Trust sanctuary in Des Moines, Iowa (http://africa.reuters.com/odd/news/usnN16285101.html). In August 2008, at the International Primatological Congress held in Edinburgh, Scotland, there was a symposium on invasive research on great apes, one of the first of its kind ever held at this prestigious meeting. This important gathering came at a time when the European Union (EU) was considering Directive 86/609, which would confirm a total ban on the use of great apes and wild-caught primates in invasive research (http://ec.europa.eu/environment/chemicals/lab_animals/revision_en.htm). Soon after that, legislation was passed in Spain to protect great apes. And in November 2008, because of unrelenting and intentional abuse, Proposition 2 was passed in California by a vote of 63 percent to 37 percent to protect factory-farmed animals so that, beginning in 2015, farm animals will have the right to lie down, stand up, turn around, and extend their limbs. A *New York Times* editorial supported this legislation and urged other states to implement it (http://www.nytimes.com/2008/10/09/opinion/09thu3.html?_r=2&th&emc=th&oref=slogin&oref=slogin).

Which animals we choose to eat also presents major problems, as shown by noted author Michael Pollan in *The Omnivore's Dilemma* and *In Defense of Food*, and in Gene Baur's *Farm Sanctuary,* a superb review of the horrors of factory farming. Not only is a diet with less meat better for us and for animals, but also for the planet as a whole. In addition to extremely important ethical questions that center on the use of animals for food, there are many environmental concerns (http://www.ciwf.org.uk/resources/publications/environment_sustainability/default.aspx; http://www.ciwf.org.uk/includes/documents/cm_docs/2008/s/sustainable_agriculture_report_2008.pdf). For example, it is estimated that by 2025, 64 percent of humanity will be living in areas of water shortage. The livestock sector is responsible for over eight percent of global human water use, and seven percent of global water is used for irrigating crops grown for animal feed. Animal agriculture is responsible for 18 percent of global anthropogenic greenhouse gases (GHGs). In New Zealand, 34.2 million sheep, 9.7 million cattle, 1.4 million deer, and 155,000 goats emit almost 50 percent of greenhouse gases in the form of methane and nitrous oxide (http://www.newscientist.com/article/mg20026873.100-how-kangaroo-burgers-could-save-the-planet.html). Animals are living smokestacks (http://www.nytimes.com/2008/12/04/science/earth/04meat.html?pagewanted=2&_r=1&ref=science). According to the Swedish group Lantmannen, "Producing a pound of beef creates 11 times as much greenhouse gas emission as a pound of chicken and 100 times more than a pound of carrots" (http://www.nytimes.com/2008/12/04/science/earth/04meat.html?pagewanted=2&_r=1&ref=science).

A major concern is the high prevalence of infectious disease that results from factory farming, including streptococcus, Nipah virus, multidrug-resistant bacteria, SARS, avian flu, and other diseases (http://www.hsus.org/farm/news/ournews/factory_farming_emerging_diseases.html; https://hfa.org/factory/index.html). There is even evidence that workers who kill pigs can suffer nerve damage (http://www.iht.com/articles/2008/02/05/healthscience/05pork.php). Physicians were mystified when three patients who visited the Austin Medical Center had the same highly unusual symptoms, including fatigue, pain, weakness, and numbness and tingling in the legs and feet. But the patients had something else in common; they all worked at Quality Pork Processors, a local meatpacking plant.

Even the United Nations' Nobel Prize-winning scientific panel on climate change urged people to stop eating meat because of the climatic effects of factory farming (http://afp.google.com/article/ALeqM5iIVBkZpOUA9Hz3Xc2u-61mDlrw0Q). They suggested, "Don't eat meat, ride a bike, and be a frugal shopper—that's how you can help brake global warming." In addition they suggested, "Please eat less meat—meat is a very carbon intensive commodity . . . Studies have shown that producing one kilo (2.2 pounds) of meat causes the emissions equivalent of 36.4 kilos of carbon dioxide."

An essay in *Conservation Magazine* (July/September, 2008) titled "The Problem of What to Eat" (http://www.conbio.org/CIP/article30813.cfm) highlights the major problems:

> It turns out that many core issues such as pesticide use, soil health, and the impact of food miles are more nuanced and complicated than you might think. . . . According to a recent study by researchers at Carnegie Mellon University, foregoing red meat and dairy just one day a week achieves more greenhouse gas reductions than eating an entire week's worth of locally sourced foods. That's because the carbon footprint of food miles is dwarfed by that of food production. In fact, 83 percent of the average U.S. household's carbon footprint for food consumption

comes from production; transportation represents only 11 percent; wholesaling and retailing account for 5 percent.

It has been calculated that the carbon footprint of meat-eaters is almost twice that of vegetarians (http://www.nowpublic.com/environment/love-mother-earth-slash-carbon-footprint-going-veggie; http://news.sg.msn.com/lifestyle/article.aspx?cp-documentid=1647349). Commercial meat production clearly is not sustainable, according to the most often quoted definition from a United Nations Report, as development "meeting the needs of the present without compromising the ability of future generations to meet their own needs" (United Nations, 1987, p. 1). Further, all definitions of sustainable development ignore the lives of animals.

Dissecting animals as part of educational practices is also being questioned. Schools at all levels around the world are banning this practice not only because of ethical issues, but also because non-animal alternatives are as good or better for reaching educational goals (http://www.interniche.org/). More than 30 published studies show that alternatives such as computer software, models, and transparencies are at least as likely as dissection to achieve the intended educational goals. Technological advances, such as imaging that allows students to view the nervous system at any level, to rotate the image, to make certain layers opaque and others transparent, to cut away certain layers, and to repeat these operations in reverse, add an overwhelming advantage to these alternatives.

Educators around the world agree. In Gujarat, India, Bhavnager University has replaced the annual use of more than 3,000 animals with non-animal alternatives, Israel banned vivisection in schools in 2003, and in March 2008, the Faculty of Zoology at Tomsk Agricultural Institute in Russia ended the use of animals for dissection (http://www.vita.org.ru/), even as Russian President and *Time* Magazine's person of the year Vladimir Putin admitted to having harassed rats when he was young (http://www.time.com/time/magazine/europe/0,9263,901071231,00.html).

Medical schools in the United States are swapping pigs for plastic (http://www.nature.com/news/2008/080507/full/453140a.html). In this essay it was noted that while doctors used to try out their surgical skills on animals before being allowed to work on patients, now only a handful of medical schools in the United States still have animal labs. Live-animal experiments were on the curriculum in 77 of 125 medical schools in 1994, but now it is thought that only 11 of 126 schools still use them, and this trend is being followed around the globe. By February 2008, all American medical schools had abandoned the use of dog labs for teaching cardiology (http://www.nytimes.com/2008/01/01/health/research/01dog.html?_r=1&scp=1&sq=belloni%20dogs&st=cse&oref=slogin). For more information on alternatives, see these Web sites: www.aavs.org, www.idausa.org/campaigns/dissection/undergradscience.html, and www.petakids.com/disindex.html. Francis Belloni, dean at New York Medical College, has said that "the use of animals was not done lightly and had value," but added that students would "become just as good doctors without it" (*New York Times* article cited above).

The debate about the use of non-animal alternatives continues. On the one hand, Roberto Caminiti, chair of the Programme of European Neuroscience Schools (htpp://fens.mdc-berlin.de/pens) has argued that it will never be possible to replace animals in research (*Nature*, 2009, p. 147). Caminiti avoids any discussion of the numerous non-animal alternatives that are available, many of which are being used successfully by many of his colleagues. On the other hand, Bill Crum of the Centre for Neuroimaging Sciences at the Institute of Psychiatry at King's College London counters Caminiti as follows:

> To my mind, there is a moral inconsistency attached to studies of higher brain function in nonhuman primates: namely, the stronger the evidence that nonhuman primates provide excellent experimental models of human cognition, the stronger the moral case against using them for invasive medical experiments. From this perspective, "replacement: should be embraced as a future goal." (http://www.nature.com/nature/journal/v457/n7230/pdf/457657b.pdf)

It is clear that people who are interested in animal rights and animal welfare are involved in an ever-growing social movement, and the time has indeed come to move forward proactively, and not merely reactively, to educate, and to raise consciousness. In March 2006, I gave a lecture at the annual meeting of the Institutional Animal Care and Use Committees in Boston. I was received warmly, and the discussion that followed my lecture was friendly, even though some in the audience were a bit skeptical of my unflinching stance that certain animals feel pain and a wide spectrum of emotions. After my talk, a man approached me and informed me that he was responsible for enforcing the Animal Welfare Act at a major university. He admitted that he'd been ambivalent about some of the research that's permitted under the act and, after hearing my lecture, he was even more uncertain. He told me that he'd be stricter in enforcing the current legal standards, and work for more stringent regulations. I could tell from his eyes that he meant what he said, and that he understood that the researchers under his watch would be less than enthusiastic about his decision. But he needed someone to confirm his intuition that research animals were suffering, and that the Animal Welfare Act was not protecting them. I was touched and thanked him. Then he put his head down, mumbled, "Thank you," and walked off. In September 2008, I learned that he had recommended that I be invited to a conference about enriching the lives of laboratory animals. Although I would like to see research with lab animals phased out entirely and the animals moved to a sanctuary, this is a first step in raising awareness that laboratory animals cannot be given what they need, and that there are non-animal alternatives that are as good or better. Over the past few years, in my extensive travels around the world, I've learned that many of my colleagues now agree that animal welfare often isn't good enough (Bekoff, 2008b).

The work on behalf of ending some laboratory uses of animals stems from the pioneering efforts of Henry Spira, founder of Animal Rights International (see Singer, 1998). In the 1970s, working from his small apartment in New York City, Spira and his grassroots organization were responsible for having federal funding pulled from a project in which researchers at the American Museum of Natural History performed surgery on cats' genitals and pumped them full of various hormones to see how the mutilated cats would behave sexually. Spira also formed the Coalition to Abolish the Draize Test, a test that involves using rabbits to test eye-makeup. The Draize test is torture, and rabbits, who have very sensitive eyes, suffer immensely. By 1981, the cosmetics industry itself awarded $1 million to Johns Hopkins University's School of Hygiene and Public Health to establish the Center for Alternatives to Animal Testing. Most cruelty-free products trace their history back to Spira's tireless and unflagging efforts to stop animal abuse.

ASKING ANIMALS WHAT THEY WANT

In this revised edition of the *Encyclopedia of Animal Rights and Animal Welfare,* there is considerably more information on animal cognition and animal emotions and sentience. These are among the hottest areas in a field called cognitive ethology (Bekoff, 2006, 2007b; Bekoff and Pierce, 2009), or the study of animal minds. This information was used in a very novel study by renowned ape language researcher Sue Savage-Rumbaugh. One way to find out what animals want is to ask them and then write a paper with them, as Savage-Rumbaugh did. She coauthored a paper for the *Journal of Applied Animal Welfare Science* (JAAWS) with the bonobos she studied for years, Kanzi Wamba, Panbanisha Wamba, and Nyota Wamba (http://www.informaworld.com/smpp/content~content=a788000924~db=all~order=page). Because Sue and the bonobos had two-way conversations, these amazing beings using a keyboard with symbols (lexigrams), and she could actually ask the bonobos questions and record their responses (http://www.myhero.com/myhero/hero.asp?hero=sue_savage_rumbaugh; http://www.iowagreatapes.org/media/releases/2008/nr_10a08.php). She also notes,

> Although it is true that I chose the items listed as critical to the welfare of these bonobos and facilitated the discussion of these particular items, I did not create this list arbitrarily. These items represent a distillation of the things that these bonobos have requested repeatedly during my decades of research with them.

Sue discovered that these were the items the bonobos agreed were important for their welfare:

1. Having food that is fresh and of their choice
2. Traveling from place to place
3. Going to places they have never been before
4. Planning ways of maximizing travel and resource procurement, for example, obtaining food
5. Being able to leave and rejoin the group, to explore, and to share information regarding distant locations
6. Being able to be apart from others for periods of time
7. Maintaining lifelong contact with individuals whom they love
8. Transmitting their cultural knowledge to their offspring
9. Developing and fulfilling a unique role in the social group

10. Experiencing the judgment of their peers regarding their capacity to fulfill their roles, for the good of the group

11. Living free from the fear of human beings attacking them

12. Receiving recognition, from the humans who keep them in captivity, of their level of linguistic competency and their ability to self-determine and self-express through language

Clearly, eating well, having the freedom to move about and to have time alone, being stimulated by novelty, being an active member of a social group, being appreciated for the beings they are, and living free from fear, all figured into the bonobos' assessment of what they needed in captivity. Enriched and challenging social and physical environments were important to them, as they would be to most animals who find themselves living in situations where their options are limited. This sort of preference testing could be used on a wider array of species, and in this way they can tell us what they want and need. In doing this we can make "good welfare" better.

EVERY INDIVIDUAL CAN MAKE
A DIFFERENCE: WE'RE WIRED TO BE KIND

The first annual Kindness Index, introduced today by Best Friends Animal Society, finds that most Americans, in addition to loving their pets, believe overwhelmingly that they have a moral obligation to protect animals. They are also adamant about passing these values on to their children . . . The major discovery of the poll is that far more people than we imagined really want better lives for animals, and they're prepared to help. We simply have to create the opportunities. (http://www.bestfriends.org/aboutus/pdfs/061906%20Kindness%20Index.pdf)

I believe that at the most fundamental level our nature is compassionate, and that cooperation, not conflict, lies at the heart of the basic principles that govern our human existence . . . By living a way of life that expresses our basic goodness, we fulfill our humanity and give our actions dignity, worth, and meaning. (His Holiness The Dalai Lama, "Understanding our Fundamental Nature")

Human beings are wired to care and give . . . and it's probably our best route to happiness. (Psychologist Dacher Keltner: http://www.nytimes.com/2008/11/27/us/27happy.html?scp=3&sq=dacher&st=cse)

Despite everything we read about competition and nastiness, most research nowadays supports what University of California psychologist Dacher Keltner claims. Humans are wired to care and to give, and it makes us feel good to help others. We're also learning that egalitarianism has been a force in shaping many human societies (Bekoff and Pierce, 2009), so it should be natural that we all work for a science of unity that respects other animals and cherishes the beautiful and magical webs of nature.

We need to replace mindlessness with mindfulness in our interactions with animals and the earth. Nothing will be lost and much will be gained. We can never be too generous or too kind. Surely we will come to feel better about ourselves if we know deep in our hearts that we did the best we could, and took into account the well-being of the magnificent animals with whom we share earth, the awesome and magical beings who selflessly make our lives richer, more challenging, and more enjoyable than they would be in their absence. Doesn't it feel good to know that there are animals out there who we have helped, even if we cannot see, hear, or smell them? Doesn't it feel good to know that we did something to help the earth, even if we do not see the fruits of our labor?

If we forget that humans and other animals are all part of the same world, and if we forget that humans and animals are deeply connected at many levels of interaction, when things go amiss in our interactions with animals, and animals are set apart from and inevitably below humans, it is certain that we will miss the animals more than they will miss us. The interconnectivity and spirit of the world will be lost forever, and these losses will make for a severely impoverished universe. As Paul Shepard wrote:

> There is a profound, inescapable need for animals that is in all people everywhere, an urgent requirement for which no substitute exists. This need is no vague, romantic, or intangible yearning, no simple sop to our loneliness or nostalgia for Paradise . . . Animals have a critical role in the shaping of personal identity and social consciousness . . . Because of their participation in each stage of the growth of consciousness, they are indispensable to our becoming human in the fullest sense.

To conclude, here are ten overlapping reasons why we all need to be concerned with animal rights and animal welfare, why we need to do better, and why we need to increase our compassion footprint (Bekoff, 2008a, 2010):

> Animals exist and we share Earth with them
> This land is their land, too
> Animals are more than we previously thought
> We have become alienated from animals
> We need to mind animals and look out for one another
> We are powerful and must be responsible for what we do to animals
> What we're doing now doesn't work
> "Good welfare" isn't good enough
> We all can do something to make the world a more compassionate and peaceful place for animals and for ourselves
> We need to increase our compassion footprint

I offer these reasons to stimulate discussion, not because they're the only reasons why we need to examine the concept of animal welfare and treat animals with more respect and dignity, but because reflecting on these and perhaps other reasons will force us to be more responsible for what we do to animals and help to increase our compassion footprint. Some people worry that more attention to animals means less attention for needy humans, but this is a baseless concern. Many people who work for animals also work for humans. In addition to working for animals, I work with many children's groups, senior citizens, and prisoners, and I also sponsor a young girl in Uganda so that she receives medical care and an education. Compassion begets compassion, and seamlessly crosses species. I truly feel that it will be much easier to live in a world where ethical choices are commonplace and compassion is the name of the game, rather than in a world where we ignore others' lives. I hope you agree.

Studying nonhuman animals is a privilege that must not be abused. We must take this privilege seriously. Although the issues are very difficult and challenging, it does not mean they're impossible to address. Certainly we cannot and must not let animals suffer because of our inability to come to terms with difficult issues or to accept responsibility for how we treat them. Questioning the ways in which humans use animals will make for more informed decisions about animal use. By making such decisions in a responsible way, we can help to ensure that in the future we do not repeat the mistakes of the past, and that we move toward a world in which humans and other animals may be able to share peaceably the resources of a finite planet.

I believe that we are born to be good, and there is hope for the future when we come to realize that the competitive survival of the fittest mentality is not who we really are or have to be. It's not really a dog eat dog world, because dogs don't eat other dogs. Being kind and good must also include cultural pluralism in the diverse and often tough world in which we live. And we need to constantly remind ourselves that we live in a more-than-human world, as philosopher and master magician David Abram reminds us. Goodness and kindness will allow us to do what needs to be done to heal the conflicts we have with other animals and amongst ourselves. Now is the time to tap into our innate goodness and kindness to make the world a better place for all beings, creating a paradigm shift that brings hope and life to our dreams for a more compassionate and peaceful planet. The essays in this encyclopedia contain the information that is needed to make the best and most enduring compassionate choices.

Further Reading

Abram, D. 1996. *The Spell of the Sensuous: Perception and Language in a More-Than-Human World.* New York: Pantheon Books.

Baur, G. 2008. *Farm Sanctuary: Changing Hearts and Minds About Animals and Food.* New York: Touchstone.

Bekoff, M. 2006. *Animal Passions and Beastly Virtues: Reflections on Redecorating Nature.* Philadelphia: Temple University Press.

Bekoff, M. 2007a. *Animals Matter: A Biologist Explains Why We Should Treat Animals with Compassion and Respect.* Boston: Shambhala.

Bekoff, M. 2007b. *The Emotional Lives of Animals.* Novato, CA: New World Library.

Bekoff, M., ed. 2007c. *Encyclopedia of Human-Animal Relationships: A Global Exploration of Our Connections with Other Animals.* Westport, CT: Greenwood Publishing.

Bekoff, M. 2008a. "Increasing our compassion: The animals' manifesto." *Zygon* (*Journal of Religion & Science*), 43, 771–781.

Bekoff, M. 2008b. "Why 'good welfare' isn't 'good enough': Minding animals and increasing our compassionate footprint." *Annual Review of Biomedical Sciences*, 10, T1-T14. http://arbs.biblio teca.unesp.br/viewissue.php.

Bekoff, M. & Pierce, J. 2009. *Wild Justice: The Moral Lives of Animals.* Chicago: University of Chicago Press.

Bekoff, M. 2010. *The Animal Manifesto: Six Reasons for Expanding Our Compassion Footprint.* Novato, CA: New World Library.

Clubb, R., Rowcliffe, M., Lee, P., Mar, K. U., Moss, C. & Mason, G. 2008. "Compromised survivorship in zoo elephants." *Science*, 322, 1649.

Dawn, K. 2008. *Thanking the Monkey: Rethinking the Way We Treat Animals.* New York: HarperCollins.

His Holiness the Dalai Lama. 2002. "Understanding our fundamental nature." In *Visions of Compassion: Western Scientists and Tibetan Buddhists Examine Human Nature*, 66–80. New York: Oxford University Press.

Pollan, M. 2006. *The Omnivore's Dilemma.* New York: Penguin Press.

Pollan, M. 2008. *In Defense of Food.* New York: Penguin Press.

Salem, D. J. & Rowan, A. N., eds. 2007. *The State of the Animals IV, 2007.* Washington, DC: Humane Society Press.

Shepard, P. 1996. *Traces of an Omnivore.* Washington, DC: Island Press.

Singer, P. 1998. *Ethics Into Action: Henry Spira and the Animal Rights Movement.* Lanham, MD: Rowman & Littlefield.

United Nations. Report of the world commission on environment and development. 42/187. December 11, 1987.

Williams, E. & DeMello, M. 2007. *Why Animals Matter: The Case for Animal Protection.* Amherst, NY: Prometheus Books.

Marc Bekoff

A

ABOLITIONIST APPROACH TO ANIMAL RIGHTS

The abolitionist approach to animal rights seeks to provide both a deontological theory (a theory of moral obligation) concerning the moral status of nonhumans, and a practical approach to animal advocacy. The central tenets of the abolitionist approach are that animal use should be abolished and not merely regulated because animal use cannot be morally justified, and that veganism is a baseline moral principle and should be the primary focus for animal advocacy. The abolitionist approach squarely and unequivocally rejects all forms of welfarism, which maintains that the central goal of animal advocacy is to regulate animal exploitation to make it more humane and regard veganism as an optional way of reducing suffering and not as a fundamental moral tenet or a central focus of advocacy.

There are animal advocates who disagree with the abolitionist position as described but who nevertheless use the term abolitionist to characterize their views. The central characteristic of the new welfarism, which is the prevalent approach to animal ethics promoted by large animal advocacy organizations in North America, South America, and Europe, is that the abolition of animal use is the long-term goal of animal advocacy, but that welfarist regulation of the treatment of animals is the most efficient way of moving incrementally toward that abolition. The abolitionist approach described here rejects this view.

Because large animal organizations adopt a traditional welfarist or new welfarist approach to animal ethics, they are understandably hostile to the abolitionist perspective. The abolitionist movement, currently developing as an international phenomenon, is one that has emerged largely as a grassroots endeavor of advocates who have little or no connection to any of the large animal organizations. Abolitionists are often part of Internet communities that provide social support and discussion of theoretical and practical issues.

Abolitionism and Animal Welfare

The abolitionist approach rejects animal welfare as a general matter for both theoretical and practical reasons. As a theoretical matter, all forms of welfare assume that nonhuman animals have a lesser moral value than humans, a notion extant in animal welfare theory from its emergence in 19th-century Britain. Although welfarists such as Jeremy Bentham and John Stuart Mill argued that animals deserved to be included in the moral community and given at least some legal protection, they did not oppose the continued use of animals by humans. According to the welfarists, although animals were sentient, they did not have an interest in not being used by humans because they were not self-aware and did not have an interest in continued existence. That is, animals lived in the present and were not aware of what they lost when we took their lives. They did not have an interest in not being used; they only had an interest in being treated gently. These and related views about the supposedly superior mental characteristics of humans led Bentham, Mill, and other early welfarists to regard animals as having less moral value than humans. This position is represented in contemporary animal welfare theory by Peter Singer, the leading figure. Singer, like the original welfarists, argues that most animals do not have any interest in continuing to live.

English philosopher and economist John Stuart Mill was an early advocate of legal protection for animals. (Library of Congress)

There probably are significant differences between the minds of humans and those of nonhumans, given that human cognition is so closely linked to symbolic communication which, with the possible exception of nonhuman great apes, nonhumans do not use. There is, however, no reason to maintain that any cognitive differences mean that animals have no interest in continuing to exist. To say that any sentient being is not harmed by death begs the question and is, in any event, decidedly odd. After all, sentience is not a characteristic that has evolved to serve as an end in itself. Rather, it is a trait that allows the beings who have it to identify situations that are harmful and that threaten survival. Sentience is a means to the end of continued existence. Sentient beings, by virtue of their sentience, have an interest in remaining alive; that is, they prefer, want, or desire to remain alive. Therefore, to say that a sentient being is not harmed by death denies that the being has the very interest that sentience serves to perpetuate. This would be analogous to saying that a being with eyes does not have an interest in continuing to see or is not harmed by being made blind.

The fact that the minds of humans differ from those of nonhumans does not mean that the life of a human has greater moral value any more than it means that the life of a human who is normal has greater moral value than the life of a mentally disabled person, or that the life of an intelligent person has greater moral value than that of a normal but less intelligent one. Although the differences between humans and animals may be important for some purposes, they are completely irrelevant to the morality of treating animals as human resources, even if we do so humanely. The abolitionist position maintains that we are obligated to accord every sentient being the right not to be treated as a resource.

The abolitionist approach does not support the idea that some species of nonhumans, such as nonhuman great apes, are more deserving of moral or legal protection than other species on the ground that the former are more similar to humans. With respect to being treated as a human resource, all sentient beings—human and nonhuman—are equal.

The abolitionist approach also rejects animal welfare on practical grounds. Animals are property; they are defined as economic commodities with only extrinsic or conditional value. To the extent that we protect animal interests, we do so only when it provides a benefit—usually an economic benefit—to humans. As a result, the protection of animal welfare is, for the most part, very limited. Regulation does not decrease animal suffering in any significant way, and it does not decrease demand by making animal exploitation more expensive. On the contrary, welfare reform generally increases production efficiency so that it becomes cheaper to produce animal products. To the extent that a welfare regulation imposes any sort of additional cost on animal production, that added cost is *de minimis*. Moreover, welfare reform makes the public feel more comfortable about using animal products, and perpetuates rather than discourages animal exploitation. There is absolutely no empirical evidence that animal welfare reform will lead to abolition or to significantly decreased animal use.

Abolitionism and Veganism

Although veganism may represent a matter of diet or lifestyle for some, ethical veganism is a profound moral and political commitment to abolition on the individual level and extends not only to matters of food, but to the wearing or use of animal products. Ethical veganism is the personal rejection of the commodity status of nonhuman animals and the notion that animals have less moral value than do humans. Indeed, ethical veganism is the *only* position that is consistent with the recognition that, for purposes of being treated as a thing, the lives of humans and nonhumans are morally equivalent. Ethical veganism must be the unequivocal moral baseline of any social and political movement that recognizes that nonhuman animals have inherent or intrinsic moral value and are not resources for human use. Ethical vegans believe that we as people will never even be able to see the moral problem with animal use as long as we continue to use animals. We will never find our moral compass as long as animals are on our plates, or on our backs or feet, or in the lotions that we apply to our faces.

Animal advocates who claim to favor animal rights and to want to abolish animal exploitation, but continue to eat or use animal products, are no different from those who claimed to be in favor of human rights but continued to own slaves. Moreover, there is no coherent distinction between flesh and dairy or eggs. Animals exploited in the dairy or egg industries often live longer, are treated worse, and end up in the same slaughterhouses as their meat counterparts. There is as much if not more suffering and death in dairy or egg products than in flesh products, but there is certainly no morally relevant distinction between or among them.

The most important form of incremental change on a social level is creative, non-violent education about veganism and the need to abolish, not merely to regulate, the institutionalized exploitation of animals. Veganism and creative, positive vegan education provides a practical and incremental strategy, both in terms of reducing animal suffering now, and in terms of building a movement in the future that will achieve more meaningful legislation in the form of significant prohibitions of animal use.

Rather than embrace veganism as a clear moral baseline, welfarists promote flexible veganism or consuming with conscience, which they see as one way to reduce suffering, along with welfarist reforms that they promote as reducing suffering. That is, welfarists restrict the scope of animal ethics to suffering; anything that arguably reduces that suffering, including being what Peter Singer calls a conscientious omnivore, represents a morally defensible position. Putting aside that welfare reforms do not result in significant protection of animal interests, the welfarist position on veganism reflects the view that animal use is itself not morally problematic, which assumes that animal life is of lesser value than human life.

Abolitionism and Single-Issue Campaigns

The abolitionist approach promotes the view that veganism and creative, non-violent education about veganism are the primary practical and incremental approaches that should be pursued. In addition to rejecting campaigns that seek to make animal exploitation more humane, the abolitionist approach generally regards single-issue campaigns, such as those involving foie gras or fur garments, as problematic because they reinforce the view that certain forms of exploitation are worse than others. For example, the anti-fur campaign implicitly and often explicitly characterizes fur as involving some greater degree of exploitation than does, say, wool or leather. But any such characterization would be inaccurate. Both wool and leather are every bit as morally objectionable as fur in terms of the suffering involved and the fact that, irrespective of any differences in suffering, all three forms of clothing involve killing animals for human purposes. Foie gras is no worse than other animal foods.

Abolitionism and Domesticated Nonhumans

The abolitionist position maintains that if we recognize that nonhuman animals should not be treated as resources, the appropriate social response would be to stop bringing domesticated nonhumans into existence. We should care for those whose existence we have caused or facilitated, but we should not cause more to come into existence.

Representative Web sites are:

Animal Rights: The Abolitionist Approach: www.AbolitionistApproach.com

Vegan Freak: Being Vegan in a Non-Vegan World: www.veganfreak.com

Further Reading
Francione, Gary L. 2000. *Introduction to animal rights: Your child or the dog?* Philadelphia: Temple University Press.

Francione, Gary L. 2008. *Animals as persons: Essays on the abolition of animal exploitation.* New York: Columbia University Press.

Francione, Gary L. and Anna E. Charlton. 2008. "Animal advocacy in the 21st century: The abolition of the property status of nonhumans," in T. L. Bryant, R. J. Huss, and D. N. Cassuto (eds.), *Animal law in the courts:*

A reader (St. Paul, MN: Thomson/West, 2008), 7–35.

Torres, Bob. 2007. *Making a killing: The political economy of animal rights.* Oakland, CA: AK Press.

Gary L. Francione
Anna E. Charlton

AFFECTIVE ETHOLOGY

Affective ethology refers to the behavioral study of one's affective states, emotions, feelings. Research toward animal emotions has been overshadowed for many years by scientific taboo, but over the last decade interest in animal emotions has gained increasing attention. Affective ethology is important for our treatment of animals, as the question of whether animals can experience feelings like pain, fear, joy and happiness is at the core of discussions on animal welfare and animal ethics.

An important root of the taboo goes back to the Cartesian school of thought. The seventeenth-century French philosopher René Descartes stated that only humans have souls, and therefore they are the only beings that can reason and feel, whereas animals are merely complex machines, which only *appear* to think or feel (Margodt, 2007). Two centuries later, Charles Darwin argued that humans and animals are not radically different, but rather related. Humans and other animals have a common ancestry and share mental characteristics. Darwin brought a range of behavioral information together in support of feelings such as fear, anger, pleasure, and love in animals (Darwin, 1872, 1890).

Behaviorism—another major root—denies the possibility of studying animal minds. It reacted against unfounded 19th-century claims regarding animal minds, such as stories about mice cooperating to cross rivers on floats of dried cow-dung, carrying mushrooms filled with berries as provisions (in Romanes, 1882). Behaviorism's goal was to have psychology accepted as a serious science, and argued it should discard consciousness and instead focus on the prediction and control of behavior. This taboo on considering animal consciousness was broken in an unprecedented way by primatologist Jane Goodall during the 1960s with her study of wild chimpanzees. Her descriptions of chimpanzees tickling, chasing and laughing, and of infant chimps being depressed after losing their mothers, only make sense within the context that they have feelings, minds, personalities (Goodall, 1971).

During the 1970s, Donald Griffin coined the term cognitive ethology for the study of behavior suggestive of consciousness and thinking in animals (Griffin, 1976). Referring to many studies, he emphasized the versatility of animal minds (e.g. in solving problems) and their rich communication systems. Though Griffin had to endure a lot of criticism, cognitive ethology has gained considerable support among ethologists.

Since the 1990s, several books have raised a variety of arguments and observations in support of animals' experiencing emotions, thus picking up a thread started by Darwin 120 years before (see Masson & McCarthy, 1994; Bekoff, 2000; Balcombe, 2006; Bekoff, 2008). These works indicate that the scope of affective ethology is no less varied as that of cognitive ethology.

How did the notion of affective ethology arise? Gordon Burghardt argued that cognitive ethology isn't appropriate to indicate the study of private experiences in animals (1997), but that animal minds are broader than the cognitive sphere; they also relate to affective and motivational aspects. However, Burghardt did not really suggest an alternative name for the phenomenon. In 2004, I proposed the terms affective ethology and motivational ethology, and naming the behavioral study of private experiences ethology of mind. This discipline then comprises three sub-disciplines, namely cognitive, affective and motivational ethology (Margodt, 2004). In 2007 I was contacted by the Hungarian philosopher László Nemes, who also suggested the notion of affective ethology. Indeed, this seems to confirm that this idea logically follows from Donald Griffin's suggestion regarding cognitive ethology and the existing field of affective neuroscience.

In recent years, the interest in animal emotions has increased due to developments in affective neuroscience. Noninvasive brain imaging techniques such as PET (positron emission tomography) and fMRI (functional magnetic resonance imaging) scans allow for the detection of changes in regional blood flow related to emotional reactions, and may lead to unprecedented comparisons between human and nonhumans (see Davidson, Scherer & Goldsmith, 2003).

A large variety of emotions remain to be studied in animals belonging to a wide range of species. In addition to behavioral observations in the wild and in captivity, carefully designed experiments allow further exploration of the emotional world of animals. Jaak Panksepp and colleagues showed that rats have a stronger preference for being tickled (rapid finger movements at their undersides) than being petted (gently stroked on the back).

Tickled rats expressed seven times more 50-kHz chirps—typical for playful situations—than petted rats. They also ran four times as quickly to a human hand, and repeatedly hit a bar to signal that they wanted to be tickled, whereas they almost never pressed a bar to signal that they wanted to be petted (see Balcombe, 2006).

Affective ethologists will be challenged by other scientists who are skeptical about emotions in animals. A leading critical voice is that of Oxford University zoologist Marian Stamp Dawkins, who argues that it remains logically possible that emotional behavior is not accompanied by any feelings in animals. Statements about what animals feel can only be personal views, not something grounded in hard facts (Dawkins, 2000). The debate on animal emotions thus remains ongoing.

Affective ethology also has a second meaning, apart from the study of affective behavior. It also implies that ethologists have to undertake their research on the animals they study in a *caring* way. Harry Harlow studied depression in primates by separating infants from their mothers and isolating them for months or even years in tiny steel chambers, which he called Pits of Despair (Blum, 1994). His research methods may have been most effective, but they were ethically highly questionable. It may be expected that the more the field of affective ethology grows, the stronger will be the calls to care for the welfare interests of sentient, feeling beings.

See also Animal subjectivity; anthropomorphism; anthropomorphism—critical; consciousness, animal; sentience and animal protection; sentience and animal protection; sentientism; Whales and Dolphins, Sentience and Suffering

Further Reading

Balcombe, J. 2006. *Pleasurable kingdom: Animals and the nature of feeling good.* New York: Macmillan.

Bekoff, M. (ed.) 2000. *The smile of a dolphin: Remarkable accounts of animal emotions.* London: Discovery Books.

Bekoff, M. 2007. *The emotional lives of animals: A leading scientist explores animal joy, sorrow, and empathy—and why they matter.* Novato, CA: New World Library.

Blum, D. 1994. *The monkey wars.* New York and Oxford: Oxford University Press.

Burghardt, G. M. 1997. Amending Tinbergen: A fifth aim for ethology. In Mitchell, R. W., Thompson, N.S., & Miles, H.L. (eds.), *Anthropomorphism, anecdotes, and animals* (254–276). Albany: State University of New York Press.

Darwin, C. 1872, 1890. The expression of the emotions in man and animals. Second edition, edited by Francis Darwin. In Barrett, P. H., & Freeman, R. B. (eds.) (1989). *The works of Charles Darwin, Vol. 23.* New York: New York University Press.

Davidson, R. J., Scherer, K. R., & Goldsmith, H. H. (eds.) 2003. *Handbook of affective sciences.* Oxford: Oxford University Press.

Dawkins, M. S. 2000. Animal minds and animal emotions. *American Zoologist, 40,* 883–888.

Goodall, J. 1971. *In the shadow of man.* London: Book Club Associates.

Griffin, D. R. 1976. *The question of animal awareness: Evolutionary continuity of mental experience.* New York: Rockefeller University Press.

Margodt, K. 2004. *The moral status of great apes: An ethical and philosophical-anthropological study.* Unpublished PhD thesis, Ghent University.

Margodt, K. 2007. Sentience and cognition: Descartes, René. In Bekoff, M. (ed.), *Encyclopedia of human-animal relationships: A global exploration of our connections with animals* (1305–1306). Westport, CT: Greenwood Press.

Masson, J. M., & McCarthy, S. 1994. *When elephants weep: The emotional lives of animals.* London: Jonathan Cape.

Romanes, G. J. 1882. *Animal intelligence.* London: Kegan Paul, Trench & Co.

Koen Margodt

ALTERNATIVES TO ANIMAL EXPERIMENTS IN THE LIFE SCIENCES

Within education and training in biology, medicine and veterinary medicine, animals often play a central role in laboratory practical classes. Alive, they are used in experiments to illustrate physiological and pharmacological principles, and for acquisition of a range of clinical and surgical skills. They are also killed for their tissue and organs, and so that students can perform dissections in anatomy classes. Tens of millions of animals—perhaps more—are used for these purposes each year around the world.

Animals suffer harm in various forms during capture, breeding and incarceration, and suffer pain and injury during experiments. These are sometimes conducted without anesthetic, and with lasting negative impact on the individual animal, if he or she survives. Killing is obviously also a serious form of harm, because the most significant freedom that each individual animal has—his or her life—is denied.

Dissecting Convention

In this conventional, harmful use of animals, the relationship between the animal and the student is clearly a negative one. This reality is not what most students are expecting when they choose to study the nature and processes of life (through biology), or train to heal people or animals (through medicine). Harmful animal use is a counter-intuitive practice, and creates a learning environment that is not conducive to effective acquisition of knowledge, skills and responsible attitudes.

The limitations of harmful animal use and the advantages of new approaches are illustrated by the many published studies comparing conventional methods with alternatives. In almost all cases, the alternatives are shown to be equivalent or superior in terms of student and trainee performance. Moreover, assessing how effectively teaching objectives are met requires an identification of a broader range of teaching objectives beyond the standard, and must address the negative messages of the hidden curriculum. These include the lessons that instrumental use of animals is acceptable, and that compassion, respect for life, and ethical concerns as a whole are unimportant—or even obstacles to effective education and training.

Awareness, Objection and Innovation

Some students may even choose not to study the life sciences at the university level because of an awareness of the harm caused to animals in many classes. This results in a loss to the related professions of some of the most sensitive and critical-thinking students. Desensitization of students who do enter these classes is a damaging consequence of harmful animal use, and self-aware students may recognize this psychological process. Students who find that practices are against their ethical positions or religious beliefs may face academic or psychological penalties from teachers if they challenge the status quo. However, informed and responsible conscientious objection can be a powerful catalyst in resolving ethical conflicts in education and in implementing progressive teaching methods, clearly illustrating the intersection of animal rights and civil rights.

Despite the inertia of convention, the replacement of harmful animal use with other methods has been gaining momentum around the world. Progressive, humane alternatives have now fully replaced animal experiments and dissections in a growing number of university departments. Technological innovation, particularly the development of multimedia software and its potential to support the learning process, has played a major role in this ongoing revolution. The economic advantages of using alternatives, and the broader social and cultural changes in favor of ethical treatment of animals, also contribute.

Types of Alternatives

Alternatives, therefore, are progressive learning tools and teaching approaches that can replace harmful animal use or complement existing humane education. They include non-animal learning tools as well as alternative approaches that are neutral or beneficial to individual animals. Often developed by teachers themselves, and typically used in combination, alternatives include:

Mannequins and Simulators Life-like mannequins can support effective training of clinical skills such as taking blood, intubation, and the management of critical care scenarios. The perfusion of ethically sourced organs in advanced simulators allows for realistic surgery practice from student to professional level. By allowing repeated practice, these alternatives enable students and trainees to gain the confidence and competence necessary to work with real patients.

Multimedia Software and Virtual Reality (VR) Visualization and understanding of anatomical structure and function can be enhanced through high-resolution images, video clips, and animations available in multimedia software. Virtual labs can illustrate the interplay between complex phenomena and related symptoms, and support the development of problem-solving skills. In true virtual reality (VR), clinical skills and surgical procedures can be practiced in a highly immersive environment, and even the sense of touch—haptics—can be simulated. Just as an airline pilot trains using flight simulators in order to be fully versed in all likely scenarios, so must all students and professionals who will be working with patients achieve the required level of mastery. Simulations can help guarantee this.

Ethically Sourced Animal Cadavers and Tissue All future veterinarians will require hands-on experience with animals and animal tissue. The use of ethically sourced cadavers and tissue is an alternative to the killing of animals for dissection and surgery practice. The term ethically sourced refers only to cadavers or tissue obtained from animals who have died naturally or in accidents, or who have been euthanized due to terminal disease or serious injury. Body donation programs can provide cadavers in an ethical way.

Clinical Work with Animal Patients Student access to clinical learning opportunities could be significantly increased in order to replace animal experiments and to better prepare students for their professions. A progressive approach to learning veterinary surgery might involve mastering basic skills using non-animal alternatives, then using ethically sourced cadavers for experience with real tissue,

and finally performing supervised work with animal patients. Shelter sterilization programs are an important potential resource; students can observe, assist and then perform castrations and spays. The clinic can also teach many other skills that the lab cannot, such as post-operative care and supporting the recovery of patients, reflecting a growing awareness that caring is a clinical skill.

Student Self-Experimentation For further experience of the whole, living body, the consenting student is an excellent experimental animal, particularly for physiology classes. The intense involvement and self-reference of such practical classes makes them highly memorable and supports effective learning.

In Vitro Labs The rapid development and uptake of in vitro technology in research and testing needs to be supported by student familiarity with the technique. Animal tissue and cells used for in vitro practical classes can be sourced ethically, and within some biology practical classes, the use of animal tissue can be replaced directly with plant material.

Field Studies Students may study animals in a laboratory setting as a model for nature, or they may face invasive or otherwise harmful interactions with wild animals. However, biology is not just experimentation, nor does its study require harm. Studying animals within their natural environment can be a particularly rewarding alternative.

The use of the above replacement alternatives illustrates the potential of humane education to transform a negative relationship between students and animals into a positive one.

Efforts to Offer Alternatives

The International Network for Humane Education (InterNICHE) works with teachers to introduce alternatives, and with students to support freedom of conscience. Resources developed by InterNICHE to catalyze change include: the multi-language book and database *From Guinea Pig to Computer Mouse* (2003), which presents case studies, information on curricular design and assessment, and details of over 500 alternatives; several Alternatives Loan Systems or libraries of mannequins, simulators, and software; the Humane Education Award, an annual grant program to support the development and implementation of alternatives; the information-rich Web site www.interniche.org; and InterNICHE conferences, outreach visits, and training around the world.

Alternatives to harmful animal use are possible for all practical classes within the life science disciplines. In many departments, the word alternative may not even be used because these are increasingly becoming the standard teaching approaches—and in some cases examples of best practice—often backed by laws and regulations stating that alternatives should be used wherever possible. The multiple positive impact of alternatives means that replacement is to the benefit of students, teachers, animals, the life sciences, and society itself. Further effort is required to replace the remaining harmful animal use internationally, but increasing success with the implementation of alternatives illustrates how science and ethics can indeed be fully compatible.

See also Dissection in Science and Health Education; Dissection, Student Objections to

Further Reading

Jukes, N., Chiuia, M., eds. 2003. *From guinea pig to computer mouse: Alternative methods for a progressive, humane education,* 2nd ed. Leicester, UK: InterNICHE.

Jukes, N., Martinsen, S. 2008. Three's a crowd: The 1R of replacement for education and training. In Proceedings of the 6th World Congress on Alternatives and Animal Use in the Life Sciences. *AATEX 2008;14(Special Issue)*:291–293.

Martinsen, S. 2008. Training the animal doctor: Caring as a clinical skill. In Proceedings of the 6th World Congress on Alternatives and Animal Use in the Life Sciences. *AATEX 2008;14(Special Issue)*:269–272.

Nick Jukes

ALTERNATIVES TO ANIMAL EXPERIMENTS: REDUCTION, REFINEMENT, AND REPLACEMENT

The concept of alternatives, or the Three Rs—reduction, refinement, and replacement of laboratory animal use—first appeared in a book published in 1959 entitled *The Principles of Humane Experimental Technique.* The book, written by two British scientists, William M. S. Russell and Rex Burch, was a report of their scientific study of humane techniques in laboratory animal experiments, commissioned by the Universities Federation for Animal Welfare (UFAW). In this book, Russell and Burch hypothesized that scientific excellence and the humane use of laboratory animals were inextricably linked, and proceeded to define in detail how both of these goals could be achieved through reduction, refinement, and replacement of animal use. Russell and Burch's work had relatively little impact upon the scientific community for almost two decades. In 1978, physiologist David Smyth conducted a survey on the Three Rs for the Research Defense Society in England and wrote the book *Alternatives to Animal Experiments,* in which he used the term *alternatives* to refer to the Three Rs. Thereafter, for those familiar with the concept, the Three Rs have become interchangeable with the word *alternatives*. In some circles, however, the word alternatives is understood to signify only replacement. Hence, in order to avoid possible misinterpretations, one of the Three Rs should precede the term alternative when discussing specific methods (reduction alternative, refinement alternative, or replacement alternative).

Definition of the Three Rs

A reduction alternative is a method that uses fewer animals to obtain the same amount of data or that allows more information to be obtained from a given number of animals. The goal of reduction alternatives is to decrease the total number of animals that must be used. In fact, reduction means better experimental design. Much progress has been made in reducing the number of animals required for product safety testing. This is partially due to the development of substantial databases as well as to the use of non-animal methods such as cell culture to prescreen for potential harmful effects. Most companies try to obtain as much information about their products as possible before they test them in animals. This has led to a large reduction in animal use.

In doing research, scientists can decrease the number of animals they use by appropriate experimental design of their experiments and by more precise use of statistics to analyze their results. Researchers can also reduce the number of experimental animals by using ever-evolving cellular and molecular biological methods. These systems are sometimes more suitable for testing hypotheses and for gaining substantial information prior to conducting an animal experiment.

Refinement alternatives are methods that minimize animal pain and distress, enhance animal well being, or use animals considered to be lower on the phylogenic scale. An important consideration in developing refinement alternatives is being able to assess the level of pain an animal is experiencing. In the absence of good objective measures of pain, it is appropriate to assume that if a procedure is painful to humans, it will also be painful to animals. Refinement alternatives include the use of analgesics and/or anesthetics to alleviate any potential pain.

Animals can also experience distress when they are unable to adapt to changes in their environment, such as might be caused by frequent handling or by experimental procedures. Refinement alternatives, such as properly-taught handling techniques that decrease distress, can significantly contribute to the welfare of laboratory animals. Animal welfare may also be enhanced by enriching the environment of the animals during the times when they are not undergoing experimental procedures. Such enrichment can range from placing species-appropriate objects for play and exploration in animal cages to group housing of social species.

Replacement alternatives are methods that do not use live animals, such as in vitro systems. The term in vitro literally means "in glass," and refers to studies carried out on living material or components of living material cultured in Petri dishes or in test tubes under defined conditions. These may be contrasted to in vivo studies, or those carried out in the living animal. Certain tests that were once done in live animals, such as pregnancy tests, have been completely replaced by in vitro tests. Other types of in vitro systems include the use of human cells in culture or human tissue obtained from surgeries and other medical procedures. In addition to replacing animals, these studies can directly provide valuable information about humans, which cannot be obtained from some animal models.

Other examples of replacement alternatives are mathematical and computer models, use of organisms with limited sentience such as invertebrates, plants and micro-organisms, and human studies, including the use of human volunteers, post-marketing surveillance and epidemiology.

The Future of the Three Rs

The Three Rs of reduction alternatives, refinement alternatives, and replacement alternatives are seen as mainstream concepts through which scientists can achieve optimal scientific goals while taking the maximal welfare of animals into consideration. In doing so they are seen by many to be the middle ground where scientists and animal welfare advocates can meet to reconcile the interests of human health and animal well-being. Those interested in promoting the Three Rs have begun a series of World Congresses on Alternatives and Animals in the Life Sciences, the first of which took place in Baltimore, Maryland in 1993 and the sixth in Tokyo, Japan in 2007. These meetings provide a forum for scientists to participate in dialogues with the animal protection community

to focus not on the differences between the two groups, but on opportunities for collaborative efforts and shared concerns. Acknowledgment and implementation of the Three Rs will ensure that the only acceptable animal experiment is one that uses the fewest animals and causes the least possible pain or distress, is consistent with the achievement of a justifiable scientific purpose, and is necessary because there is no other way to achieve that purpose.

The issues of pain and distress are the focus of most laws pertaining to animal use in biomedical research. However, an overriding consideration is that the general public accept that animals have intrinsic value, and this recognition is a significant consideration in how animals can be used in biomedical research.

See also Toxicity Testing and Animals

Further Reading

Animal Welfare Information Center and Universities Federation for Animal Welfare, 1995.

Balls, M., Goldberg, A. M., Fentem, J. H., Broadhead, C. L., Burch, R. L., Festing, M.F.W., et al. 1995. The Three Rs: The way forward. *Alternatives to Laboratory Animals* 23:838–866.

Environmental Enrichment Information Resources for Laboratory Animals: 1965–1995. Birds, cats, dogs, farm animals, ferrets, rabbits, and rodents. U.S. Department of Agriculture, Washington, D.C.

Experimental Design and Statistics in Biomedical Research. 2002. *ILAR Journal,* Vol. 43.

Gardner, R. and Goldberg, A. M. 2008. Proceedings of the 6th World Congress on Alternatives and Animals in the Life Sciences, Tokyo, 2008.

Goldberg, A. M. and Thomas Hartung. 2006. *Scientific American* (Nov. 2006).

Russell, W.M.S., Burch, R. L. 1959. *The principles of humane experimental technique.* London: Methuen. (Reprinted by the Universities Federation for Animal Welfare, UK: Potters Bar, Herts, 1992).

Smyth, D. 1978. *Alternatives to animal experiments*. London: Scolar Press.

Zurlo, J., Rudacille, D., Goldberg, A.M. 1994. *Animals and alternatives in testing—History, science and ethics*. New York: Mary Ann Liebert.

Joanne Zurlo and Alan M. Goldberg

THE AMERICAN SOCIETY FOR THE PREVENTION OF CRUELTY TO ANIMALS (ASPCA)

The American Society for the Prevention of Cruelty to Animals, or the ASPCA as it is known, was the Western Hemisphere's first humane society and was founded by Henry Bergh on April 10 1866. Shortly after its founding it served as the inspiration and model for the formation of SPCAs and humane societies across the country.

Bergh was the son of a wealthy New York City shipbuilder who enjoyed travel and the theater. While serving as a diplomat in St. Petersburg, Russia he was inspired to dedicate the rest of his life to the protection of animals. On his return trip to the United States he stopped in London to meet with representatives of the Royal Society for the Prevention of Cruelty to Animals to learn how their organization functioned.

Shortly after his return to New York City, he organized a meeting of influential business and political leaders at Clinton Hall on February 8 1866. Bergh gave a speech enumerating the many terrible deeds done to animals, the important role that animals played, and the need for a society to protect them. The original charter for the ASPCA listed the names of many prominent New Yorkers, including Horace Greeley, members of the Rockefeller family, and the mayor of New York City. Just nine days after the charter was granted by the New York State Legislature, Bergh convinced the legislature to pass an anti-cruelty law that gave the new society the authority to enforce it.

From the very start the ASPCA was active in publicizing the plight of animals and intervening on their behalf. One of the first cases that Bergh and the new ASPCA brought before the courts was that of a cart driver beating his fallen horse with a spoke from one of the cart wheels. This event would eventually be depicted in the seal adopted by the ASPCA, showing an avenging angel rising up to protect a fallen horse.

Within the first year, Bergh and the ASPCA would address many of the same questions that continue to occupy the efforts of his successors at the ASPCA and

Henry Burgh, angered at seeing horses mistreated on the streets of New York, founded the American Society for the Prevention of Cruelty to Animals in 1866. (AP Photo)

other humane societies including the treatment of farm animals, dogfighting, horses used to pull trolleys, turtles transported for food, and vivisection.

Recognizing the difficulty of coordinating the efforts of a far-ranging national organization, Bergh encouraged and helped others to start independent SPCAs across the country. The ASPCA became the model for hundreds of others societies, many of them using a variation of the SPCA name, the charter, and even the seal. The first such society was founded in 1867 in Buffalo, New York and included Millard Fillmore, C. J. Wells and William G. Fargo among it supporters. Boston, San Francisco, and Philadelphia soon followed.

Bergh's aggressive tactics soon earned him a host of enemies. The carting and transportation companies that depended on horses, butchers, dogfighters, and gentlemen's fox hunting organizations soon sent up an outcry that the ASPCA was interfering with their business and affairs. By 1870 Bergh and the ASPCA were hard pressed to defeat efforts to limit its charter and weaken the anticruelty laws.

The issues in these early years were frequently played out in the pages of the newspapers. Stories about the ASPCA's arrests, court cases and rescues of animals were given great attention. In addition, Bergh wrote many letters to the papers to explain the actions of the ASPCA and to point out problems that needed to be addressed. The newspapers were soon in the middle of a long feud between two of America's most famous men, Henry Bergh and P. T. Barnum. Bergh would attack Barnum on the care provided for the animals in his menagerie and performing in his shows. Barnum would defend his practices and use the publicity from the dispute to attract even larger crowds. Over time, Barnum would become a grudging admirer of Bergh and the work of the ASPCA, eventually helped to form an SPCA in Connecticut.

In 1873, Henry Bergh and the ASPCA's attorney, Elbridge Gerry, helped to rescue a young girl from an abusive home. The "Mary Ellen case" would lead to the myth that Bergh had claimed she deserved at least the same protection provided for animals. While the myth was unfounded, the case did, however, lead to the formation of the Society for the Prevention of Cruelty to Children, and the movement for child protection.

The ASPCA helped to change the way that Americans thought about animals. The organization also helped to introduce a number of innovations that provided for their care and protection. Bergh helped to design and introduce an ambulance for horses, and promoted an early version of the clay pigeon instead of live pigeons as a target for shooters. Further innovation continued into the 1950s, when the ASPCA helped with the design and implementation of equipment for the humane slaughter of animals for food.

Its hands-on services in New York City would grow to include an animal hospital and animal shelters. For one hundred years, from 1894 to 1994, the ASPCA would provide animal control services for the City of New York. During this time, hundreds of thousands of animals would be rescued by ASPCA ambulances, treated in clinics, sheltered, and placed in new homes whenever possible. Before the ASPCA assumed the animal control duties for New York City, unwanted dogs were drowned in an iron cage lowered into the river. During the following century, methods employed to euthanize unwanted dogs and cats would evolve from the use of gas, to decompression chambers, and ultimately to sodium pentobarbital injection. At the same time the promotion of responsible care of companion animals, including spaying and neutering, helped to reduce the numbers of animals euthanized by 99 percent.

In August 1996, the ASPCA negotiated with the University of Illinois to acquire the National Animal Poison Control Center. This is the nation's only 24/7 animal poison control center, staffed full-time by specialists in veterinary toxicology. Staff will typically answer over 125,000 calls from veterinarians and members of the public, providing expert advice for dealing with exposure to various toxins. In 2007 the ASPCA Animal Poison Control Center (APCC) found itself at the center of the largest pet food recall in history. Beginning in February 2007, pets around the country were getting sick after eating one of what turned out to be a wide range of pet food brands. By the middle of March, Menu Foods recalled over 60 million packages of cuts and gravy-type foods from over 100 brands. Veterinary toxicologists at the APCC were in regular contact with veterinarians at the ASPCA's Bergh Memorial Animal Hospital, comparing the information that they were collecting from veterinarians and members of the public across the country with cases being treated at the hospital. They were able to provide substantial information to the veterinary profession and the public on the symptoms to look for, and aggressive treatment protocols for animal that had eaten the food. Eventually it was discovered that Menu Foods had purchased wheat gluten from China that had been adulterated with melamine and cyanuric acid to enhance its nitrogen/protein profile. When mixed into the pet food to help thicken the gravy, and then consumed by the pets, the melamine and cyanuric acid would react in the kidney to form crystals that would block kidney function, sickening and killing the animals.

In 2007 the ASPCA launched a vigorous community-based program called ASPCA Mission: Orange. The focus of the effort was to develop community collaborations to address issues that put companion animals in the designated communities at risk. The first group of cities included Austin, Texas, Spokane, Washington, Tampa, Florida, Gulfport-Biloxi, Mississippi, and Philadelphia, Pennsylvania. The ASPCA provided financial grants and staff leadership to evaluate the unique needs of each community and work with community leaders to establish programs to address the issues that put animals at risk.

In 2006 the ASPCA expanded its national anti-cruelty training and support programs. In 2007 that included the addition of veterinary forensic services, including the nation's only mobile Crime Scene Investigation unit dedicated to animal cruelty cases. This service proved invaluable during the investigation of dogfighting charges against professional football quarterback Michael Vick. ASPCA staff assisted federal authorities in the investigation, including examining the remains of dogs found on the site of a suspected dogfighting and breeding operation. When Vick and his co-defendants pleaded guilty to federal charges, the ASPCA was called upon to provide behavioral expertise to evaluate the dogs seized during the investigation, and make recommendations for their disposition. Approximately 50 dogs were evaluated, and all but one were found suitable for placement in either foster care or sanctuary facilities.

The ASPCA continues as one of the world's largest humane societies. It still operates animal hospitals and shelters in New York City, and its humane law enforcement agents enforce the anti-cruelty laws in New York State. The ASPCA also promotes education and legislative activities that fulfill the original mission described for the organization by its founder Henry Bergh, ". . . to provide effective means for the prevention of cruelty to animals throughout the United States."

See also Animal Protection: The Future of Organized Activism; Royal Society for the Prevention of Cruelty to Animals (RSPCA) History

Further Reading

Franz, William C. 1980. Bergh's War: The first crusade for animal rights. *The Elks Magazine* (October 1980).

Lane, M. and S. Zawistowski. 2008. *Heritage of care.* Westport, CT: Praeger.

Loeper, John J. 1991. *Crusade for kindness: Henry Bergh and the ASPCA.* New York: Atheneum.

Pace, Mildre Mastin. 1995. *Friend of animals.* Ashland, KY: The Jesse Stuart Foundation. (Original publication Charles Scribner's Sons, New York, 1942).

Steele, Zulma. 1942. *Angel in a top hat.* New York: Harper & Brothers Publishers.

Turner, James. 1980. *Reckoning with the beast.* Baltimore: Johns Hopkins University Press.

Stephen L. Zawistowski

AMPHIBIANS

Many biologists today are concerned by evidence that populations of amphibians around the world are declining and the welfare of amphibians is seriously affected in their natural habitats by human-caused environmental deterioration. Because the skin of amphibians is not readily resistant to water loss, most species are restricted to streams and ponds or to moist terrestrial and arboreal habitats. The moist skin of amphibians may also make them more vulnerable to injurious ultraviolet rays and chemical pollution than other groups of vertebrates with better skin protection. There is general concern that major global changes in the environment may be specifically injuring amphibian populations throughout the world. For example, ultraviolet (UV) radiation is harmful to humans, and the middle part of the spectrum (UV-B) is particularly dangerous. Recent evidence has shown that the eggs of some species of frogs and toads are very sensitive to UV-B, with high mortality within egg clutches exposed to this radiation. This raises fears that the current reduction in the ozone layer around the earth may subject amphibians to increased levels of UV-B.

There are three groups of amphibians: caecilians, salamanders, and frogs. Caecilians are earthworm-like amphibians that occur in aquatic and terrestrial habitats in Asia, Africa, and America. Little is known about their biology. Therefore, populations may or may not be declining.

About 400 species of salamanders occur in Asia, Europe, North America, and northern South America. Some species are entirely aquatic, living in streams, rivers, or ponds. Other species are semi-aquatic or consist of aquatic larvae and terrestrial adults, while yet others are strictly terrestrial, inhabiting burrows in the soil, or strictly arboreal. The arboreal species, though less well studied, are probably suffering from deforestation in Central and northern South America. Adult males and females of terrestrial species are territorial, defending feeding areas under rocks and logs, and they are aggressive toward some other species of *Plethodon* that appear to be declining. Terrestrial salamanders may not be greatly affected by UV-B or by airborne pollution, due to the buffering influence of the soil.

Streamside salamanders live in habitats that are flushed by flowing water, and thus they too may be relatively protected from airborne pollution, such as acid rain, but not necessarily from UV-B. The salamanders that may be most affected by pollution and UV-B are those that either live in ponds as adults or breed in ponds and produce aquatic larvae. If worldwide changes in the environment are occurring, the welfare of pond species might be most at stake.

Vance T. Vredenburg, a researcher from the University of California, Berkeley, weighs a mountain yellow-bellied frog caught in a pond in the Sierra Nevada near Ebbetts Pass, California. Vredenburg has been studying the decline of the mountain yellow-legged frog. (AP Photo/Rich Pedroncelli)

About 4,000 species of frogs occur throughout North and South America, Europe, Asia, and Australia. They inhabit arboreal, terrestrial, semi-aquatic, and aquatic habitats. As with the salamanders, considerable attention has been focused on pond-breeding species with regard to the injurious effects of pollution (such as acid rain) and UV-B radiation.

Because of the decline of numerous species of amphibians in nature, scientists who study amphibians in the laboratory have had to reevaluate the ethics of using large numbers of individuals in research or in teaching. For example, a biologist who wishes to conduct an experiment can often estimate just how many frogs or salamanders are needed to obtain significant results; that biologist can then collect or purchase just the minimum number of animals needed to perform the experiment effectively. In the laboratory, animals can often be housed in individual containers, thus reducing the potential for mortality caused by the spread of infections and contaminants. Another tactic used by laboratory biologists is to cycle the same frogs or salamanders through a series of experiments, rather than obtaining a different set of animals for each individual experiment. This is not always possible when, for instance, surgery is required, but cycling animals among behavioral or ecological experiments is often feasible.

Concern about amphibians takes two basic forms: concern about their welfare in nature and, given the decline of once-abundant species, the treatment of these animals in the laboratory. More and more species are being listed as threatened or endangered, and these designations should help to improve awareness and reduce local human-induced impacts on their populations. Such restrictions will

18 | *Animal Body, Alteration of*

also limit the number and kinds of species that can be used in biological research.

See also Reptiles

Further Reading

Blaustein, A. R. 1994. Amphibians in a bad light. *Natural History Magazine* 103(10): 32–39.

Buchanan, B. W., and R. G. Jaeger. 1995. Amphibians. In B. E. Rollin and M. L. Kesel (eds.), *The experimental animal in biomedical research,* vol. 2 (Boca Raton, FL: CRC Press), 31–48.

Duellman, W. E., and L. Trueb. 1994. *Biology of amphibians*. Baltimore: Johns Hopkins University Press.

Mathis, A., R. G. Jaeger, W. H. Keen, P. K. Ducey, S. C. Walls, and B. W. Buchanan. 1995. Aggression and territoriality by salamanders and a comparison with the territorial behavior of frogs. In H. Heatwole and B. Sullivan (eds.), *Amphibian biology,* vol. 2, *Social behavior* (Chipping Norton, Australia: Surrey Beatty and Sons), 633–676.

Stebbins, R. C., and N. W. Cohen 1995. *A natural history of amphibians*. Princeton: Princeton University Press.

Zug, G. R. 1993. *Herpetology: An introductory biology of amphibians and reptiles*. New York: Academic Press.

Robert G. Jaeger

ANIMAL BODY, ALTERATION OF

People have been adorning and modifying their bodies for thousands of years, most likely since we first evolved as humans. All societies everywhere physically alter their bodies in an attempt to meet cultural standards of beauty as well as their religious and social obligations. In addition, since the earliest farmers first domesticated animals around 10 thousand years ago, humans have been modifying, and sometimes adorning, the animal body as well. Here we refer to the physical alteration of animal bodies through selective breeding, surgery, tattooing, branding, genetic modification, cloning, and other practices.

Since the first animals were domesticated for food, labor, and their skins, domesticated animals have changed in a whole host of ways, both behaviorally and physically. Natural selection has favored those traits that made individual species, and individual animals, good prospects for domestication—lack of fear, curiosity, relatively small size, and gregariousness, for example—making the earliest domesticates look and behave differently from their wild relatives.

Of course once humans began selectively breeding their animals (and killing those whose bodies or temperaments were unwelcome) in order to emphasize or discourage certain traits, the animals changed even further, resulting today in animals who are, for the most part, smaller (yet fleshier), more brightly colored, with shorter faces, rounder skulls, and more variations in fur and hair type as well as ear and tail appearance. They also became tamer, friendlier, and more dependent on the humans who cared for them.

As farmers and, later, show breeders, learned more about the inheritance of traits, animal breeders began selectively breeding their animals for more specific characteristics, such as overall size, fur and wool color or texture, ear and tail shape, and more. Termed artificial selection by Darwin, selective breeding has led to the creation of hundreds of breeds of dogs, one of the most intensively bred animals in the world. Using dogs as an example, breeds were created in order to fulfill human desires. Some breeds were created to retrieve ducks during a hunt, others were created to herd sheep, and still others were created to race.

With the advent of industrial methods of food production in the 20th century, changes in livestock breeds accelerated. To produce the most meat in the shortest amount of time, animal agribusiness companies bred farm animals such as pigs and chickens to grow at unnaturally rapid rates. These changes have been encouraged by new developments in agricultural science aimed at improving the productivity of food animals. For example, U.S. beef cattle are routinely administered hormones to stimulate growth, and to increase milk yield, producers often inject dairy cows with hormones.

Since the early part of the 20th century, farmers have been experimenting with creating new livestock breeds, via careful cross-breeding, in order to maximize size, fat composition, productivity, or other traits. Since the development of artificial insemination and the ability to freeze semen, cattle farmers are able to more selectively breed their prized bulls and cows to replicate the traits of the parents.

The pet and show industries, too, rely on artificial selection (and today, following the livestock industry, artificial insemination) to create breeds of animals with favorable (to humans) traits. Recent years have seen an escalation in the varieties of dogs, cats, and other companion animals being developed in order to appeal to discriminating consumers.

While early breeds of dogs were created to highlight working traits, recent breeds have been geared more toward aesthetics. On the other hand, since cats are not working animals, most cat breeds have been created for aesthetic purposes, with an eye toward color, size, fur type, tail, ear, and body type. The result is hundreds of breeds of dogs, and dozens of breeds of cats, rabbits and other species, all bred by large and small breeders to sell through the pet industry. Another result is a whole host of health problems associated with these breeds. Dogs in particular are at risk for problems associated with the odd proportions in body, legs, and head that are bred into many of the breeds. Even without the specific genetic defects associated with certain dog or cat breeds, many modern breeds of dog or cat are unable to survive without close human attention. While dependency has been bred into domestic animals since the earliest days of domestication, it has accelerated in recent years with the production of animals such as Chihuahuas, *WHO are physically and temperamentally unsuited to survival outside of the most sheltered of environments.

Another form of artificial selection refers to breeders' emphasis on deleterious traits in the breeding process. Japanese Bobtails (cats with a genetic mutation resulting in a bobbed tail), hairless cats, and Scottish Folds (who have folded down ears) are examples of this type of breeding. More disturbing are cats that go by the name of Twisty-Cats, or Kangaroo Cats, all of whom have a genetic abnormality which results in drastically shortened forelegs or sometimes a flipper-like paw rather than a normal front leg, and who are being selectively bred by a handful of breeders.

Genetic manipulation of animals represents a new scientific development that has irreversibly changed animal bodies. Because pigs, beef cows, and chickens are created for one purpose—food consumption—their genes have been altered in a whole host of ways to suit that purpose, resulting in, for example, pigs engineered to have leaner meat, tailor-made to suit a more health-conscious consumer.

Genetically engineered animals are also becoming more popular among scientists who experiment on or test animals. Genetically modified mice and rats are especially popular, allowing researchers to study the ways that genes are expressed and how they mutate. Genetic engineering has even found its way into the pet world, with the production of a new hypoallergenic cat (selling for 12–28 thousand dollars), created by manipulating the genes that produce allergens.

In terms of reproduction, cloning animals is the wave of the future, allowing humans the greatest level of control over animal bodies. Thus far, the livestock industry has been most active in the use of cloning, reproducing prized breeder animals in order to ensure higher yields (in meat, wool, etc.) by cloning only very productive animals, but cloning is found in the vivisection and pet industries as well. Laboratory scientists are also cloning mice, rabbits and other laboratory animals in order to ensure that the animals used in research are genetically identical, and to control for any imperfections. In the pet world, cloning has been less successful, but a handful of companies today either offer cloning (for cats; dogs have not yet been cloned) or tissue-freezing services for those animals which cannot yet be cloned.

Another way that animal bodies have been changed is through surgical procedures. Because the control of animal reproduction is critical to keeping domestic animals, castration has been used for thousands of years to ensure that undesirable animals cannot breed, or to increase the size or control the temperament of certain animals. Castration methods include banding (in which a tight band is placed around the base of the testicles, constricting blood flow and eventually causing the scrotum to die and fall off after about two weeks), crushing (this method uses a clamping tool called a Burdizzo, which crushes the spermatic cords) and surgery (in which the testicles are removed from the scrotum with a knife or scalpel). In the 20th century, with the keeping of companion animals rising in popularity, surgical techniques to remove the uterus and ovaries of female animals were developed, and spaying is now an extremely common surgery for companion animals, although it is very rarely performed on livestock. Castrated animals are often referred to by different names than intact males, using names such as *ox, bullock* or *steer* for cattle, *barrow* for pig, *wether* for sheep, and *gelding* for horse.

Other forms of surgical modification have also been common for years, particularly in livestock and purebred companion animals. For example the last century has seen a number of procedures performed on livestock as a result of the close confinement necessitated by factory farm production. The debeaking of hens (amputating, without anesthesia, the front of the chicken's beak) is common in the egg industry, where chickens are so intensively confined in tiny cages that they may attack each other due to stress and overcrowding. Even in situations where livestock is not as closely confined, farmers often remove body parts. One mutilation that's increasing in popularity is tail-docking of dairy cows, in which producers amputate up to two-thirds of the tail, usually without painkillers. Cattle are often dehorned, and sheep often have their tails removed (usually via banding, also without anesthesia).

In the pet breeding world, companion animals undergo surgical procedures in order to make them conform to the artificial requirements of the breed. Breed standards demand that certain dogs, for example, must have their tails docked, their ears cropped, or both. In addition, many companion animals today experience surgical procedures which are used to control unwanted (by humans) behavior. Some people, for example, have their dogs de-barked (by cutting their vocal cords) in order to reduce barking, and many cat owners elect to have their cats declawed (which involves amputating the front portions of a cat's toes) in order to prevent harm to their furniture.

Identifying animals in order to determine ownership is important in the livestock, animal science, and pet worlds. Branding is the oldest form of marking ownership on animal bodies, and has been used since the ancient Greeks, Egyptians and Romans marked both cattle and human slaves with iron brands. Still popular amongst cattle ranchers today, brands are used to prevent theft, to identify lost animals, to mark ownership, and to identify individual animals. Some horses are branded as well, either because they are very expensive or, in the case of some wild American horses, because they are federally protected. Today, freeze brands (which freeze, rather than burn the skin), ear tags, tattoos, and microchips are often used instead of brands for livestock, laboratory animals, and companion animals.

While much less common than the above forms of modification, animals are also, occasionally, subject to tattooing, piercing, or hair dying not for practical purposes, but for aesthetic reasons. The most common form of adornment for animals is found in the show and pet dog worlds, where long-haired breeds of dogs have their hair professionally cut and styled, often with ornaments like barrettes and other accessories. Poodles, in particular, are expected to have a certain look which must be maintained via often rigorous grooming.

In the United States, in the heyday of the circus and carnival sideshow, tattooed families were a popular sideshow attraction, and they often included a tattooed dog. Today, some people involved in the body modification community pierce or tattoo their own pets, although most tattooists and piercers do not appear to condone these procedures (which, after all, do hurt). Here, as with people tattooing or piercing themselves, the tattoos are ostensibly marks of individuality (although they likely reflect the owner's personality more than the dog's) and, like branding, marks of ownership as well.

Some people also dye their animals' hair, usually for a special event. Feed stores around the country routinely sell dyed chicks and baby bunnies for Easter, for example, and some pet owners dye their own animals' fur for holidays like St. Patrick's Day, either with commercially produced pet fur dye, or products like food coloring. Finally, in recent years, evocative photos of "painted cats" began to appear, which showed cats with intricate designs painted on their bodies. While the photos turned out to be Photoshopped, they continue to circulate on the Internet, inciting awe, outrage, and interesting discussions regarding what humans can or should do with animals.

Further Reading

Clutton-Brock, Julia. 1987. *A natural history of domesticated mammals.* Cambridge: Cambridge University Press.

DeMello, Margo. 2007. *Encyclopedia of body adornment: A cultural history.* Westport, CT: Greenwood Publishing.

Silver, Burton and Heather Busch. 2002. *Why paint cats: The ethics of feline aesthetics.* Berkeley: Ten Speed Press.

Williams, Erin and Margo DeMello. 2007. *Why animals matter: The case for animal protection.* Amherst, NY: Prometheus Books.

Margo DeMello

ANIMAL LIBERATION ETHICS

At the core of animal liberation ethics is an argument from consistency directed against the contemporary view of egalitarianism. This view claims that all human beings are equal whatever their gender, race, or psychological traits, such as intelligence, skills, and sensitivity. It rejects the view that the members of a particular biological group may be discriminated against because they belong to that group, and it considers ethically offensive the idea that the intellectually less endowed, the disabled, small children, or the elderly may be routinely taken advantage of by more rational or autonomous human beings. Thus neither biological characteristics nor particular psychological properties over and above sentience are important for equal treatment. If we are ethically required to treat like cases alike, as ethicists since Aristotle have urged, then the moral status of members of other species should be the same as the moral status of members of our own species at a similar psychological level. This means giving basic rights to most of the individual animals whom humans use as means to their ends.

Animal liberation ethics, which became important in the 1970s, was perceived as subversive to received theory and practice. In response to its challenge, defenders of humanism—the view that human lives and interests should always be given greater weight than nonhuman lives and interests—offered a number of objections. They claimed that humans have special duties toward their closest kin; that, in contrast with race, species differences corresponded with significant differences; that it is not possible to have rights without the capacity to claim them; that it is not possible to have rights without the capacity to have duties; and even that nonhuman animals, lacking verbal language, have no conscious interests that need to be taken into consideration. Such objections can be rebutted. To begin with, the notion of closest kin can be used to justify discrimination against members of the human species as well as members of other species. Even if race does not correspond with significant differences, gender does. Also, we grant basic rights to small children, although they certainly cannot claim them or have duties. Finally, the theory of evolution has wiped out the traditional notion of fixed, totally distinct essences; since Darwin, the idea of differences in kind rather than in degree between us and all other animals is unlikely. Even the (highly controversial) appeal to the potential for becoming full rational beings in order to draw a line between human infants and nonhuman animals at a similar mental level overlooks the fact that there are human beings whose mental disabilities cannot be reversed.

All things considered, those who argue against speciesism believe that there is no argument for discrimination between members of different species that could not be used as an argument for discrimination among humans. Justifications for equality cannot be accepted only up to a point and then arbitrarily rejected. In highlighting the arbitrariness of the humanist position, animal liberation ethics not only seeks to protect nonhuman beings, but also challenges the direction

Masked animal activist holds a monkey who was once used for laboratory experimentation. (Animal Aid)

and basis of much of Western moral thinking.

From this perspective, the request to remove other animals from the realm of *things* in order to include them in our own moral community, and the goal of dismantling the social institutions and practices that are based on their exploitation for human ends, are part and parcel of that slow but steady process of enfranchisement which has until now marked what we call moral progress.

See also Animal Rights; Animal Rights, Abolitionist Approach; Animal Rights Movement, New Welfarism; Animal Welfare and Animal Rights, A Comparison; Evolutionary Continuity; Speciesism

Further Reading

Cavalieri, Paola, and Peter Singer (eds.). 1993. *The Great Ape Project: Equality beyond humanity* London: Fourth Estate.

Cavalieri, Paola, and Will Kymlicka. 1996. Expanding the social contract. *Etica & Animali* 8: 5–33.

Regan, Tom. 1983. *The case for animal rights.* Berkeley: University of California Press.

Sapontzis, Steve F. 1987. *Morals, reason, and animals.* Philadelphia: Temple University Press.

Singer, Peter. 1990. *Animal liberation.* New York: New York Review of Books.

Singer, Peter. 2005. *In defense of animals. The second wave.* Oxford: Blackwell.

Paola Cavalieri

ANIMAL MODELS AND ANIMAL WELFARE

Animals serve in laboratories as models of human biology and medicine. Controversy exists as to whether such use is scientifically sound; if it is not, then it would not be ethically justifiable. Furthermore, there is controversy over whether animal use would be justifiable even if the science is good. Consumers can try to avoid cosmetics and some other products that have been safety-tested on animals. In contrast, few medicines are developed without the use of animals as models.

Animal modeling is more complex than it at first would appear. Animals cannot be thought of as miniature people, identical in every way but size and language. Not even humans' closest relatives, the great apes, can be seen as substitute people. Rather, animals must be carefully chosen to model some particular aspect of human biology—not the *whole* of human biology. Data extrapolated from animals must be interpreted in this limited context. Over-interpretation of animal data invites criticism.

A bewildering array of animal species is pressed into service as models.

Rodents, rabbits, and primates may first come to mind, but horses, roundworms, fruit flies, zebra fish, songbirds, and many, many others model some aspect of human biology. Woodchucks, for instance, are susceptible to a virus similar to human hepatitis-B. Leprosy can be produced in nine-banded armadillos but in few other animals. Labrador retrievers develop a hip dysplasia that resembles human osteoarthritis. Squid nerve axons transmit nervous signals much as human nerves do. These are but a few of the thousands of ways in which animals are used as models for normal and diseased human biology.

Cell and tissue culture have made many uses of animals obsolete. After all, why use an animal as a model when actual human cells can be grown in the lab and studied? Typically, scientists use tissue-culture systems to study events at the cellular and subcellular levels. For example, tissue culture is used to study which types of cells HIV, the human immunodeficiency virus, is capable of infecting, and what events occur in the cell that eventually kill it. But when research requires studying the interaction of many different cells and tissues, such as how the immune system fails to protect the brain from the effects of the AIDS virus, or how medications will affect this, then scientists may turn to whole-animal models.

Many factors influence the choice of model. Animal welfare requires scientists to consider using less-sentient species when possible (such as fruit flies instead of mice or monkeys). Cost considerations push scientists to choose smaller animals with shorter life cycles for many studies. Data may be most easily obtained and analyzed from smaller, simpler organisms than from larger ones; thus, zebra fish are chosen for studies of organ development,

because the embryo is largely transparent and develops in an egg outside of the mother's body. On the other hand, larger size is sometimes required, such as when surgeons develop new techniques by using pigs. To best interpret data in light of what is already known, scientists will often choose the animal models most common to their fields, whether that original association was somewhat arbitrary (such as use of rats rather than hamsters in psychology experiments) or based on unique biological attributes (such as studies of vitamin C in guinea pigs, one of the few non-primate mammals to require vitamin C the way humans do). Increasingly, research requires knowing an animal's genetic makeup, so well-studied and easily modified species, especially mice, zebra fish, and fruit flies, have become more widely used.

It is controversial just how useful animal models are. Certainly no drug is marketed in the United States without having been studied in animals. Is this because there is always a biological need to use animal data to develop drugs, or simply because the law requires animal safety data to be submitted before a drug can be licensed?

Models may be classified in many ways. This essay looks at three broad categories of animal models: testing for product development, skills development, and induced and spontaneous models of disease.

Animal Models in Safety Testing

Using animals to test the safety of drugs, cosmetics, and environmental chemicals is what most people think of— and criticize—first. The crudest version of safety testing is to apply a compound to an animal—either acutely at high doses or at lower doses over a longer period of time—and watch for reactions. The reactions may be eye irritation, rashes, fetal deformities, cancers, or other toxicities. To move beyond a simple Safe/Not Safe determination, scientists developed more measurable practices, such as the Draize test for eye irritation or the Lethal Dose 50 (LD50) test. Both are ways of quantifying how much of a compound leads to what degree of injury. Both are still used today, though less than in the past.

This crude approach to safety testing may be criticized both for the suffering it inflicts on animals and for how reliably this information, especially the quantitative information, really applies to people. There is no guarantee that a compound that causes cancer in mice will do so in people, or that one that is safe in mice will not cause human disease. Moreover, small animals have much faster metabolisms and may have variations in the enzymes that process chemicals, so the *amount* of compound that is safe or dangerous may be different for a mouse than it would be for a person.

Safety testing docs not usually generate truly new biological information and, for this reason, replacing animals in safety testing is a more realistic goal than replacing them in original research. The Center for Alternatives to Animal Testing was started in 1981, at Johns Hopkins University, to develop alternatives to these types of testing. It is necessary that animal alternatives, such as development of an artificial skin to replace guinea pigs in testing for contact irritation, be evaluated scientifically and validated as useful before regulatory agencies, such as the Food and Drug Administration or the Environmental Protection Agency, will consider them an acceptable replacement for animal studies.

Animal Models for Skills Development and Teaching

Animals have been used in classrooms for years. They have been used to teach students anatomy or to demonstrate physiological functions, such as how the heart beats. In addition, animals have been used to allow human and veterinary surgeons-in-training to develop their manual skills before working on actual patients. For surgeon training, dogs and pigs have often been chosen, because their size more closely approximates a human patient than that of other animals. For microsurgical training—such as learning to repair blood vessels or nerves—smaller animals, such as rats, are often used. For most such training, healthy animals are used and then euthanized at the end of the training session.

As with safety testing, animal models are still used, but their classroom use has been decreased, because of greater reliance on alternative methods and models. Many medical schools have phased out animal use during the four-year MD curriculum, though animals may still be used in advanced surgical-residency training. Many veterinary schools allow students to opt out of classes that would require medically unnecessary surgeries to be conducted on healthy laboratory animals. Human and veterinary surgeons-to-be can acquire many of the basic skills of cutting tissues and placing stitches via the use of artificial (plastic, foam, etc.) models—and through practice on the cadavers of animals euthanized for other purposes. An imitation rat has been marketed for teaching microsurgical skill.

Surgical research differs from surgical training. Although large (humanlike) sizes may be useful for surgical training, they are less relevant in researching surgical concerns such as organ-transplant rejection, surgical infection, or healing processes. For these studies, rats and genetically modified (transgenic) mice are commonly used.

Animal Models of Disease

There is no field of human medical, surgical, or psychiatric research that does not include some use of animal models. Animals are used to study the normal, healthy biology relevant to disease processes, as well as to study the diseases themselves. How animals are used depends on what the scientist is trying to learn. For example, a scientist may cause a cancer in an animal by implanting some cancer cells into the animal's body. This will tell little about what causes cancer in people (people don't typically get cancers from transplanted cancer cells), but may be useful in studying some approaches to treating already-developing cancers. Conversely, a study on how influenza is transmitted may yield valuable information for preventing an epidemic, but may tell nothing about how to treat the infection once it has developed.

Spontaneous animal models of human disease are those that develop more or less naturally, possibly by genetic mutation. A mutation in the gene that codes for the molecule dystrophin, for example, knocks out that molecule's function, resulting in Duchene muscular dystrophy in dogs, people, and mice. Though the mutations arise naturally in these species, they are then continued through selective breeding. In this ways, colonies of dogs or mice with a predisposition to muscular dystrophy were developed for research. Other spontaneous animal models for study can include infections that develop and spread in wild, pet or food animal populations, or through accidents, injuries and poisonings that occur outside of the laboratory. Laboratory animals may be maintained into older age to study the conditions known to naturally arise in geriatric animals of a particular species or strain.

Induced animal models are those that start with healthy animals and then cause a disease in those animals in the laboratory. Cancers may be induced by exposure to chemicals, by irradiation, or through transplant of tumor cells. Infections may be caused by directly exposing an animal to a virus, bacteria, prion, or fungus. Psychiatric conditions may be caused by manipulating an animal's environment, by subjecting an animal to shocks or other stressors, or through injection of chemicals. Some conditions may be caused by surgically altering an animal, say, by creating an abnormal blood flow through an organ or removing some organ or gland entirely, such as in early studies of diabetes mellitus that involved removal of the pancreas from previously healthy laboratory dogs.

An increasingly active approach to animal modeling is through genetic modification of mice, zebra fish, rats, and other species. Genes may be introduced that will cause disease. One example is the "oncomouse," developed at Harvard University, into which a cancer-promoting oncogene was inserted, making the animal more prone to cancers. The opposite is to "knock out" a gene to cause disease: removing a functional gene, such as one that encodes a cell's insulin receptors, leading to diabetes in those mice unable to respond to their body's own insulin. Even more complicated is the ability to turn added genes on, or to knock out a gene's function, at any point in an animal's life, simply by adding a special chemical to the diet that the gene responds to.

Negative models, in which an animal fails to respond as a human might, can also be useful. Scientists find value in studying the small number of animals capable of being infected with HIV and susceptible of developing an AIDS-like condition. But there can also be reason to study animals (most apes and monkeys, for example) that are resistant to that virus, in order to figure out why they are resistant and to see what lessons that might hold for preventing human HIV infections.

No matter the animal model, none is a perfect replica of human health or disease. Those models that involve animal sickness or death—as most of them do—must be chosen only when a scientist is convinced no other method will answer important biological questions.

See also Alternatives to Animal Experiments in the Life Sciences

Further Reading

LaFollette, H., & Shanks, N. (1996). *Brute science: Dilemmas of animal experimentation.* London: Routledge.

Quimby, F. (2002). Animal models in biomedical research. In J. G. Fox, et al. (Eds.), *Laboratory Animal Medicine* (1185–219). Academic Press: New York.

Rowan, A. N. (1984). *Of mice, models, and men: A critical evaluation of animal research.* Albany: State University of New York Press.

Zurlo, J., Rudacille, D., & Goldberg, A. M. (1994). *Animals and alternatives in testing: History, science, and ethics.* New York: Mary Ann Liebert.

Larry Carbone

ANIMAL PROTECTION: THE FUTURE OF ACTIVISM

From its institutional beginnings in the second half of the 19th century to the period of the second World War, the humane movement focused on developing a worldwide network of societies for the prevention of cruelty to animals, which provided direct care of animals and a range of other services, pushed the passage of basic anti-cruelty laws in the United States and many other countries, promoted humane education as an instrument of childhood socialization, and advanced the notion that cruelty to animals is the sign of a socially maladapted personality. For the most part, however, these focus areas centered on the regulation or improvement of individual behavior, and organized animal protection achieved more limited gains in its efforts to confront cruelty by corporate or institutional actors.

The latter part of the 20th century witnessed a surge in worldwide activism on behalf of animals, with a more concentrated focus on institutional forms of cruelty and a commitment to changes in policy to address these large-scale contributors to animal mistreatment. Few people would quarrel with the idea that cruelty to animals is a serious matter. The difficulties come in applying anti-cruelty principles to legal, institutional uses of animals which, however abusive or harmful, have a wide array of corporate and political defenders.

As we examine the current state of the humane movement, it is obvious that we are situated in an odd and even contradictory place in history. There are more people and organizations devoted to helping animals, and extraordinary participation in pet keeping, wildlife watching, and other expressions of kinship or identification with animals—all of which manifest a deep appreciation and love for them. Yet, there is also more exploitation than ever—from staged animal fights to puppy mills, from trophy hunting to factory farms, from exotic pets to bush meat, from animal testing to tiger farming. Each industry, from animal cloning to Internet hunting, has its built-in defense mechanisms and its innovative means of exploitation.

With the rise of powerful new economies in China and other Pacific Rim nations, where humanitarian concerns hold little or no influence within institutional or cultural traditions, there are enormous challenges ahead for the movement. These problems are compounded by the lack of a free press in some nations and the absence of non-governmental organizations to drive reforms. Moreover, in a world beset by so many other pressing social and political concerns—war, resource scarcity, pandemics, and global financial crises—we face powerful competition for attention, capital, and human resources.

One lesson from animal protection's past is the need to establish humane, animal-friendly values permanently within relevant institutions of government and civil society. Within schools of social work and education, at veterinary and medical colleges, in wildlife and agricultural sciences departments, and in law schools, the movement must work to see that animal welfare concerns are sustained. The same is true for law enforcement and environmental protection agencies, and international regulatory bodies where animal welfare issues surface.

The humane movement must also reinforce the case for animal protection by continuing to draw the connections between cruelty to animals and other pressing social concerns. With the spread of disease and the danger of pandemics threatening humankind, we need a serious international campaign to stem the exotic animal trade and the cockfighting culture, wherever they thrive. With the metastasis of domestic violence undermining our families and communities, we need to ensure that people make the connection between cruelty to animals and interpersonal violence. With adulterated animal products finding their way into school lunch programs and other commodity programs sponsored by the federal government, we need to underscore the urgency for reform in food production and food policy.

Expanding the definition of corporate social responsibility to include animals, and shifting consumer preferences and corporate behavior toward cruelty-free or more humane choices, will be the key to many positive changes for animals. Recent developments in the farm animal welfare sector have validated this principle, as growing numbers of consumers opt for non-factory farm products and companies increasingly shift their purchasing preferences to less intensive production practices like cage-free or crate-free livestock. One of our great challenges will be to translate these trends to China and other developing nations, since global capitalism often migrates to areas that lack adequate regulatory standards. Humane values will not necessarily take hold in other markets solely because they have taken hold in the United States.

Throughout the world, farm animal welfare is inextricably bound to a broader debate over food and its relationship to public health, environment, energy use, and national security. In addition to animal protectionists, advocates for food reform, public health, small-scale farming, anti-hunger, and smart energy are also pressing for change. The humane movement is part of the larger pattern of growth for organics, sustainable agriculture, locavorism (eating only what grows locally), flexitarianism (a semi-vegetarian diet with occasional meat consumption), vegetarianism, and other manifestations of conscious eating. Animal agriculture is the subject of unprecedented scrutiny and criticism, in the wake of high-profile exposes (e.g., the HSUS investigation into a Southern California slaughter plant called the Hallmark Meat Company), and major reports from the FAO (*Livestock's Long Shadow*) and the Pew Commission (*Putting Meat on The Table: Industrial Farm Animal Production in America*), and widely viewed or read treatments of the issue (e.g., *Supersize Me, Food, Inc., Fast Food Nation,* and *The Omnivore's Dilemma*).

With the passage of successful ballot initiatives on farm animal welfare in Florida (2002), Arizona (2006), and California (2008), organized animal protection has become a catalyst for public debate about factory farming, while forcing industry to abandon some of the most controversial intensive confinement practices. These victories, especially the passage of Proposition 2 in California, have reordered political perceptions of this issue, signaling to lawmakers that there is a dominant sentiment in the public for animal welfare and a new paradigm in food production. The younger generations of Americans will grow up with a new sensibility about the basic treatment of farm animals, regulatory bodies will be charged with ensuring their welfare as new laws are passed, and the entire landscape of opportunity in this sector of humane work will be transformed.

Innovation, technological or otherwise, as a continuing force for good or ill to animals, is another hallmark of our age. Genetic engineering, however, can cut both ways. It can make it possible to prevent suffering by precluding the birth of male chicks in the egg industry, or advance humane population control through the mechanism of immunocontraception. On the other hand, it promises to open up the prospects for replicating several thousand monogenic disorders in laboratory animals, perhaps leading to their expanded use in biomedical research, and for increased emphasis on cloning and the propagation of transgenic animals, with attendant suffering and health problems.

On the unambiguously positive side, innovations in the marketplace are making it easier to reduce our impacts on animals. Soy- or wheat-based meat facsimiles, in vitro testing in the cosmetics and household products industry, and synthetic and natural fiber clothing all provide a pathway for alternatives to animal use, without requiring sacrifice or any reduction in our quality of life.

Clearly, in the face of global trade and capitalism, the humane movement must expand its reach to address problems in developing nations. Many animal issues, such as testing, animal agriculture, and the fur trade, necessarily transcend national boundaries, while others such as companion animal overpopulation and wildlife protection, present imposing challenges in nations where animal care and control entities and wildlife protection agencies are weak or lacking entirely. Through direct aid, training, and improved worldwide enforcement of international wildlife treaties, the United States and other affluent nations can and must extend themselves in support of animals in need and help build local and regional capacity to address these problems.

In the United States, high-profile cases of animal abuse or tragedy have raised consciousness about our responsibilities to other creatures. The abandonment of pets during Hurricane Katrina (2006), the pet food adulteration scandal (2007), and the Michael Vick dogfighting conviction (2007) all revealed a widespread intensity of feeling and regard for companion animals. The Hallmark Meat Company/Westland scandal and the passage of Prop 2 showed that such concerns could extend to animals raised for food. All of these situations and their outcomes are part of an emerging consensus that animals matter and that we must do better in our dealings with all species.

Of the current range of threats, it is climate change—now finally finding its place on the geopolitical agenda—that poses a macro-level threat to animals. The Nobel Prize-winning Intergovernmental Panel on Climate Change (IPCC) has predicted that without immediate and meaningful action to reverse the warming trend, 15–37 percent of plant and animal species will be extinct by 2050. Climate change is already adversely affecting animals around the globe: Diseases are more frequently emerging and spreading to new areas; rising air and sea temperatures are damaging critical habitats and threatening species that rely on these habitats for survival; and increasing numbers of extreme weather events are displacing or killing unprecedented numbers of farm animals, companion animals, and wildlife.

Other human-caused threats to the environment, such as habitat destruction and the pollution of freshwater and ocean habitats, also threaten the lives of animals, and require the animal protection movement to align itself more frequently with environmental advocates.

How ever great the threats, there are also more opportunities for animal protection to make tremendous gains in the years to come, taking advantage of the tremendous popular interest in animal welfare, the depth of popular understanding and affection for animals, and a growing appreciation for the principle that the fate of humanity is bound up with that of other species.

Wayne Pacelle

ANIMAL REPRODUCTION, HUMAN CONTROL

For animals who live their lives directly under the control of humans, one of the most important forms of influence that we exert is the control of the animals' reproduction and family relationships. This control is exerted in order to achieve the number and the type of animals to meet various human requirements for food, work, commerce, entertainment, research, or companionship. Humans have created highly specialized breeds within animal species, some of which could never have occurred naturally, which have particular qualities such as a defined size, shape, color, strength, ability to win races, or capacity to produce large quantities of meat, milk, and eggs.

The physical and sometimes psychological characteristics of animals kept directly under human control are selected not by the evolutionary pressure of the environment but by the needs and choices of humans. The reproductive choices that animals would normally make for themselves are made instead by humans. Human influence extends to when the animals breed, which animals breed and which do not breed, how many young are produced, in what physical and social environment, what social relationships exist between parent and offspring, and how the genotype and phenotype of the animals may be changed. The widespread use of reproductive technologies such as artificial insemination (and increasingly frequently, embryo transfer and possibly cloning) means that one highly-valued bull, for example, can be the biological father of hundreds of thousands of calves on several continents, altering and reducing the gene pool of the entire breed. The widespread use of one selected pedigreed dog for breeding can have a dramatic impact on the appearance, and possibly health, of the breed as a whole.

Young domestic animals are often removed from their mothers at a much younger age than would be the case in nature, and some have no contact with their mothers at all. Naomi Latham and Georgia Mason have recently reviewed numerous scientific studies showing that

maternal deprivation leads to abnormal behavior that is indicative of stress and has a profound effect on the mental and physical health of young animals.

Farmed Animals

Human control of farmed animal reproduction has led to very large increases in the production of meat, milk and eggs, with productivity increasing most steeply over the last 35 years.

Selective breeding by humans has specialized domestic cattle into those used for producing milk (dairy cows) and those used for producing meat (beef cattle). Dairy cows have been specialized to put most of their physiological effort into producing milk in their very large udders, and tend to be thin animals. The amount of milk produced for human use

Scientists and park rangers move a tranquilized elephant cow after she was darted in Kruger National Park, South Africa. After tranquilizing the animals a team of scientists examine them to see if a contraception program can limit population growth. (AP Photo)

by specialized dairy cows (such as the Holstein breed, which now dominates in developed countries, and is increasingly being exported to developing countries) is about 10 times what a calf would need. The highest yielding dairy cows now produce about 5,500 gallons (or 10,000 kg) of milk a year or more. The average milk yield per cow is eight times higher in North America than in developing countries, where specialized breeds may still be a minority. Beef cattle, in contrast, have been bred to put most of their physiological effort into fast growth and heavy musculature. This over-specialization has welfare impacts for cattle of both dairy and beef breeds.

In herds of wild and feral cattle that scientists have studied, adult females would normally have one calf and one yearling with them, and family bonds often continue when the offspring have reached adulthood. A calf would normally suckle for at least eight months or until the next calf is born, and the herd's calves often stay together in a crèche guarded by the herd. But commercial dairying also requires that the calf be separated from its mother a few days after birth, breaking the emotional bond that has formed between them. The calves are then reared away from their mothers and, if they are reared for veal production in veal crates, they are reared in isolation from others of their kind. (The use of veal crates for calves has been prohibited in the European Union since January 2007, on grounds of animal health and welfare. Phase-outs or bans have been enacted in Arizona and California and agreed to by some major North American food companies.) Because dairy breeds are selected for high milk production, not for muscle, the male calves of dairy breeds are often considered useless for beef production in

developed countries and may be shot at birth. If the cow fails to become pregnant again soon enough, she is considered economically worthless and is likely to be sent to slaughter.

Highly specialized dairy cows have such high physiological demands on their bodies that they are likely to suffer from painful lameness, mastitis, and low fertility. Often they are worn out and in poor health after having produced only two or three calves, compared to traditional breeds of cows that can last for 15 lactations. In this sense the breeding strategy adopted by humans, which in the short term produces high milk yield from a cow, is also costly from the point of view of creating healthy and long-living cows. Specialized beef breeds have a different problem; the most heavily muscled beef cows, such as the Belgian Blue breed, often require surgery in order to give birth.

Equally dramatic changes have been made in the control of the reproduction of commercial pigs (hogs). Wild and feral pigs live in small groups of a few sows and their litters. When she is about to give birth, a sow walks away from the herd and builds a nest of grass, sticks and leaves to cover herself during birth and suckling for the first couple of weeks. The mother and piglets then join the rest of the herd and the piglets become integrated into the group gradually. Sows wean their piglets gradually at up to 16 to 17 weeks of age.

The aim of commercial pig farming is to rear and sell the maximum number of piglets per sow per year, with a steady supply throughout the year. Maximizing production means control of the sow during pregnancy, birth, lactation and weaning, and severely restricting her natural behavior. Nearly all sows, at least in developed countries, are artificially inseminated. In order to monitor and control the sow during pregnancy, she may be kept in a sow stall (gestation crate), a narrow stall which prevent her from turning around or even lying down easily. (Sow stalls/gestation crates are prohibited in the European Union from 2013 onward, on the grounds of animal health and welfare. Phase-outs or bans have also been enacted in Australia, Florida, Arizona, Oregon, California, and agreed by some major North American food companies.)

Commercial sows have large litters of around 12 piglets, compared to around 4–6 piglets produced by their ancestor, the wild boar. In the search for productivity, selective breeding has created sows that are very large compared with their many tiny piglets, making it more likely that some of the piglets may get crushed to death when the sow accidentally lies on them. To try to solve this problem, most sows are kept in farrowing crates when they give birth and are suckling their piglets. These are narrow stalls that prevent the sow from turning around and prevents the piglets from coming any closer to her than to be able to reach her teats. In order to reduce the time to the sow's next pregnancy, the piglets are weaned and removed from their mothers at a time when naturally they would still be suckling and they are still very dependent on their mother socially. In Europe they are removed around 3–4 weeks of age; in North America this can be done as early as two weeks of age.

Sows have not lost their very strong motivation to build a nest, and make the same movements to try to do so even in a bare farrowing crate. Piglets have not lost their need for their mothers. Abrupt early weaning and mixing with unfamiliar pigs stresses the piglets and results in a high incidence of diarrhea and other disease.

Dan Weary and David Fraser at the University of British Columbia observed that in the first few days after weaning the piglets call constantly for their mothers.

In natural conditions, a hen builds a hidden nest and lays a small clutch of eggs, then stops laying and incubates the clutch. The mother communicates with her chicks even before hatching, and after hatching she spends her time protecting and teaching her chicks for several weeks. In commercial production, hens lay around 300 eggs continuously during a year. Chicks are reared in tens of thousands from eggs incubated in hatcheries, without ever seeing a parent bird.

The human selection of chickens, by specializing the birds into laying breeds and meat breeds, has caused biological anomalies on perhaps the largest scale yet known in human uses of animals. Laying hen breeds have very little breast muscle development, the muscle needed for meat production. In commercial hatcheries, the just-hatched chicks of laying breeds are separated by sex and the male chicks are killed at one day old (approximately 368 million per year in North America and 416 million a year in the EU25, according to statistics collected by the UN's Food and Agriculture Organization).

The economics of large-scale meat chicken farming depends on the chickens' speed of growth, their quantity of breast muscle, and their efficiency at converting food into muscle. The application of breeding technology to developing commercial hybrid chickens during the period since the 1960s has resulted in chickens designed to grow at a speed that puts them just on the edge of biological viability, typically to the age of five to seven weeks, when they are ready for slaughter. Recent research by Toby Knowles and his colleagues at Bristol University has found that nearly 30 percent of fast-growing meat chickens become moderately or severely lame. These birds are normally unable to reach adulthood in good health unless their food intake is severely restricted, because their skeletal and heart development cannot keep up with their growth rate if they are allowed to eat as much as they want. Human control of chicken breeding in the service of human needs has thus created animals that can be seen either as maximally productive or alternatively as biologically unviable and even, in the case of male layer chicks, commercially worthless.

Companion and Sports Animals

Human intervention has created large numbers of breeds and types of dogs, cats, horses, and other animals that have been kept for use or cooperation with people in work, in sport, and for companionship. Many of these are classed as pedigrees, and their breeding is highly controlled in order to produce traits that people see as desirable. As with food animals, this can often conflict with the health and welfare of the animals. In addition, since the animals have been bred with only one function in mind, any animal who fails to look right or perform to the highest standard is in danger of being rejected or even destroyed at an early age. Critics believe that pedigree breeding contributes to the already severe welfare and social problems caused by surplus and unwanted dogs, cats and horses.

Approximately 400 dog breeds have been created so far by humans over hundreds of years, all of them believed to be descended from the grey wolf. Modern dog breeds include extremes of size and shape very far removed from the wolf ancestor. Dogs were bred to have short legs to chase animals underground, to have strong jaws for guarding or fighting, to be large and strong for hunting large animals. Even in modern urban society, where nearly all dogs are kept as companions rather than for work, people still appear to prefer dogs of defined breeds. In Europe, typically three-quarters of the dogs owned are pedigree dogs, sometimes called purebred dogs, rather than mongrels.

In most modern societies, the most important characteristics of dogs are their appearance rather than their working ability, behavioral characteristics, or personality. Dog breeds have been refined and defined into breed standards by the breed societies and Kennel Clubs of the world. New breeds are still being designed for the requirements of modern urban life, such as tiny teacup dogs as accessories for celebrities, and hairless dogs bred for people suffering from allergies. The enthusiasm for dog breeding is often driven by competitive dog shows, which are often criticized for encouraging breeders to select for features that damage welfare. Examples include flat faces and short noses (such as for the bulldog and Pekingese, as well as the Persian cat), which make breathing difficult and increase the risk of heart problems; legs that are too short in proportion to the back (such as for the dachshund), increasing the risk of painful spine problems; loose skin and skin folds on face and body (such as the Shar Pei), leading to irritating and painful dermatitis between the folds; ears and hair that are too long, which may prevent dogs from keeping themselves clean without human help; and very long hair covering the eyes that may make a dog timid or defensive.

The emphasis on breed standards and breed purity can give the impression that pedigree dogs are in some sense of higher quality than dogs of a thoroughly mixed breed, but that is far from being the case. Veterinarians are aware that certain dog breeds have a much greater risk of inherited or breed-related disease than the general dog population. When breeders strive to perfect an ideal dog type or develop a new breed, two serious problems can arise. These are inbreeding, and the development of breed standards that call for unnatural and inappropriate body conformations. Inbreeding is almost inevitable in breeds that have only a relatively small number of dogs, and for numerically large and established breeds it is common to use only a fraction of the dogs for breeding, in order to maintain the desired appearance. Inbreeding (sometimes called line-breeding) decreases the genetic diversity of the breed and increases the effect of deleterious, often recessive, gene mutations.

Many breeds, including Labrador and golden retrievers, German shepherds and Rottweilers, suffer from high incidences of hip and elbow dysplasia (disorders of bone growth that lead to painful arthritis and lameness). Between a quarter and a third of the world's dog breeds have inherited eye diseases, including painful and blinding conditions such as glaucoma and degeneration of the retina. Several breeds that carry the piebald or merle genes for coat color have inherited deafness. These conditions can be disabling and lead to euthanasia. Recently it has become clear that a high proportion of the popular Cavalier King Charles Spaniel breed, in addition to being at high risk of heart disease, suffer from a mismatch between the shape of the brain and the shape of the skull, caused by breeding for a particular head shape. This results in a very painful neurological condition known as syringomyelia, which causes the dogs to scratch their necks continually and sometimes scream with pain. The massive head size of bulldogs means that puppies often have to be born by caesarean section. And dogs bred for certain behaviors such as herding, guarding, or chasing can be frustrated by the restrictions of modern urban living conditions, with resulting behavior problems.

A positive development is that both professionals and the public are now debating how our animal breeding practices impact animal rights. The UK's Kennel Club, in response to criticism, has announced a reform of breed standards to remove the worst features that cause ill health and disability. The revised standard for the Pekingese, for example, requires the dog to have a defined muzzle. In dairy cow breeding, breeders claim to be paying more attention to traits that improve health, rather than only selecting for high production. These initiatives have the potential to improve welfare, although it is too early to predict how effective they will be. Unfortunately the human desire to design animals for our own convenience remains a powerful force. Whatever viewpoint is taken, the evidence must make us question to what extent intervention operates to the benefit of the animals.

Further Reading
Advocates for Animals. 2006. *The Price of a pedigree: Dog breed standards and BREED-related illness.* Edinburgh, UK: Advocates for Animals. Download at http://www.advocatesforanimals.org/content/view/264/580/

Dybkjær, L. (ed.). 2008. Early weaning. Special issue of *Applied Animal Behaviour Science* 110(1–2): 1–216.

Gough, A. and A. Thomas. 2004. *Breed predispositions to disease in dogs & cats.* Oxford, UK: Blackwell Publishing.

Keeling, L. J. and H. W. Gonyou, eds. 2001. *Social behaviour of farm animals.* Wallingford, UK: CABI Publishing.

Knowles, T. G. et al. 2008. Leg disorders in broiler chickens: Prevalence, risk factors and prevention. *PLoS ONE* 3(2): e1545. doi:10.1371/journal.pone.0001545.

Latham N. R. and G. J. Mason. 2007. Maternal deprivation and the development of stereotypic behaviour. *Applied Animal Behaviour Science* 110: 84–108.

McGreevy P. D. and F. W. Nicholas. 1999. Some practical solutions to welfare problems in dog breeding. *Animal Welfare*, 8, 329–341.

Rauw, W. M., E. Kanis, E. N. Noordhuizen-Stassen, & F. J. Grommers. 1998. Undesirable side effects of selection for high production efficiency in farm animals: A review. *Livestock Production Science*, 56, 15–33.

The Associate Parliamentary Group for Animal Welfare. 2007. *The welfare of greyhounds: Report of the APGAW enquiry into the welfare issues surrounding racing greyhounds in England.* APGAW. Download at http://www.scribd.com/doc/15726306/Report-of-APGAW-Inquiry-Into-the-Welfare-of-Greyhounds

Weary, D. M. and D. Fraser. 1997. Vocal response of piglets to weaning: Effect of piglet age. *Applied Animal Behaviour Science*, 54, 153–160.

Jacky Turner

ANIMAL RIGHTS

Two opposing philosophies have dominated contemporary discussions regarding the moral status of nonhuman animals: (1) animal welfare (welfarism) and (2) animal rights (the rights view).

Animal welfare holds that humans do nothing wrong when they use nonhuman animals in research, raise them to be sold as food, and hunt or trap them for sport or profit, if the overall benefits of engaging in these activities outweigh the harms these animals endure. Welfarists ask that animals not be caused any unnecessary pain and that they be treated humanely.

The animal rights view holds that human utilization of nonhuman animals, whether in the laboratory, on the farm, or in the wild, is wrong in principle and should be abolished in practice. Questions about how much pain and death are necessary miss the central point. Because nonhuman animals should not be used in these ways in the first place, any amount of animal pain and death is unnecessary. Moreover, unlike welfarism, the rights view maintains that human benefits are altogether irrelevant for determining how animals should be treated. Whatever humans might gain from such utilization (in the form of money or convenience, gustatory delights, or the advancement of knowledge, for example) are and must be ill gotten.

While welfarism can be viewed as utilitarianism applied to animals, the rights view bears recognizable Kantian features. Immanuel Kant was totally hostile toward utilitarianism, not because of what it implies may be done to nonhuman animals, but because of its implications regarding the treatment of human beings. To the extent that one's utilitarianism is consistent, it must recognize that not only nonhuman animals may be harmed in the name of benefiting others; the same is no less true of human beings.

Kant abjured this way of thinking. In its place he offered an account of morality that places strict limits on how individuals may be treated in the name of benefiting others. Humans, he maintained, must always be treated as ends in themselves, never merely as means. In particular, it is always wrong, given Kant's position, to deliberately harm someone so that others might reap some benefit, no matter how great the benefit might be.

The rights view takes Kant's position a step further than Kant himself. The rights view maintains that those animals raised to be eaten and used in laboratories, for example, should be treated as ends in themselves, never merely as means. Indeed, like humans, these animals have a basic moral right to be treated with respect, something we fail to do whenever we use our superior physical strength or general know-how to inflict harm on them in pursuit of benefits for ourselves.

Among the recurring challenges raised against the rights view, perhaps the two most common involve (1) questions about where to draw the line and (2) the absence of reciprocity. Concerning the latter, critics ask how it is possible for humans to have the duty to respect the rights of other animals when these animals do not have a duty to respect our rights. Supporters of the rights view respond by noting that a lack of such reciprocity is hardly unique to the present case; few will deny that we have a duty to respect the rights of young children, for example, even while recognizing that it is absurd to require that they reciprocate by respecting our rights.

Concerning line-drawing issues, the rights view maintains that basic rights are possessed by those animals who bring a unified psychological presence to the world—those animals, in other words, who share with humans a family of cognitive, attitudinal, sensory, and volitional capacities. These animals not only see and hear, not only feel pain and pleasure, they are also able to remember the past, anticipate the future, and act intentionally in order to secure what they want in the present. They have a biography, not merely a biology.

Where one draws the line that separates biographical animals from other animals is bound to be controversial. Few will deny that mammals and birds qualify, since both common sense and our best science speak with one voice on this matter. Moreover, new evidence concerning fish cognition and behavior is leading some philosophers and scientists to recognize the psychological complexity of these animals.

Line-drawing issues to one side, the rights view can rationally defend the sweeping and, indeed, the radical social changes that recognition of the rights of animals involves—the end of animal model research and the dissolution of commercial animal agriculture, to cite just two examples.

See also Animal Liberation Ethics; Animal Welfare and Animal Rights, A Comparison

Further Reading

Armstrong, Susan and Richard Botzler, eds. 2003. *The animals ethics reader*. London and New York: Routledge.

Carl Cohen and Tom Regan. 2003. *The animal rights debate*. Lanham, MD: Rowman and Littlefield.

Dunayer, Joan. 2004. *Speciesism*. Derwood, MD: Ryce Publishing.

Francione, Gary. 1995. *Animals, property and the law*. Philadelphia: Temple University Press.

Franklin, Julian H. 2006. *Animal rights and moral philosophy*. New York: Columbia University Press.

Midgley, Mary. 1983. *Animals and why they matter*. Athens: University of Georgia Press.

Pluhar, Evelyn. 1995. *Beyond prejudice: The moral significance of human and nonhuman animals*. Durham, NC: Duke University Press.

Regan, Tom. 1983. *The case for animal rights*. Berkeley: University of California Press.

Regan, Tom. 2001. *Defending animal rights*. Urbana: University of Illinois Press.

Regan, Tom. 2003. *Animal rights, human wrongs: An introduction to moral philosophy*. Lanham, MD: Rowman and Littlefield.

Regan, Tom. 2004. *Empty cages: Facing the challenge of animal rights*. Lanham, MD: Rowman and Littlefield.

Rollin, Bernard. 1992. *Animal rights and human morality,* rev. ed. Buffalo, NY: Prometheus Books.

Singer, Peter, ed. 1986. *In defense of animals.* Walden, MA: Blackwell Publishing.

Singer, Peter. 1990. *Animal liberation.* New York: New York Review of Books.

Singer, Peter, ed. 2006. *In defense of animals: The second wave.* Walden, MA: Blackwell Publishing.

Sunstein, Cass R. and Martha C. Nussbaum, eds. 2004. *Animal rights: Current debates and new directions.* Oxford: Oxford University Press.

Taylor, Angus. 2003. *Animals and ethics: An overview of the philosophical debate.* Peterborough, ON: Broadview Press.

Wise, Steven. 2000. *Rattling the cage: Toward legal rights for animals.* New York: Perseus Publishing.

Zamir, Tzachi. 2008. *Ethics & the beast: A speciesist argument for animal liberation.* Princeton, NJ: Princeton University Press.

Tom Regan

ANIMAL RIGHTS MOVEMENT, NEW WELFARISM

Until the 1970s, the prevailing approach to animal ethics was represented by the animal welfare position. This position holds that it is acceptable to use animals for human purposes, but recognizes a moral and legal obligation to regulate our treatment of animals to ensure that it is humane and that we do not impose unnecessary suffering on them. The welfarist approach was challenged in the 1970s by the emergence of the animal rights position, which rejects welfarism on theoretical grounds (even humane animal use cannot be justified morally) as well as practical grounds (regulation simply does not work and fails to protect animal interests). The rights position proposes that recognizing the moral significance of nonhuman animals requires that animal exploitation be abolished and not merely regulated.

New welfarism is a term that describes an approach to animal ethics that is characterized by a recognition of the limitations of traditional animal welfare but an unwillingness to embrace the rights/abolitionist approach, and the consequent promotion of some improved version or theory of welfare reform. There are several versions of new welfarism, including the following three.

Welfare as a Means to Abolition

Many new welfarists believe they seek the abolition of animal exploitation as a long-term goal but advocate the improved regulation of animal use in the short term as the means to achieve the abolition (or significant reduction) of animal use by gradually raising consciousness about the moral significance of nonhuman animals. Although this position has been promoted by many of the large animal organizations in North America, South America, and Europe, it has both theoretical and practical problems.

As a theoretical matter, if our use of animals is not morally justifiable, promoting more humane exploitation as a means to the end of abolition raises a serious issue. For example, if we believe that any form of pedophilia is morally wrong, we cannot, consistent with that position, campaign for humane pedophilia. In the struggle against human slavery in the United States, many of those who favored abolition refused to campaign for the reform of slavery because they considered reform as inconsistent with the basic moral principle that slavery was an inherently unjust institution. Similarly, the promotion of more humane animal use is inconsistent with the idea that we cannot justify animal use in the first instance.

As a practical matter, this first version of the new welfarist position—that improved protection for animal interests in the short term will eventually lead to abolition—is problematic for at least three reasons. First, the history of animal welfare regulation makes clear that, because animals are property, animal welfare regulation is, as a practical matter, incapable of providing any significant protection for animal interests in the short term. Welfarist regulation generally protects animal interests only to the extent that there is an economic benefit to humans.

Second, there is no evidence that making exploitation more humane leads to the abolition of that exploitation. Indeed, the contrary appears to be true. We have had animal welfare laws for nearly 200 years, and yet we now exploit more animals in more ways than at any time in the past. To the extent that animal welfare reform raises consciousness about animals, it merely reinforces the notion that animals are things that we are entitled to use as long as our treatment of them is humane, and facilitates the continued acceptance of exploitation which is characterized as meeting that standard.

Third, the phenomenon of new welfarism has resulted in a curious partnership between those who claim to endorse animal rights and institutional animal exploiters who claim to seek mutually acceptable welfare reforms, which the former believe will lead to abolition and the latter believe will further reassure the public that animal treatment is at a morally acceptable level. But because animals are property, these reforms are necessarily limited to minor changes in animal treatment that, in many cases, actually improve animal productivity and increase producer profit. Animal advocates have, in effect, become advisers to institutional exploiters and have helped them to identify certain practices that are not cost effective. To the extent that welfare reforms result in any benefits to animals, these benefits are offset by the fact that exploiters can point to the support of animal advocates, which in turn promotes the continued social acceptance of animal exploitation. Indeed, many large animal advocacy organizations actively promote animal products that supposedly have been produced in a humane manner. Such promotion may actually increase consumption by those who had stopped eating animal products because of concerns about treatment, and will certainly provide a general incentive for continued consumption of animal products.

This first version of new welfarism presents the false dichotomy that, even if we embrace abolition as the ultimate goal, we have no choice but to pursue welfarist regulation in the short term, because that is the only realistic strategy, given that animal use will not be abolished any time soon. Putting aside that welfarist regulation does not significantly protect nonhumans in the short term and does not lead to abolition in the long term, this position neglects other strategies that are arguably not only more consistent with a theory that rejects animal use as immoral, but are also more effective as a practical matter in reducing demand for animal products and in building a political movement that will support abolitionist measures. The rights/abolitionist approach focuses on veganism as a moral baseline and prescribes incremental social and political change primarily through creative, nonviolent vegan education.

Many new welfarists, however, reject veganism as a moral baseline. They maintain that it is more practical to support welfarist reform and to promote animal uses that are more humane. But this approach reinforces the prevailing view that animal use is morally acceptable if treatment is humane, and it makes veganism appear to be a radical or extreme response to animal exploitation, which is counterproductive to the goal of abolishing animal use.

Peter Singer and Animal Welfare

The second form of new welfarism is the position advocated by Peter Singer. Although Singer is often characterized as an animal rights advocate he, like Jeremy Bentham (1748–1832), is a utilitarian who maintains that normative matters are determined only by consequences, and he rejects the concept of moral rights for humans and nonhumans alike. Singer agrees with Bentham that sentience is the only characteristic required for animals to be morally significant, and that no other characteristic, such as rationality or abstract thought, is needed. Singer maintains that we should apply the principle of equal consideration and should treat animal interests in essentially the same way that we would treat the similar interests of a human, and not discount or ignore those interests on the basis of species alone. But, also like Bentham, Singer regards most nonhumans as living in a sort of eternal present that precludes their having an interest in a continued existence. This position leads Singer to maintain that killing animals per se does not raise a moral problem, and so he does not challenge the property status of animals as inherently problematic.

Because Singer does not challenge the property status of nonhumans, and maintains that their use per se does not raise a moral issue, his theory is essentially a version of animal welfare. It is arguably more progressive, in that it requires that we accord greater weight to animal interests than is required under the traditional welfarist approach but, as a theoretical matter, Singer never explains how to do this and, as a matter of his individual animal advocacy, he promotes traditional animal welfare reform such as more humane slaughtering processes or larger cages for battery hens. In any event, Singer does not see animal welfare as a means of abolishing animal use, because he does not advocate abolition as a long-term goal and, therefore, he differs from the new welfarists described in the previous section. Rather, he sees animal welfare as a means to reduce animal suffering. He maintains that we can be what he calls conscientious omnivores if we take care to eat flesh and other products made from animals who have been raised and killed in a humane fashion.

If Singer is wrong in assuming that animals do not have an interest in their continued existence, then our use of animals in ways in which we do not use humans and our treatment of animals as our property *necessarily* violates the principle of equal consideration. Humans who lack the reflective self-awareness of normal adults, such as those with particular forms of amnesia, or very young children, or those with certain mental disabilities, are still self-aware and have an interest in continuing to live. There may, of course, be a difference between the self-awareness of normal adult humans and that of nonhuman animals. But even if that is the case, it does not mean that the latter have no interest in continuing to live and it does not justify treating the latter as commodities. Critics believe that Singer begs the question from the outset by maintaining that the only self-awareness that matters to having an interest in life is the sort that normal humans possess. Singer's view that if some animals, such as the great apes, have humanlike self-awareness, they are entitled to greater moral significance and legal protection than other nonhuman animals, merely perpetuates an unjustifiable speciesist hierarchy.

Moreover, even if animals do not have an interest in continued life, the application of the principle of equal consideration to issues of animal treatment is problematic in a number of respects. Any such endeavor requires that we make interspecies comparisons in order to determine whether the animal interest in question is similar to a human interest and, therefore, merits similar treatment. This sort of determination is difficult when only humans are involved. It is almost impossible when comparing members of different species. There is an understandable tendency to think that a human interest is always different and more important. In addition, assessments of similarity are particularly difficult given the property status of animals. The fact that an animal is property and has only extrinsic or conditional value automatically prejudices us against perceiving an animal interest as similar to a human one. Given the importance of property rights, it should not be surprising that many humans think that any inability to use their property as they wish is a significant deprivation that leads them to discount heavily any animal interests at stake.

The Feminist Critique of Rights

Another version of new welfarism may be found in the writings of certain feminist theorists who assert that rights are patriarchal and reinforce hierarchies, and that we must therefore move beyond rights to develop an ethic of care for our relationship with nonhumans. Those who adopt this view reject universal rules, such as an absolute prohibition on the use of animals as human resources, in favor of using values such as love, care, and trust to guide our use and treatment of animals in particular situations.

Although rights certainly have been used to establish and reinforce a variety of morally odious hierarchies, rights are certainly not inherently patriarchal. Instead, a right is simply a way of protecting an interest; it treats that interest as inviolable even if the consequences to others of violating it are considerable. Such normative notions are necessarily part of feminist theory in that no feminist believes that the morality of rape is dependent on a case-by-case analysis in light of an ethic of care. On the contrary, a woman's interest in the integrity of her body is correctly treated as inviolable: a woman has a *right* not to be raped.

Similarly, if nonhumans are sentient, we have no justification for ignoring the fundamental interests of those nonhumans and treating them as a resource. The feminist ethic of care does not go beyond rights, as some of these theorists maintain. Rather, it is a form of welfarist theory which, like Singer's position, seeks to accord greater weight to nonhuman interests but still preserves the hierarchy of humans who, despite what these theorists state, are accorded protection of their rights that is denied to nonhumans.

See also Abolitionist Approach to Animal Rights; Law and Animals; Utilitarianism

Further Reading

Francione, Gary L. 1996. *Rain without thunder: The ideology of the animal rights movement.* Philadelphia: Temple University Press.

Francione, Gary L. 2008. *Animals as persons: Essays on the abolition of animal exploitation.* New York: Columbia University Press.

Gary L. Francione

ANIMAL SHELTERS

See Shelters, No-Kill; Rescue Groups

ANIMAL STUDIES

Animal studies is the interdisciplinary study of human-animal relations. At times referred to as anthrozoology, animal humanities, critical animal studies, or human-animal studies, it examines the complex interactions between the worlds of humans and other animals. Several features of animal studies are emphasized in this entry.

First, animal studies is an emerging discipline and one of the fastest growing fields in the academy. The human relationship to other animals is of obvious interest and concern to a great many people. The popularity of companion animals, nature videos, animal-focused ecotourism, bird-watching, animal art, and social movements to protect wild and domestic animals are but a few examples. Animal studies is both root and fruit of this interest and concern. Overall, the field seeks to understand, and in some instance critique and revise, how humans relate to nonhumans in a more-than-human world.

The growth to date of animal studies is akin to that of other forms of social problems research that, because of the complexity of the issues and the need for interdisciplinary collaboration, evolve from subfields of others disciplines into a discipline of their own. The emergence and institutionalization of environmental studies and women's studies are models that animal studies scholars point to when describing this process.

Currently animal studies is in a pre-disciplinary phase. One can find it as an official or de facto subfield represented through courses, research and/or special interest groups in a variety of disciplines. These included interdisciplinary fields (e.g. environmental studies and geography), the social sciences (e.g. anthropology, political science, psychology, and sociology), as well as the arts and humanities (e.g. history, literature, philosophy and religious studies). There are an increasing number of journals (e.g. *Anthrozoos*; *Humanimalia, Society and Animals*), book series (e.g. Brill, Temple University Press, Columbia University Press), international societies, and online networks (e.g., the International Society of Anthrozoology; H-Animal, Animal Inventory), as well as policy institutes that make use of the fruits of this scholarship (e.g., the Institute for Society and Animals, Humane Society University). Out of this nexus, graduate degrees and undergraduate majors/minors are beginning to appear. Of particular note is the Graduate Specialization in Animal Studies at Michigan State University.

Second, animal studies emerged in response to three problematic ways of understanding animals. The first is the failure of the natural and behavioral sciences to adequately address the sentience, sapience, and agency of many animals. The second is the recognition of anthropocentrism and speciesism as prejudicial paradigms that distort our moral relationship with other people, animals, and the rest of nature. The third is a burgeoning interest in the cultural, social, and political place of animals in human societies. A clarion response to these problems was the publication of two books by Mary Midgley—*Beast and Man* (1978) and *Animals and Why They Matter* (1984). Both texts were motivated in part as responses to the ethical and scientific blinders of behaviorism, genetic determinism, and sociobiology. Midgley is arguably the field's most celebrated scholar, and her incisive critiques of ethical, philosophical, and scientific themes inspired scholars to consider the animal question as a serious subject of study.

Third, the interdisciplinary nature of animal studies produces a wealth of theories, methods, and topics. Scholars approach the field from diverse theoretical positions, ranging from empiricism and positivism, to interpretivism and critical theory. They undertake their studies using qualitative, quantitative, and mixed methodologies, and their topics touch on wild, companion, farm and research animals. While this plurality generates a vibrant dialogue that should be praised, it can also obscure fundamentally different approaches to ethics, science, and society. This is becoming something of an unacknowledged struggle for HAS, as positivists and anti-positivists begin to clash in conferences, faculty meetings, seminars, and publications. This is to be expected, as the positivist claim to undertaking value-free and objective science is discredited, and the anti-positivist alternatives represent such a diversity of theoretical and methodological points-of-departure that it is both impossible and undesirable to establish a unitary paradigm. Indeed, these paradigms are

particularly incommensurable with respect to naturalistic versus interpretive theories of science, quantitative versus qualitative methods of research, and the vision of value-free versus value-forming scholarship. These clashes have not become the primary focus of debate as of yet, but bear watching as sources of rough weather.

Fourth, like any academic field with social relevance, there is an ongoing tension between scholarship and activism. The perspectives of activists for animal welfare, protection, or rights are a source of inspiration and insight to the academy and society alike. Yet scholarship and activism are neither identical nor inseparable. Some scholars and students have precommitments to animal social movements and, for reasons of academic freedom and social relevance, this is well and good. Even so, the intellectual arm of social movements frequently engages in moral and political intransigence. So too, the academic empires some scholars attempt to build in an effort to valorize their own work is equally problematic. Dogmatism may serve academics and advocates well as they mobilize support for their positions. It is antithetical, however, to the best norms of scholarship that aspire to theoretical and methodological rigor. It is equally antithetical to the contextual realities that confront advocates on a daily basis. The trick to managing this tension is not to privilege the academy over advocacy, one concern or discourse over another, but to allow each to inform and challenge the other. We need reason and action as nuanced as the world's complexity.

Fifth, animal studies will face crucial challenges in the years ahead. One such challenge has to do with its legitimacy in academia. Despite the interest in animal studies shown by academics and the general public, the field is receiving a cool reception in many academic departments. The reasons for this vary, but are not so different from what women, minorities, and others have experienced when they too advocated for new arenas of scholarship. This opposition includes:

- hostility toward animals as a serious subject of study
- fears that interdisciplinary fields diminish students and resources for established departments
- theoretical imperialism and the distaste for upstart disciplines that do not toe the theoretical line
- advocacy concerns that a focus on the well-being of animals will detract from the well-being of humans, and
- censorship by university administrators who fear animal studies will jeopardize corporate and government sources of funding

Proponents of animal studies will have to directly face all of these concerns if their efforts to institutionalize the field are to win out over ivory tower politics.

Another challenge has to do with creating a learning community in the context of the globalization of knowledge. As noted above, animal studies draws insights from many disciplines, theories, methods, topics, and experiences. These insights are drawn not only from North America and the animal protection movement, but from places and identity groups around the globe. The globalization of animal studies will likely continue in the years ahead. This then raises questions about how academics and others learn to generate a body of knowledge that is open to a wide diversity of perspectives, without lapsing into a lazy relativism about knowledge or moral norms. Grappling with the problem of relativism—and its opposite, objectivism—will likely require an ongoing debate over the status of situated knowledge in ethics, science, and society. It will also require ongoing attention for dialogue that creates the possibility for such knowledge.

Further Reading

Baker, Steve. 2001. *Picturing the beast: Animals, identity, and representation.* Urbana: University of Illinois Press.

Balcolmbe, Jonathan P. 1999. Animals and society courses: A growing trend in postsecondary education. *Society and Animals* 7 (3): 229–240.

Jamieson, Dale. 2002. *Morality's progress: Essays on humans, other animals and the rest of nature.* New York: Oxford University Press.

Kalof, Linda. 2007. *Looking at animals in human history.* London: Reaktion Books.

Kalof, Linda and Amy Fitzgerald, eds. 2007. *The animals reader: The essential classic and contemporary writings.* Oxford: Berg.

Lavigne, David, ed. 2006. *Gaining ground: In pursuit of ecological sustainability.* Limerick, IRL: University of Limerick Press.

Lynn, William S. 2002. Canis lupus cosmopolis: Wolves in a cosmopolitan worldview. *Worldviews* 6 (3): 300–327.

Lynn, William S. 2004. Animals. In *Patterned ground: Entanglements of nature and culture,* edited by S. Harrison, S. Pile and N. Thrift. London: Reaktion Press.

Midgley, Mary. 1995. *Beast and man: The roots of human nature.* London: Routledge.

Midgley, Mary. 1998. *Animals and why they matter.* Reissue ed. Athens: University of Georgia Press.

Midgley, Mary. 2005. *The essential Mary Midgley.* New York: Routledge.

Patton, Kimberly, and Paul Waldau. 2006. *A communion of subjects: Animals in religion, science and ethics.* New York: Columbia University Press.

Philo, Chris, and Chris Wilbert, eds. 2000. *Animal spaces, beastly places: New geographies of human-animal relations.* London: Routledge.

Rollin, Bernard E. 2006. *Science and ethics.* Cambridge: Cambridge University Press.

Sax, Boria, ed. 2001. *The mythical zoo: An A-Z of animals in world myth, legend and literature.* New York: ABC-Clio.

Wolch, Jennifer, and Jody Emel, eds. 1998. *Animal geographies: Place, politics and identity in the nature-culture borderlands.* London: Verso.

Wolf, Cary, ed. 2003. *Zoontologies: The question of the animal.* Minneapolis: University of Minnesota Press.

William S. Lynn

ANIMAL SUBJECTIVITY

We care about animal rights and animal welfare because we assume that animals are able to experience their lives subjectively—that they have an individual perspective on things and can feel good or bad about them. Thus we naturally see animals as sentient, and assume that they have an inner life of some sort, that there is "something it is like to be" them, to quote the words of philosopher Thomas Nagel. For scientists working in the field of animal welfare, the problem is whether and how we can objectively assess this subjectivity, for example, what it is like to be a battery cage hen or a laboratory rat. There exists as yet no agreement between either scientists or philosophers on precisely how we should understand the subjective aspects of life, how they might relate to observable behavior, and how we might measure them. These are deep philosophical problems that we cannot expect to resolve in the near future, but that nevertheless affect the way we think about animal suffering and our responsibility to alleviate it.

This difficulty in studying how animals subjectively experience life feeds into another meaning of the term subjective, which refers to the difficulty of gaining certain, factual knowledge. Many scientists are concerned that because experience is subjective, it is not open to reliable, objective assessment, only to prejudiced, untrustworthy, subjective judgment. Many go so far as to believe that because feelings are difficult to study, they are literally hidden from view, and should be defined as internal mental states. In such a light, describing animals as happy or sad, frustrated or content, can quite easily be dismissed as the misguided anthropomorphic projection of human emotions onto nonhuman animals.

It is, however, very important that we do not confuse the two meanings of the word subjectivity. That feelings are of an inner, personal nature does not automatically imply that they are completely hidden from others and cannot be observed and investigated. It is true that, generally, you do not directly feel what someone else (human or animal) feels, but that is not to say that, with some effort, you could not perceive and understand the quality of another's experience. With appropriate criteria and assessment procedures, objective investigation of subjective experience in animals may well be possible.

Various approaches to the study of subjective experience in animals have been developed over the years. In science, one of the first and most influential ideas was to let animals vote with their feet: when given a choice of environments or situations, animals will presumably spend most of their time in the situations they like best. Another proposal was to test how hard animals are prepared to work for various kinds of reward. To gain access to litter, for example, chickens are willing to peck a key many times. Such studies indicate what animals like and value; however, they do not tell us whether animals suffer when they are deprived of what they value.

One approach is to test whether out of sight is out of mind: if animals can be shown to remember previous experiences of, say, companionship, play opportunities, or preferred foods, they may well miss these experiences when they are absent. Another approach is to test whether taking away valued goods, for example cage enrichment materials, affects how animals make decisions in learning tasks. Researchers found that such deprivation made rats more pessimistic in their attempts to solve learning tasks, indicating that the changes in their cage had affected them negatively and made them more anxious and uncertain.

Approaches such as these study the specific responses given by animals under controlled experimental test conditions. Another way of addressing what animals experience is through careful, patient study of their body language. In the way animals interact with and pay attention to their surroundings, the way that they orient their body, eyes, ears, nose, nostrils, or tail, they continuously express how they perceive and evaluate these surroundings. By learning to judge whether the animal's demeanor is relaxed, lively, confident and curious or, by contrast, tense, agitated, fearful or lethargic, we can get closer to how animals feel about the situation they are in, whether it makes them happy or distressed. Research on farm animals has shown that such judgments, if based on careful observation, have scientific validity. However, it is good to realize that if you don't know an animal well, or the species to which it belongs, it is possible to misinterpret its body language expressions. Indeed, many animals communicate in

ways that are largely inaccessible to us, for example through smell, echolocation, or kinesthetic vibration. Judging animal body language is thus a skill that takes years to develop, and relies on extensive observation of animals in a wide range of circumstances.

It is perhaps not surprising in this light that field researchers such as Jane Goodall or Cynthia Moss who spend extended periods of their lives with animals in their natural environments speak confidently of these animals' individual personalities and emotional lives. Equally, people who work and live with animals in mutual partnership, such as dog- and horse-trainers, zookeepers and pet owners, often develop an intimate acquaintance with their animals' expressive repertoire, and many have written books about how their animals communicate with them. Such understanding leads to strong bonds and friendships, which is perhaps the best evidence that animals are not just complex physical objects, but sentient subjects with a perspective of their own.

These, amongst others, are constructive and fruitful ways of studying the subjective perspectives of animals and the quality of life they enjoy or are forced to endure. The extent to which they truly prove that animals are capable of happiness and suffering remains a point of scientific debate; however, this does not mean that until we resolve this debate there can be no compelling evidence of animal suffering. That science cannot as yet explain subjective experience does not mean its existence is uncertain or unavailable for assessment. Careful description of phenomena is the start of scientific explanation, not the result, and that our judgments of animal experience have a certain open-endedness and vulnerability to misinterpretation does not mean that they are fundamentally shaky and unreliable. Such open-endedness always exists in communication with other sentient living beings, and we should accept this out of respect for these beings' autonomy. The way to deal with this uncertainty is not to be dismissive of studying the feelings of animals, but to devise better ways of communicating with them, and to study their expressions more closely.

The brain is of course a vitally important source of information for understanding the physical mechanisms that facilitate subjective experience. However, if we want to know what this experience is like, the range and diversity of experience of which animals are capable, behavior in all its richly expressive aspects, provides the best starting-point. We should enable and encourage animals to express to us how they experience their world, and we should learn to listen to them in as many ways as we can.

See also Affective Ethology; Consciousness, Animal; Whales and Dolphins: Sentience and Suffering

Further Reading
Bekoff, M. 2007. *The emotional lives of animals.* Novato, CA: New World Library.
Crist, E. 1999. *Images of animals: Anthropomorphism and animal mind.* Philadelphia: Temple University Press.
Dawkins, M. S. 1990. From an animal's point of view: motivation, fitness, and animal welfare. *Behavioral and Brain Sciences* 13, 1–61.
Gaita, R. 2004. *The philosopher's dog.* London: Routledge.
Hearne, V. 1986. *Adam's task. Calling animals by name.* London: Heinemann.
Midgley, M. 1983. *Animals and why they matter.* Athens: The University of Georgia Press.
Nagel, T. 1974. What is it like to be a bat? Reprinted in: Nagel, T. 1991. *Mortal questions,* 2nd edition, 165–181. Cambridge: Cambridge University Press.
Paul, E. S., E. J. Harding, and M. Mendl. 2005. Measuring emotional processes in animals: the utility of a cognitive approach. *Neuroscience and Biobehavioral Reviews* 29, 469–491.
Wemelsfelder, F. (2007). How animals communicate quality of life: the qualitative assessment of animal behaviour. *Animal Welfare* 16(S), 25–31.

Françoise Wemelsfelder

ANIMAL WELFARE

When dictionaries define *welfare* and *well-being,* they use phrases such as "the state of being or doing well" and "a good or satisfactory condition of existence." These phrases tell us that the welfare or well-being of animals has to do with their quality of life, but to be more precise about the meaning of the terms we must go beyond the semantic issue of how the words are used and address the value issue of what we consider important for animals to have a good quality of life.

Three main approaches to this question have emerged. Some people emphasize how animals feel. According to this view, the affective states of animals (feelings or emotions) are the key elements of quality of life. Thus a high level of welfare requires that animals experience comfort, contentment, and the normal pleasures of life, as well as being reasonably free from prolonged or intense pain, fear, hunger, and other unpleasant states. A second approach emphasizes the biological functioning of animals. According to this view, animals should be thriving, capable of normal growth and reproduction, and reasonably free from disease, injury, malnutrition, and abnormalities of behavior and physiology. A third approach considers that animals should be allowed to live in a reasonably natural manner or in a manner for which they are well suited. This view takes two slightly different forms: that animals should be in natural environments (fresh air, sunshine, natural vegetation) and that animals should be able to express their natural behavior and develop their natural adaptations. The three views of animal welfare have close parallels in the timeless philosophical debate about what constitutes a good life for humans, and they represent different values that are deeply rooted in human thought.

The three approaches to animal welfare often agree in practice. For example, allowing a pig to wallow in mud on a hot day is good for its welfare according to all three views: because the pig will feel more comfortable, because its bodily processes will be less disturbed by heat stress, and because it can carry out its natural thermoregulatory behavior.

However, there are some real differences between the three views of welfare. A pig farmer using criteria based on biological functioning might conclude that the welfare of a group of confined sows is high because the animals are well fed, reproducing efficiently, and free from disease and injury. Critics using other criteria might conclude that the welfare of the same animals is poor because they are unable to lead natural lives, or because they show signs of frustration and discomfort.

Scientific research is often very helpful in assessing animal welfare. For example, housing calves in individual stalls has many effects on their degree of movement, disease transmission, levels of stress hormones, and so on, and these measures can be studied scientifically. But which measures we choose to study in order to assess animal welfare, and how we use such measures to draw conclusions about animal welfare, involve value judgments about what we think is more important or less important for the animals. Knowledge alone cannot turn such judgments into purely factual issues. Science cannot, for instance, prove whether freedom of movement is more important or less important for animals than freedom from certain diseases.

There are also several confusing semantic issues concerning the application of the concepts of welfare and well-being to animals.

First, welfare (when it is applied to humans) has a second meaning: specifically, it is used to refer to social assistance programs (food, housing, money) designed to help vulnerable members of society. To avoid confusion between the two meanings of welfare, scientists generally use the term animal welfare to refer to the state of the animal, and use other terms (animal care, animal husbandry, humane treatment) to refer to what people provide to support a good quality of life for animals.

Second, many scientists write about a certain level of welfare and thus use the term as a kind of scale, running from high to low. Thus one might speak of poor welfare. This usage will seem strange to those who think of welfare as referring only to the good end of the scale. However, we do not have a distinctive term for the scale, and using the term welfare (or well-being) in this dual sense fills the need. A precedent is the word health, which means both (1) freedom from illness and injury, and (2) the general condition of an organism with reference to its degree of freedom from illness and injury.

Third, confusion also arises because people have tried to distinguish between welfare and well-being in various ways. One approach uses well-being to mean the state of the animal and welfare to mean the broader social and ethical issues; thus one might say that the well-being of animals is at the heart of animal welfare controversies. A second approach uses the term welfare to refer to the long-term good of the animal and the term well-being for its short-term state, especially how the animal feels. Hence a painful vaccination may enhance an animal's welfare but reduce its feelings of well-being. A third approach, often followed in Europe, uses the term welfare exclusively because it is the traditional term in ethical and scientific writing, in most legislation, and in the names of animal welfare organizations. A fourth approach, sometimes followed in the United States, uses the term well-being instead of the term welfare because welfare (in its second meaning of social assistance programs) represents a controversial issue. Finally, many people treat the two terms as synonymous, following the lead of many dictionaries. Treating welfare and well-being as synonyms is probably the simplest approach and conforms best to everyday usage of the terms, but that will not stop scholars from continuing to propose more specialized meanings.

See also Animal Rights; Utilitarianism; Pain, Suffering and Behavior

Further Reading

Broom, D. M., and A. F. Fraser. 2007. *Domestic animal behaviour and welfare,* 4th ed. Wallingford: CAB International.

Duncan, I.J.H., and M. S. Dawkins. The problem of assessing 'well-being' and 'suffering' in farm animals. In D. Smidt, ed., *Indicators relevant to farm animal welfare* (The Hague: Martinus Nijhoff, 1983), 13–24.

Fraser, D. 2008. *Understanding animal welfare: The science in its cultural context.* Oxford: Wiley-Blackwell.

Rollin, B. E. 1995. *Farm animal welfare.* Ames: Iowa State University Press.

David Fraser

ANIMAL WELFARE AND ANIMAL RIGHTS, A COMPARISON

The notion of *animal welfare* dates back far before the notion of *animal rights.* In fact, the concept of rights in their modern sense did not enter common usage until the 1700s. It was notably through the publication of *Animal Liberation* by Australian philosopher Peter Singer in 1975 that the animal liberation movement as we know it coalesced. There were several reasons for the new radical view, all of which directly influenced the content of Singer's important book: (1) using the liberation movements on behalf of blacks and women as models, the animal liberation movement rejected *speciesism* (arbitrary discrimination on the basis of species or species-characteristics) as well as racism, sexism, homophobia, and ableism; (2) advances in evolutionary biology blurred species boundaries between humans and other animals; (3) rebellions occurred within human organizations (e.g., the Royal Society for the Prevention of Cruelty to Animals' earlier support of hunting—many of its wealthy patrons were fox hunters—led to the formation of the Hunt Saboteurs Association in 1963; now fox-hunting is illegal in Britain); and (4) modern animal cruelties were documented in Ruth Harrison's 1964 book *Animal Machines,* which exposed factory farming, and in Richard Ryder's 1975 *Victims of Science,* which revealed horrors in the laboratory.

Technically, animal rights can refer to any list of rights for animals. In 1988, for example, Sweden passed a law explicitly giving animals raised for food the right to graze. Currently, though, animal rights is widely understood to refer to the idea of abolishing all use or exploitation of animals, a view reflected in Tom Regan's *The Case for Animal Rights.* Animal welfare, in comparison, is generally understood as advocating humane or kind use of animals, at minimum upholding animal well-being by prohibiting unnecessary cruelty (a common legal phrase).

In spite of this general meaning of animal welfare, there remains a spectrum of views as to what this phrase represents:

- **The animal exploiters' animal welfare.** To critics, a view represented by the reassurances of those who use animals as food, commercial, or recreational resources (e.g., factory farmers), stating that they care for animals well, which is a position that seems to many to be primarily exhibited for public relations or advertising purposes

- **Commonsense animal welfare.** The average person's typical and usually vague concern to avoid cruelty and perhaps to be kind to animals

- **Humane animal welfare.** A view that offers a more principled, deep, and disciplined stance than commonsense animal welfare in opposing cruelty to animals, often advocated by humane societies, for example. This form still does not reject most animal-exploitive industries and practices (fur and hunting are occasional exceptions, along with the worst farming or laboratory abuses)

- **Animal welfare as a misnomer for animal ill-fare.** A label, originated by David Sztybel, stating that even if efforts are made to be humane, animal exploitation is an *ill fate* overall. (In other words, imagine comforts being secured for humans who are to be eaten, skinned, vivisected, etc. This would still be a bad overall situation for these people). Along similar lines, in *Empty Cages,* Tom Regan disputes that animal welfare is really the norm in America, and Joan Dunayer, in *Speciesism,* places animal welfare in skeptical quotation marks when applied to the industrial uses of animals

- **Utilitarian animal welfare.** A view championed by Peter Singer, which would seek to minimize suffering overall, while possibly accepting, for example, some types of medical vivisection, but not the wearing of furs by affluent urbanites

- **New welfarism.** An approach that Gary Francione characterizes as recognizing the limitations of traditional animal welfare but one that is unwilling to embrace the animal rights abolitionist approach, resulting in the consequent promotion of a new or improved theory of welfare reform

- **Animal welfare/animal rights views that do not clearly distinguish between the two.** For example, psychologist and philosopher Richard Ryder subscribes to both ideas, although he is a complete abolitionist regarding animal use. Both animal welfare and animal rights, he says, are concerned with the suffering of others, and he evidently does not see the value of using the term to distinguish abolitionists from non-abolitionists who are still humanitarians

In general, it is possible to consider animal welfare and animal rights using a common frame of reference. We can envision animal rights as championing the full protection of all of animals' vital interests. Animal welfarists, by contrast, generally agree that only some interests should be protected (e.g., avoiding unnecessary suffering, although not avoiding premature death). Also, protection of interests usually occurs to a lesser degree in the case of animal welfarists as compared to animal rightists (e.g., humans generally have more freedom of movement than animals confined for industrial purposes).

See also Animal Rights Movement, New Welfarism

Further Reading

Carson, Gerald. 1972. *Men, beasts, and gods: A history of cruelty and kindness to animals.* New York: Charles Scribner's Sons.

Dunayer, Joan. 2004. *Speciesism.* Derwood, MD: Ryce Publishing.

Finsen, Lawrence, and Susan Finsen. 1994. *The animal rights movement in America: From compassion to respect.* New York: Twayne.

Jasper, James M., and Dorothy Nelkin. 1992. *The animal rights crusade: The growth of a moral protest.* New York: Free Press.

Regan, Tom. 1983. *The case for animal rights.* Los Angeles: University of California Press/.

Regan, Tom. 2004. *Empty cages: Facing the challenges of animal rights.* New York: Roman & Littlefield.

Ryder, Richard D. 1989. *Animal revolution: Changing attitudes towards speciesism.* Oxford: Basil Blackwell.

Singer, Peter. 1975. *Animal liberation.* New York: Avon Books.

Sztybel, David. 2006. The rights of animal persons. *Journal for Critical Animal Studies* 4 (1): 1–37.

David Sztybel

ANIMAL WELFARE: ASSESSMENT

Assessment of animal welfare requires knowledge about the biology and psychology of animals—their needs and preferences, their responses to how they are treated, their perceptual and mental abilities, and their emotional states. This knowledge allows us to better understand how animals perceive the impact of housing and management on their health and welfare, and hence helps us to make more informed decisions about animal welfare issues.

A central role of *animal welfare science* is to provide this information. However, measuring the biological and psychological state of nonhuman animals is scientifically challenging. In particular, the subjective emotional experiences of animals, such as pain, fear, and pleasure, that lie at the heart of most people's concerns about animal welfare, are inherently private and therefore very difficult to assess. Many scientists and philosophers argue that we may never know whether nonhuman species have conscious experiences, let alone measure what they might be, and some contend that we should therefore only assess welfare by investigating whether the animal's biological functioning appears normal or impaired in some way. Others argue that we need to develop measures that, although indirect, may be useful proxy indicators of subjective emotional states in animals, and some researchers believe that these states can be assessed directly. Despite these differing views, animal welfare scientists have developed a number of methods to assess welfare that can be usefully split into two main approaches: the welfare indicators approach (or what animals do), and the motivational priorities approach (or what animals want).

The *welfare indicators* approach involves measuring behavior, physiology, and physical state in order to get an idea of how animals respond to the ways in which they are treated. For example, abnormal or damaging behavior, chronic changes in the functioning of physiological stress systems, suppression or alteration of immune function, increased susceptibility to disease, and physical damage may all indicate that the animal's welfare is impaired. Those who believe that welfare can only be measured by assessing the biological functioning of an animal tend to use these types of indicators. However, other researchers, assuming that many of the species under our care *are* capable of subjective experiences, also use these measures as proxy indicators of subjective suffering. The premise here is that, as appears to be the case in humans, changes in behavior and physiology may reflect emotional experiences including pain, fear, anxiety, and frustration. Although we cannot be certain about this, and to some it may smack of anthropomorphism, it is arguable that as long as we are interested in the subjective experiences of animals, the only model species we can refer to is the human being. Humans are able to provide linguistic reports on their emotional states, which can then be related to accompanying behavioral and physiological changes, and these can be used as proxy measures of such states in animals. Researchers following this approach are currently developing a range of new welfare indicators that may more closely reflect emotional states in animals. These include vocalizations, qualitative ratings of posture and behavioral expression, and changes in decision-

making behavior. Some of these indicators may reflect positive emotions as well as negative ones. There is currently a paucity of such indicators, but increasing scientific and political interest in the idea that we should not only be minimizing poor welfare, but also actively enhancing quality of life, means that this is a growing research area. There is also increasing interest in the development of welfare indicators that can be used in the field, for example on farms. The task here is to identify indicators which, although only measured at one point in time (e.g., during a farm visit) and for a subset of animals in a population, provide a reliable and valid representation of welfare.

One significant challenge that the *welfare indicators* approach still faces is how to combine information on a variety of indicators to provide a single measure of the animal's welfare. Solutions to this problem remain an important goal of animal welfare science. Similar problems exist for scientific attempts to specify absolute cut-off points at which welfare becomes unacceptable. The problem here is in identifying conditions where welfare is agreed to be good and acceptable that can act as standards against which other conditions can be compared. An obvious suggestion is to take the animal in its natural environment as the baseline condition. However, for many domestic species, it is difficult to identify what a natural environment actually is, and in most environments that we might call natural, animal welfare is far from perfect. Animals living in the wild are often under threat from starvation, extreme temperatures, injury, and predation and, in many cases, it would seem inappropriate to use measures of their behavior or physiology in the wild as benchmarks for defining acceptable welfare in animals under our care.

The *motivational priorities* approach may offer one possible solution. This approach examines how the animal values different features of its environment. Scientists have developed ways of measuring how hard animals will work to get access to resources such as food, shelter, or companions. They have shown that animals will continue to maintain access to the same amount of certain resources even if they have to work very hard for them. In the same way, the extent to which animals work to avoid things can also provide valuable information about how aversive or damaging these are. This information can be used in designing animal housing from first principles to provide what animals want and omit what they don't want. It may even be used to examine animal preferences for two existing systems, thus allowing the animal to express an overall decision about its welfare which precludes the need to assimilate many welfare indicators into one final welfare score. However, the approach has its own problems, including the fact that animals don't always choose what's best for them, conclusions are limited to the resources that are tested (the animal may be choosing the lesser of two evils), and it is difficult to decide at exactly what level of work a resource becomes important enough for it to be considered an essential feature of the captive animal's environment.

The scientific assessment of animal welfare has much to offer in terms of informing us about how animals perceive their environments and what they do or do not want. A combination of *welfare indicators* and *motivational priorities* may be the best way of assessing welfare, and a few studies have taken this combined approach. Scientific information can be used to argue that the welfare of animals

kept in one way is better or worse than that of animals kept in a different way. More generally, it can inject some much needed knowledge about the animals' perceptions of the ways in which they are managed into debates about animal welfare.

Further Reading

Boissy, A., G. Manteuffel, M. B. Jensen, R. O. Moe, B. Spruijt, L. J. Keeling, et al,. 2007. Assessment of positive emotions in animals to improve their welfare. *Physiology and Behavior* 92: 375–397.

Botreau, R., M. Bonde, A. Butterworth, P. Perny, M.B.M. Bracke, J. Capdeville et al. 2007. Aggregation of measures to produce an overall assessment of animal welfare. Part 1: A review of existing methods. *Animal* 1: 1179–1187.

Broom, D. M. 1991. Animal welfare: Concepts and measurements. *Journal of Animal Science* 69: 4167–4175.

Dawkins, M. S. 1990. From an animal's point of view: Motivation, fitness and animal welfare. *Behavioral and Brain Sciences* 13: 1–61.

Manteuffel, G., B. Puppe, and P. C. Schön. 2004. Vocalization of farm animals as a welfare measure. *Applied Animal Behaviour Science* 88: 163–182.

Mason, G., and M. Mendl. 1993. Why is there no simple way of measuring animal welfare? *Animal Welfare* 2: 301–319.

Mendl, M. 2001. Assessing the welfare state. *Nature* 410: 31–32.

Paul, E. S., E. J. Harding, and M. Mendl. 2005. Measuring emotional processes in animals: The utility of a cognitive approach. *Neuroscience and Biobehavioral Reviews* 29: 469–491.

Wemelsfelder, F., T.E.A. Hunter, M. T. Mendl, and A. B. Lawrence. 2001. Assessing the 'whole animal': A free-choice profiling approach. *Animal Behaviour* 62: 209–220.

Whay, H. R., D.C.J. Main, L. E. Green, and A.J.F. Webster. 2003. Assessment of the welfare of dairy cattle using animal-based measurements: Direct observations and investigation of farm records. *Veterinary Record* 153: 197–202.

Michael Mendl

ANIMAL WELFARE: COPING

Substantial challenges to animal functioning, include those resulting from pathogens, tissue damage, attack or threat of attack by a predator or another individual from the same species, other social competition, complexity of information processing in a situation where an individual receives excessive stimulation, lack of key stimuli such as social contact cues or a teat for a young mammal, lack of overall stimulation, and inability to control interactions with the environment. Hence potentially damaging challenges may come from the environment outside the body, for example, many pathogens or causes of tissue damage, or from within it, for example, anxiety, boredom or frustration which come from the environment of a control system. Systems that respond to or prepare for challenges are coping systems, and *coping means having control of mental and bodily stability.*

Coping attempts may be unsuccessful, in that such control is not achieved, but as soon as there is control, the individual is coping. Systems for attempting to cope with challenge may respond to short-term or long-term problems, or sometimes to both. The responses to challenge may involve activity in parts of the brain and various endocrine, immunological, or other physiological responses, as well as behavior. However, the more that we learn about these responses, the clearer it becomes that these various types of response are interdependent. For example, not only do brain changes regulate bodily coping responses, but adrenal changes have several consequences for brain function, lymphocytes have opioid receptors and a potential for altering brain activity, heart-rate changes can be used to regulate mental state and hence further responses. It is often combinations of difficulties that make coping difficult. This is true for all species of animals. The methods of coping that are used may help with several problems at once. For example, many emergency responses require more energy than normal to allow the animal to utilize skeletal muscle more efficiently, make the heart pump faster, and reduce response time. Such general physiological methods of trying to cope are usually combined with one or more of a variety of physiological responses that are specific to the effect that the environment is having upon the animal. Hence if it is too cold, the animal may raise its hair, shiver, and reduce blood supply to peripheral parts of the body, but in extreme circumstances, adrenal responses are involved as well.

Coping methods may be behavioral and mental as well as physiological, and vary from very active responses to some hazards to passive responses in which the individual minimizes movement. The initial responses to a situation may be largely automatic, but if these are not effective, other changes may be brought about that affect the mental state of the individual. Some coping systems include feelings as a part of functioning, for example, pain, fear and the various kinds of pleasure, all of which are adaptive. Bad feelings which continue for more than a short period are referred to as suffering. Other high- or low-level brain processes and other aspects of body functioning are also a part of attempts to cope with challenge. In order to understand coping systems in humans and other species, it is necessary to study a wide range of mechanisms including complex brain functioning, as well as simpler systems. Investigations of how easy or difficult it is for the individual to cope with the environment, and of how great is the impact of positive or negative aspects of the environment on the individual, are investigations of welfare. If, at some particular time, an individual has no problems to deal with, that individual is likely to be in a good state, including good feelings, and indicated by body physiology, brain state, and behavior. However, an individual may face problems in life that are such that it is unable to cope with them. Prolonged failure to cope results in failure to grow, failure to reproduce, or death. The individual is said to be stressed, and welfare is poor. A further possibility is that an individual faces problems but, using its array of coping mechanisms, is able to cope but only with difficulty and usually also with bad feelings. The greater the difficulty in coping, the worse the welfare.

Further Reading

Broom, D. M., ed. 2001. *Coping with challenge: Welfare in animals including humans.* Berlin: Dahlem University Press.

Broom, D. M., and A. F. Fraser. 2007. *Domestic animal behaviour and welfare,* 4th ed. Wallingford: CABI.

Broom, D. M., and K. G. Johnson. 2000. *Stress and animal welfare.* Dordrecht: Kluwer.

Monat, A., and R. S. Lazarus, eds. 1991. *Stress and coping,* 3rd ed. New York: Columbia University Press.

Donald M. Broom

ANIMAL WELFARE: FREEDOM

Freedom means the possibility to determine actions and to make responses, and has been thought of by many philosophers, including Immanuel Kant, as necessary for a good life in sentient individuals. However, freedom to seek pleasure without concern for all consequences is wrong, and there are few freedoms or rights which would be accepted as valid under all circumstances. The right to free speech can cause great harm to certain individuals and hence can be morally wrong, as can the right or freedom to drive a car as fast as you wish, to carry a gun, or to select the sex of your offspring. In the same way, social animals are constrained by their relationships with others such that specification of individual freedoms can sometimes be erroneous. The socially competent pig or dog is not free to do as he or she chooses. The safer argument when evaluating what comprises a moral action is to consider the obligations of the actor.

One of the approaches that has been adopted when attempting to ensure that the welfare of animals is good is to list the freedoms that should be provided for. The idea of specifying the freedoms that should be given to animals was put forward in the Brambell Committee Report, which was presented to the Government of the United Kingdom in 1965. The freedoms were defined as freedom from: (1) hunger and thirst; (2) discomfort; (3) pain, injury, or disease; (4) fear and distress; and (5) the freedom to express normal behavior by providing sufficient space, proper facilities, and company of the animal's own kind.

This list of freedoms has been a useful general guideline, but animal welfare science has progressed rapidly since that time and there is now good evidence for the needs of most domestic species. The needs are identified by strength-of-preference studies and research identifying the extent of poor welfare if it is not possible to fulfill the needs. There is now little point in listing the freedoms, because the species needs are a much more accurate way to decide upon what should be provided to ensure good welfare.

Further Reading
Broom, D. M. 1988. Needs, freedoms, and the assessment of welfare. *Applied Animal Behaviour Science* 19: 384–386.
Broom, D. M. 2003. *The evolution of morality and religion*. Cambridge: Cambridge University Press.
Broom, D. M., and A. F. Fraser. 2007. *Domestic animal behaviour and welfare*, 4th ed. Wallingford: CABI.
Webster, J. 1995. *Animal welfare: A cool eye towards Eden*. Oxford: Blackwell.

Donald M. Broom

ANIMAL WELFARE: RISK ASSESSMENT

Many people who are interested in the assessment of animal welfare want to use a science-based approach. Indeed, such science-based analyses will likely become the major way in which future legislation will be formulated. This trend is already obvious in the European Union (EU). Risk analysis is one way to quantitatively study animal welfare. Risk analysis comprises three parts: risk assessment, risk management, and risk communication. This essay deals only with risk assessment.

Risk assessment is a systematic, scientifically based process to estimate the probability of exposure to a hazard, and the magnitude of the adverse welfare effects (that is the consequences in terms of severity) of that exposure. The aim is to analyze the risk of animal suffering, that is, poor welfare, in a quantitative or semi-quantitative fashion depending on the type of data available. Conversely, a similar approach could be used to make an assessment of the likelihood of good welfare.

Hazards are identified as events or circumstances occurring in an animal's life that may result in adverse effects for an individual animal. For example, concrete floors may result in lameness for a dairy cow; lack of space may lead to stereotypic behavior for a captive animal; crowding of fish during capture or grading may lead to scale loss and other superficial injuries; misuse of a captive bolt (also called a cattle gun, which is used to stun cattle prior to slaughter) may cause pain by not inducing immediate unconsciousness. It may also be possible to characterize the hazard more precisely in order to define its quality or quantity in some way (e.g., the nature of concrete floor surface, power of the captive bolt, exact size of floor area).

The consequences of being exposed to the defined hazard are analyzed in terms of the intensity and duration of the adverse effect(s) being suffered by an individual animal. The combination of intensity and duration (severity) is then expressed in some way as the magnitude of the severity.

The likelihood of the severity occurring is also assessed or calculated depending on the quality and type of data available, and assessors are asked to give maximum, minimum, and most likely incidence.

While the above considerations refer to an individual, it may be necessary to know how commonly it happens, that is, the exposure of a population of animals to the hazard, for example, in the national herd, or in a trading area such as the United States or European Union.

The data are also analyzed for their degree of reliability or uncertainty/certainty, as information may vary from a metanalysis at one extreme (low uncertainty/high certainty) to little published evidence, that is, scarce or no data available, or evidence provided in unpublished reports, or few observations and personal communications, or experts' opinions which vary considerably (high uncertainty/low certainty).

Data can be also confounded by the degree of biological variation. For example, there is good data to show that animals that have had their pancreas removed will develop diabetes, that is, there is a high degree of certainty (or low uncertainty) and the biological variation is probably zero. However, the time it takes for fish to die from anoxia on ice, or the chance of a dairy cow becoming lame on a concrete surface is neither 100 percent nor zero, and any average will depend on biological and other factors leading to a range due to biological variation; this can also possess a high degree of certainty. Overall uncertainty associated with exposure to a hazard is recaptured by measuring the maximum and minimum estimates of the most likely value of the proportion of the population exposed, and those that suffered the adverse effect.

Risk pathways are helpful to identify precisely what hazards an animal may be exposed to and to make sure that multiple hazards are covered. For example, in abattoirs it would be important to look at pre-slaughter gathering in lairages, as well as the methods of stunning and killing and bleeding out.

Finally, a risk score is calculated applying either numbers obtained from the data (a quantitative risk assessment), or allocating numerical scores to bands of data, for example, no, low moderate or severity (say from 0–3) (a semi-qualitative risk assessment). Uncertainty is reflected in the range of values obtained from maximum, minimum, and most likely.

The methodology used does not give a precise numerical estimate of the risk attributed to certain hazards. However, the output can be used to rank problems and identify areas of concern, as well as guidance for future research. The methodology does not take into account interactions between factors and assumes linearity in the scores. These assumptions cannot be tested. When the risk scoring is semi-quantitative, as it always is for welfare assessments, the figures are not on a linear scale, and so a risk score of 12 cannot be interpreted as twice as important as a risk score of 6.

The risk assessment approach is the first to compare the severity of procedures and environments to which animals are exposed in a mathematical way. Alternatives can be compared, for example, within and between different systems of husbandry, between different breeds or strains, between different methods of slaughter, mutilations, or breeding. In this way risk scores can be used to prioritize risk management in a trading area, for example, by passing legislation. It will also be useful to prioritize research funding.

Hazard Analysis and Critical Control Points: After a risk assessment has been carried out, it may be possible to identify particular points in the risk pathway that can be used to monitor stages of the process which involve exposure to specific hazards that jeopardize animal welfare, and for controls to be put in place.

Further Reading

European Food Safety Authority (EFSA): http://www.efsa.europa.eu/EFSA/ScientificPanels/efsa_locale-1178620753812_AHAW.htm (See EFSA Web site for opinions that comment on welfare.)

David Morton

ANIMAL-ASSISTED THERAPY

Animals' ability to motivate and bring comfort and joy to people's lives can be harnessed to enhance the quality of a person's life. Dramatic evidence of benefits in specific cases inspires people to incorporate animals into institutional settings, and some make regular visits with a companion animal to a facility. It is not surprising that the practice of using animals for activities, therapy, and in education has developed faster than the scientific knowledge of efficacy, the animals' needs, and the educational curricula for health professionals.

Seeing a sick or depressed child come alive in the presence of a dog, or an elderly person with Alzheimer's disease emerge from a silent cloud when a cat approaches motivates countless people to participate as volunteers in animal-assisted activities, therapy, and education. Although such programs are staffed by volunteers in virtually every community, the promise of animals helping people remains to be broadly mainstreamed in medical settings as a common intervention for treatment. Standards for veterinary screening and oversight of the animals is an essential aspect of an integrated plan for therapy that uses animals.

People's enjoyment of animals, along with growing evidence for the healthful effects of contact with companion animals, has facilitated the expanding practice of incorporating animals in interventions. Therapeutic uses of animals range from brief visits to full-time partnership. In the United States, since the 1980s, dogs and cats have been brought to visit in nursing homes and hospitals. Full-time service dogs also began to be placed with persons using wheelchairs, providing them with therapeutic companionship and comfort 24 hours a day, and normalizing their lives in the community.

Assistance dogs are now specially trained to offer specific assistance in a growing range of tasks. Dogs offer emotional support and companionship that presumably are of greater importance than the instrumental tasks they perform. Dogs assist people with their personal needs, including giving warning and assistance with epileptic seizures, warning of hypoglycemic episodes, and calming during episodes of mental illness. Whether the animal is a short-term visitor brought by a handler or a full-time assistance animal, they have the potential to be beneficial.

Another common animal-assisted intervention that requires professional supervision, physical infrastructure, extensive care and management for the horses, and ongoing assistance from several helpers, is equine assisted therapy, or hippotherapy, in which the movement of horseback riding is used to offer muscular and postural stimulation and motivate riders in their learning and classroom activities. Since a treatment team is required for working with horses, special organizations address this form of therapy, including the North American Riding for the Handicapped Association (NARHA).

The Animal and Handler

Many applications of animal-assisted activities, therapy, and education use an animal that is brought by a handler to serve another person who can benefit. The handler may complete special instruction courses and take the animal for training and screening in order to be well-prepared to visit institutions and other settings. An advantage of this system is that the person offered the visitation is spared the responsibility of oversight and care of the animal. Someone who is institutionalized, in hospice, or in medical recovery often is not prepared to assume responsibility for an animal's care, but can still benefit from occasional visits with an animal.

When the person is vigorous and healthy enough to oversee and provide most of the fulltime care for the animal, it may be more beneficial for the same person to be the handler and also receive the benefits. Assistance dogs provide a fulltime therapeutic relationship. Dogs may be specially bred and extensively trained over a couple of years, as with guide and service dogs, or the training may be conducted over a shorter period of time, as with hearing and seizure dogs. Psychiatric service dogs are a new development, where the handler arranges for the training, usually with a companion animal that is already on hand. The handler may have a physical or mental disability and still assume the major responsibility for the dog's care. Dogs placed with people in wheelchairs have been termed service dogs, and are prepared similarly to guide dogs, with special breeding, puppy raisers, and extensive training. As the applications of assistance dogs have broadened, the designation of service, guide, and hearing dogs has often converged with the term assistance dogs; however, the nomenclature is not entirely consistent.

Legislative protection permits an assistance animal for people with a disability that interferes with their ability to perform the activities of daily living. Regulatory language allowing public access may use the term service or assistance animal, and the terminology has become less specific and more overlapping. The lack of any system for governmental or regulatory certification, paired with the personalized training of dogs to address specific needs of the person, results in a continuing expansion of the special roles of dogs.

The Welfare of the Animal

Most animal-assisted interventions employ dogs or horses. Both of these species benefit when handlers are knowledgeable about their basic needs and veterinary guidance is available. Dogs readily take to partnership with their human companions. Most breeds of dogs used are those that were specifically shaped to assist humans in particular tasks. When a breed that is well-suited for the expected tasks is selected, a dog given suitable experiences and training has a high probability of becoming a successful partner. Virtually all dogs welcome the handler, enjoy walks, and are expressive, loyal, and attentive—all traits that are highly valued by people who spend time with dogs. The subtle attentiveness of dogs to humans is now well documented, showing that dogs respond to the gaze, pointing, or yawning of a human. Thus, a natural compatibility arises between the dog that likes working as a partner and the handler who feels appreciated and loved by the dog.

Horses offer inspiring partnerships that can be highly motivating as an intervention. The safety concerns and the challenges of managing such a large animals require that a number of people be involved in providing equine-assisted therapy. The welfare of horses has been well studied, and information is available on methods of training, husbandry and transport.

Adequate curricula to assure the appropriate application of animal-assisted interventions, as well as the welfare of the animals involved, are only now coming to be available. In the United States, optional programs for certifying handlers and their animals have focused on preparing volunteers and their animals.

There have been few educational avenues for health professionals to gain coursework and practical experience in applying animal-assisted interventions and learning about the animals' needs. Practitioners have been self-taught and sought out their own path of study and experience. A recent development is the establishment of the International Society for Animal-Assisted Therapy, which accredits educational programs designed to prepare health professionals.

For the past couple of decades, practitioners of animal-assisted therapy have begun with traditional educational programs in the health professions and have had to develop their own techniques for animal-assisted interventions, building them upon their own health disciplines. To address this curricular gap, the International Society for Animal-Assisted Therapy now offers an accreditation process, with stated requirements and a detailed review process. Already, two institutions are accredited to offer instruction in Animal-Assisted Interventions, the Institute for Social Learning with Animals, in Germany, and the Institute for applied Ethology and Animal Psychology, in Switzerland. Applications from other countries are forthcoming. These programs accept students from a variety of health professions and offer flexibility for enrollees to focus on a specific area of interest for their internship and special project, such as equine-assisted therapy or animal-assisted pedagogy. These programs will be of value to those working in a wide range of settings, and with a range of species of animals. They also expand the curricular materials available for practitioners.

In the United States, the University of Denver offers a special emphasis on animal-assisted interventions within the social work program. The Bergin University of Canine Studies in Santa Rosa, California, provides undergraduate and graduate instruction focused on assistance dogs. These recent developments signal an accelerated emphasis on bolstering the number of professional opportunities in animal-assisted interventions.

Further Reading

Delta Society, accessed on December 15, 2008: http://www.deltasociety.org/Volunteer AboutAbout.htm

Fine, Aubrey H., ed. 2006. *Handbook on animal-assisted therapy: Theoretical foundations and guidelines for practice*, 2nd ed. Boston: Elsevier/Academic Press.

Hart, Lynette A. 2006. "Community context and psychosocial benefits of animal companionship." In Fine, Aubrey H., ed., *Handbook on animal-assisted therapy: Theoretical foundations and guidelines for practice*, 2nd ed. Boston: Elsevier/Academic Press, 73–94.

International Society for Animal-Assisted Therapy, accessed on December 15, 2008: www.aat-isaat.org

Melson, Gail. 2001. *Why the wild things are: Animals in the lives of children*. Cambridge: Harvard University Press.

North American Riding for the Handicapped Association (NARHA), accessed on December 15, 2008: http://www.narha.org/.

Psychiatric Service Dog Society, accessed on December 15, 2008, http://www.psychdog.org/.

Serpell, James. 1991. "Beneficial effects of pet ownership on some aspects of human health and behaviour." *Journal of the Royal Society of Medicine* 84 (Dec.): 717–720.

Lynette A. Hart

ANIMALS IN SPACE

Before human beings ventured into space, American and Russian scientists launched animals with the aim of testing both their rocket engineering and the living conditions of the environment(s) which they would eventually encounter (e.g., the effects of weightlessness and risk of sun radiation). Once human missions began, astronauts typically also took animals with them so as to conduct further biological experiments, a practice that continues to this day at the International Space Station.

Insects and animals launched on either orbital or suborbital flights with little chance of survival have included chimps, dogs, monkeys, cats, rabbits, mice, rats, turtles, frogs, spiders, bees, crickets, silkworms, fruit flies, ants, and fish. Many of them never returned. Over the years many countries have issued stamps, bubblegum cards, and even cigarette packets commemorating both those that did and those that did not make it back to Earth; such acts could potentially benefit animal welfare by promoting awareness, but it is far from clear that this has ever been the aim behind them.

With the exception of a jar of fruit flies which was successfully flown 106 miles above the earth and parachuted back in 1947, the first five animals to be sent into space—collectively known as the Albert Series—all boarded V-2 Blossoms from White Sands, New Mexico between 1948 and 1950. Albert I was a rhesus monkey who, on June 11 1948, was launched with virtually no publicity or documentation. Three days later another lab monkey, Albert II, reached an altitude of 83 miles, but died upon impact. In August that same year, an anesthetized mouse, Albert III,

was the first astronaut to return alive. He was followed, in December 1949, by Albert IV, a monkey who died on impact after a successful flight. In May 1950 the last of the Alberts, a mouse named Albert V, was launched; this mouse survived impact, having been photographed in flight. Next came the animal astronauts of the Aerobee missile flights, launched from Holloman Air Force Base, New Mexico. First up, on September 20 1951, were Yorick, another monkey, and his 11 co-passengers, all mice. The 236,000-foot missile flight was successful, and Yorick became known as the first monkey to survive spaceflight. On May 22 1952 he was followed by Patricia and Mike (two Philippine monkeys) and Mildred and Albert (two mice), who were all placed in different positions (the last two inside a drum where they could float weightlessly) and shot up 36 miles at an average speed of 2,000 mph, in order to test various effects of rapid acceleration. They were all recovered safely by parachute.

Meanwhile, back in the USSR, Soviet scientists began experimenting on rats, mice, rabbits and, eventually, dogs. The latter were chosen with the ultimate aim of designing a human space cabin (monkeys were thought to be too fidgety). Between 1951 and 1952, the Soviets launched at least nine stray female dogs (always in pairs, some dogs flew twice) on at least six of their R-1 series rockets (the precise facts are disputed among researchers). The first pair of hounds, Dezik and Tsyganka ("Gypsy"), were launched on August 15 1951 and were successfully recovered by parachute. Next came Dezik and Lisa, who in September of that year tragically died in an unsuccessful flight. The third pair, Smelaya ("Bold") and Malyshka ("Little One"), were launched successfully, despite Smelaya's brief escape on the eve of their mission. The fourth launch (carrying a pair of dogs whose names remain unknown) was as tragic as the second, but the fifth (also carrying two anonymous strays) was successful. Finally, on September 15 1951, the last pair of R-1 canines were launched successfully, though only after one of the original crew (Bobik) escaped and was replaced at the last minute by a dog found near the local canteen, which they named ZIB (the Russian acronym for "substituting for missing dog Bobik").

Four years later, on November 3 1957, Sputnik 2 famously orbited with a 13-pound stray female mongrel named Laika ("Barker," though her real name was Kudryavka, which means "little curly"). Two other dogs had been trained for this flight—Albina, who was the first back-up, and Mushka, upon whom the life support and instrumentation were tested. The dogs were all kept in increasingly smaller cages for periods of two to three weeks.

Laika's mission commemorated the 40th anniversary of the Great October Socialist Revolution (celebrated on November 7), yet shortly after the launch the Soviets caused public outrage by announcing that Laika—the first living being to orbit the earth—would almost certainly die in space because there was no recovery method, at the time, for orbital flights. For a shockingly long time, they somehow managed to persuade the world that Laika ate and barked for about a week before dying painlessly in orbit. It was not until 2002 that Dimitri Malashenkov of the Institute for Biological Problems in Moscow revealed that Laika had died from panic and/or overheating a mere five to seven hours after takeoff. The dead dog then circled the earth more than 2,500

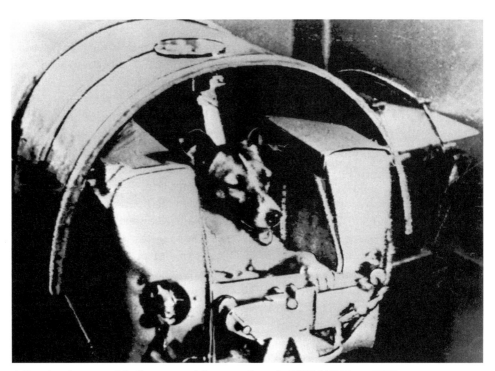

A Russian dog named Laika prepared for space launch, 1957 (AP-Photo/HO)

times before burning up in its atmosphere. In so doing she provided proof that a living organism can survive in weightlessness for a long time, thus paving the way for the human astronauts of the 1960s. Did the end justify the means? In 1988 Oleg Gazenko (one of the leading Soviet scientists involved in their animals-in-space program) announced that he regretted sending Laika into space: "The more time passes, the more I'm sorry about it. We shouldn't have done it . . . we did not learn enough from this mission to justify the death of the dog." Laika's death also enraged anti-vivisectionists in America, who were joined by anti-communists in their public expressions of outrage, an ironic turn of events given that medical researchers in the United States had previously characterized those who opposed animal experimentation as communist-led fanatics (cf. Los Angeles *Times* editorial, April 18, 1950).

On December 13 1958, a Jupiter rocket was fired, carrying a South American squirrel monkey named Gordo who, after a 15-minute, 1,800-mile flight, died when a parachute failed to open. The naval medicals concluded that his heartbeat and respiration showed that humans could survive a similar trip. However, the ASPCA complained that only inanimate objects should be used for such tests, and The British Royal Society for the Prevention of Cruelty to Animals expressed "grave concern and apprehension." Even so, Able, an American-born rhesus monkey, and Baker, a female spider monkey from South America, were launched on May 28 1959, aboard an Army Jupiter missile. They traveled 300 miles into space at speeds which at times exceeded 10,000 mph, and were recovered unharmed, becoming the first living creatures to survive a space flight. The flight lasted 15 minutes,

during nine of which the monkeys were weightless. The mission was criticized by various animal welfare groups. Able died on June 1 from the effects of anesthesia. On June 3, four mice were launched from Vandenberg Air Force Base on Discoverer 3 as part of the Corona program of US spy satellite. According to the CIA, the first try at launch failed when the mice were found dead before the flight after having eaten the Krylon that had been sprayed on their cages. The second try was delayed when scientists realized that the humidity sensor couldn't distinguish between water and mouse urine. When the rocket was finally launched, it fired into the Pacific ocean, and the back-up crew of mice died.

One of the best known space monkeys was a rhesus called Sam (an acronym for the U.S. Air Force's School of Aviation Medicine), launched on December 4 1959 with the aim of testing the launch escape system of a Mercury spacecraft. The experiment was successful, and Sam was recovered a few hours later. He lived a long and healthy life until 1982. His mate, a rhesus monkey named Miss Sam, was launched in a similarly successful test on January 21 the following year.

Soon after, the Soviet Union began testing on more dogs, including Otvazhnaya ("Brave One") and Snezhinka ("Snowflake'"), who made a successful high altitude test in 1959, accompanied by a rabbit called Marfusha. Over the next year, Otvazhnaya was to participate in five more similar experiments. A few weeks later, on July 28, Bars ("Panther") and Lisichka ("Little Fox") were launched on a Korabl Sputnik 1; both dogs died when the booster exploded. On August 19 1960, Belka ("Squirrel") and Strelka ("Little Arrow") boarded Korabl Sputnik 2, accompanied by 40 mice, two

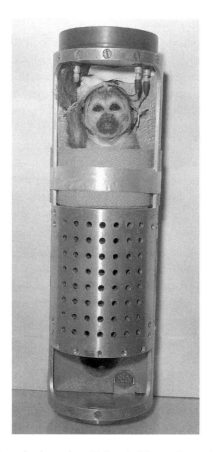

A squirrel monkey, Baker, in bio-pack couch being readied for the Jupiter AM-18 flight, launched on May 28, 1959. The Jupiter, AM-18 mission, also carried an American-born rhesus monkey, Able, into suborbit. The flight was successful and both monkeys were recovered in good condition. Able died four days after the flight and Baker died in November 1964. (NASA)

rats, a grey rabbit, and 15 flasks containing plants and fruit flies. The flight was a success, and Strelka later gave birth to six puppies one, of which was presented to President John F. Kennedy as a gift for his children.

Arguably the most famous animal astronaut of all was a four-year-old West African chimpanzee called Ham (his name was an acronym for Holloman Aero

Med) who, having been chosen from a short list of six astrochimps, on January 31 1960, donned his spacesuit and boarded the Mercury Redstone rocket at Cape Canaveral to become the first chimpanzee in space. Reaching a record speed of 5,857 mph and an altitude of over 155 miles (both due to technical problems) Ham was weightless for 6.6 minutes of his 16.5-minute flight. He landed dehydrated and fatigued to be rewarded with an apple and half an orange, but went on to live a healthy life until 1983. His body was preserved by the Smithsonian Institute, which has permanently loaned it to the International Space Hall of Fame in Alamogordo, New Mexico. Without Ham, America would not have been able to launch its first human astronaut, Alan B. Shepard, Jr., on May 5 1961, though by then the Soviets had already orbited Yuri Gagarin around the Earth for almost two hours on April 12 of that same year, following their successful Kotabl dog launches. While the Soviets had chosen dogs over monkeys for their experiments (because they fidgeted less), the Americans preferred chimpanzees over dogs because they were more similar to humans.

Some of the technical defects observed during Ham's flight were not corrected until November 1961, when a chimpanzee named Enos orbited the Earth twice. The mission plan had called for three orbits, but the flight was terminated early due to technical difficulties, which included a thruster malfunction. Without this further animal testing, John Glenn would not have been able to orbit Earth in 1962. That year, Enos was reported to have died at Holloman Air Force Base of a case of dysentery unrelated to his space travel. Equally unfortunate was Goliath, a squirrel monkey killed on November 10 1961 when the Atlas E rocket he was launched in was destroyed within 35 seconds.

On October 18 1963, the French launched and successfully recovered Felix, the first cat to make it to space. A second cat launched six days later could not be retrieved. On February 22 1966, the Soviets launched two more dogs, Veterok ("Breeze") and Ugoyok ("Little Piece of Coal"), in order to test the prolonged effects of radiation caused by the Van Allen Belt during space travel. Their 21 days in space remain the canine record to this day, surpassed only by humans in 1974. On September 15 1968, Soviet scientists launched a number of turtles, worms, flies, and bacteria on a one-week mission to orbit the Moon. The reentry capsule was successfully retrieved. A similar mission on November 10 of that same year was unsuccessful.

Between 1966 and 1969, the United States launched three missions in the Biosatellite series. The first of these carried insects and frog eggs, as well as various microorganisms and plants; it was never recovered. The second, launched in 1967, had a similar cargo, but was recalled early, while the third, launched in 1969, carried a male monkey named Bonnie. This mission's main purpose was to investigate the effect of space travel on numerous functions and abilities including behavior, cognition, and metabolism. Eight hours after his recovery, Bonnie died of a heart attack caused by dehydration.

After the successful human landing of Apollo 11 on the Moon, the use of animals in space was mainly restricted to biological experiments. Popular subjects included turtles, rabbits, fish, and insects. On July 28 1973, two spiders named Anita and Arabella were to make headline news by spinning their webs in space. Between then and 1996, the Russians launched a series of life-science missions (involving various monkeys, rats, tortoises, insects, frogs, fish, newts, and quail eggs) in cooperation with a number of countries and organizations including the Commonwealth of Independent States, the European Space Agency, and the United States. A monkey named Multik died the day after his recovery from one such two-week mission, the Bion 11, putting ethical questions relating to animal experimentation back on the agenda, and causing NASA to back out of participation in subsequent Bion missions. Other life-science experiments have included Spacelab missions (1983-present day) which experimented on both humans and animals. The environment within the animal enclosure modules used in these missions meets most of the recommendations of the *NIH Guide to Care and Use of Laboratory Animals,* with the exception of its increased ambient temperature and housing density. On April 17 1988, a record number of over 2,000 creatures accompanied the seven human astronauts of shuttle *Columbia* (STS-90) on a 16-day Neurolab mission.

The 1990s saw China launch guinea pigs and Japan launch newts, while the United States extended its menagerie to include snails, sea urchins, moths, crickets, carp, and oyster toadfish. More recently, in December 2001, 24 mice boarded the space shuttle *Endeavour* as part of an experiment on a bone-regulating protein called osteoprotegerin, while in January 2003, the space shuttle *Columbia* carried bees, ants, silkworms, and Japanese killifish, all part of various international high school projects. To this day, the United States, Russia, China, Japan, and France all continue to fly animals into space.

Further Reading

BBC. *On this day*, http://www.bbc.co.uk

Borkowski, G. L., Wilfinger, W. W., and Lane, P. K. (1996). Laboratory animals in space: Life sciences research, in *Animal Welfare Information Center Newsletter*, Vol. 6, N. 2–4, Winter 1995/1996.

Cassidy, D. & Hughes, P. (2005). *One small step: America's first primates in space*. New York: Chamberlain Bros.

Fuller, J. (2008). Why are there dozens of dead animals floating in space? 17 March 2008, HowSTuffWorks.com, retrieved 13 April 2008 from http://science.howstuffworks.com/dead-animals-in-space.htm.

Shapiro, R. N., & Teigen, P. M. (2006). *Animals as cold warriors: Missiles, medicine, and man's best friend* (United States National Library of Medicine Exhibition, http://www.nlm.nih.gov/exhibition/animals/index.html).

Tara G. (1988). *A brief history of animals in space*, http://history.nasa.gov/animals.html.

Constantine Sandis

ANTHROPOCENTRISM

The term anthropocentrism refers to any view that asserts the centrality, primacy, or superiority of human beings in the scheme of things; that claims the purpose of nature is to serve human needs and wants; or that posits the greater value of human life and interests relative to the lives and interests, if any, of nonhumans. Such views are highly characteristic of modern civilization and are frequently implicated in discussions of the world environmental crisis, the abuse of animals, and threats of species extinction. From the anthropocentric standpoint, other species—and nature as a whole—exist in a subservient relationship to our own species. This relationship may be rationalized by some kind of metanarrative, such as a story about divinely ordered creation (and humans' bearing the image of God), the great chain of being, or a putative evolutionary hierarchy, or it may merely be asserted as the natural outcome of human development and exploitative skill. In other words, the concept of human superiority may be understood in either a *de jure* (justified) or a *de facto* (happenstance) manner.

Anthropocentrism is also characterized by such terms as "homocentrism", "human chauvinism", "speciesism," and "human-centered ethics." In its crudest expression, anthropocentrism entails an outlook of the following kinds: that human interests, needs, and desires are the only ones that count; that if any life-form can be said to possess intrinsic value, only *Homo sapiens* can; that humans represent a different order of being that exists apart from nature rather than as a part of nature. Anthropocentrism is often equated with anthropomorphism, but this is an error; the two terms should be carefully distinguished.

Three main varieties of anthropocentrism can be identified:

1. **Dominionism:** Rooted in the Old Testament and in ancient Greek philosophy, this is the position that nature and individual things in nature exist only for human benefit. Dominionism is also referred to in the literature as "strong anthropocentrism," and is commonly associated with such ideas as mastery of nature and nature's possessing only instrumental or use value, and with the human species' self-glorification. Arrogance rather than humility is the mark of strong anthropocentric attitudes and behavior. Dominionists think of nature as a boundless storehouse of resources. The frontier

mentality and unrestrained development are representative modes of dominionism.

2. **Stewardship:** A milder form of anthropocentrism may also be traced to the Judeo-Christian tradition and is found in others as well. Often labeled "weak anthropocentrism" in the literature, the stewardship view is manifested in such ideas as responsible husbandry, wise management and conservation of resources, and preservation of species and natural wonders, although sometimes preservation is linked to the notion of something's being of value for its own sake. By one estimate (Butkus, 2002) there are no fewer than 26 references to stewardship in the Bible, and even the dominionist account (in Genesis 1:26–28) of how God assigns to humans their place in nature is often interpreted as a prescription to tend the Earth in a measured and loving way. Within Islam too, humans may be seen as nature's caretakers, the vice-regents of Allah for whose glory all acts are performed. And in the thought systems of Indigenous Peoples in many parts of the world, ideas of stewardship are present—for example, in the principle that the Earth is inherited from our ancestors and must be carefully looked after, in order to be passed on intact and in good health to future generations. Within stewardship, arguably, humans still matter most, but other species matter and possess noninstrumental value as well. This framework allows room for the projects of advancing biodiversity and pursuing sustainable development.

3. **Evolutionary Perspectivism:** According to this view, it is natural for each species to act as if its survival, flourishing, and reproduction are the highest goods. Given this premise, inter-species clashes are inevitable; there could not be an ecosphere as we know it without conflict and competition. Some infer from this that whatever humans choose to do in nature is simply a reflection of our species-specific behavioral repertoire, which we exhibit just as other animal types exhibit theirs. Others suggest that nature's wellbeing is not in conflict with human-centered behavior but actually coincides with an enlightened form of our species' self-interest, so that there need be no ultimate opposition between humans and the rest of nature. That is, when humans pursue their proper end, they will then act in the best interests of nature as a whole. Entomologist E. O. Wilson even argues that what he calls "biophilia" (love of life) has played a crucial role in the history of human development (see Kellert, 2003).

Many philosophers perceive anthropocentrism as a belief that, if it ever had an important function, has now outlived its usefulness and become not just outmoded, but a dangerous threat to fragile ecosystems and even to the survival of life on Earth. Others maintain that anthropocentrism is in some sense inescapable and, at a certain level of interpretation, scarcely remarkable at all. Just as spiders, if they could evaluate the world around them conceptually and articulate their thoughts in language, would be arachnicentric, so would wolves be lupucentric and cows bovicentric. How, then, could humans be other than homocentric? But while we may, and perhaps must, accept that human values and experience determine the standpoint from which we project outwards, it does not necessarily follow that overcoming or at least mitigating the more harmful effects of our anthropocentric outlook is an impossible goal. The human viewpoint is an anchoring reference to which we will always return, but this does not mean that all values must in the end be human-centered or that we must continue, in our thinking, to place ourselves above all else, at all times, at the center of significance. We should not conclude that empathy and connection with nonhuman nature are unavailable to us merely because we happen to belong to the species *Homo sapiens,* any more than we should conclude that it is beyond us to empathize and connect with fellow human beings just because we all happen to be individual, separate subjects of consciousness with our own peculiar identities. Perhaps it is not too great a step to recognize that in the natural world there are nonhuman entities and configurations possessing their own intrinsic value. How far we can and should try to extend ourselves beyond our foundational anthropocentrism, therefore, is something that cannot be decided in advance, and only time will tell how successful we might become at this or whether we might evolve into beings who can coexist with our own kind as well as with nonhuman species.

See also Animal-Human Interactions, Ecological Inclusion, Empathy for Animals; Religion and Animals

Further Reading

Agar, N. 2001. *Life's intrinsic value: Science, ethics, and nature.* New York: Columbia University Press.

Butkus, R. A. 2002. *The stewardship of creation.* Waco, TX: Center for Christian Ethics at Baylor University. www.baylor.edu/christianethics/CreationarticleButkus.pdf.

Crocker, D. A., & Linden, T. 1997. *Ethics of consumption: The good life, justice, and global stewardship.* Lanham, MD: Rowman & Littlefield.

Goldin, O., & Kilroe, P., eds. 1997. *Human life and the natural world: Readings in the history of western philosophy.* Peterborough, ON: Broadview Press.

Johnson, L. E. 1991. *A morally deep world: An essay on moral significance and environmental ethics.* Cambridge: Cambridge University Press.

Kellert, S. R. 2003. *Kinship to mastery: Biophilia in human evolution and development.* Washington, DC: Island Press.

Manes, C. 1992. "Nature and silence." *Environmental ethics,* 14, 339–350.

Passmore, J. 1980. *Man's responsibility for nature: Ecological problems and western traditions,* 2nd ed. London: Duckworth.

Steiner, G. 2005. *Anthropocentrism and its discontents: The moral status of animals in western philosophy.* Pittsburgh: University of Pittsburgh Press.

Taylor, P. W. 1986. *Respect for nature: A theory of environmental ethics.* Princeton: Princeton University Press.

Michael Allen Fox

ANTHROPOMORPHISM

Anthropomorphism is, at its most general, the assignment of human characteristics to objects, events, or nonhuman animals. Notably, belying this neutral definition is a non-neutral connotation to the word and to the phenomenon it describes. Specifically, an anthropomorphic characterization is generally held to be an erroneous one—at best, premature or incomplete, and at worst, dangerously misleading. That anthropomorphism is, further, incorrect as a description is often assumed.

Of greatest relevance to the study of animal rights and welfare are anthropomorphisms of animals as having attributes and mental states (especially emotional and cognitive states) similar to human attributes and mental states. Pets are regular subjects—a dog's low, rapid tail-wagging explained as guilt for eating a shoe, or a cat rubbing against its owner interpreted as an expression of fondness. The pain or grief of laboratory animals is often evoked by those pressing for improvements in the animals' welfare. Research in the recently developed field of cognitive ethology accumulates empirical data on precisely the kinds of mental states that anthropomorphism claims (without the backing of science): the purposes, feelings, motivations, and cognition of animals. Thus, the science and the attributions are interwoven. This is the form of anthropomorphism with which we shall primarily concern ourselves in this essay.

The Meaning of the Term

As we shall see, anthropomorphizing is generally disapproved of in describing animals. By its very definition, anthropomorphism is the misapplication to animals of words used to describe humans. Some excuse anthropomorphism as simply a form of analogy. "My dog loves that little poodle," one might say, is a claim of the presence of emotions between dogs that is analogous to those emotions in humans. In other words, the dog may not feel "love," per se, but something like love: he follows her around, he wags his tail uncontrollably when she appears, he persists in attempting to mount her . . . and so on, more or less just like human love. This is credible, although it does not exempt anthropomorphizers from criticism

on factual grounds, if the claim is without scientific support. But even if all anthropomorphisms are simply analogies relying on similarities between the target and the source, not all such analogies are anthropomorphisms. Forming analogies between humans and other animals is regularly considered nonanthropomorphic. For instance, dissection of a sheep's brain in a class on human cognition is not taken to be an anthropomorphic activity. On the other hand, the protest outside the classroom making claims about the suffering of the sacrificed sheep may be.

A Brief History of Anthropomorphism

Anthropomorphic representation appeared in Paleolithic art of 40,000 years ago, when some drawings of animals included characteristically human features; anthropomorphisms have appeared in human writings for thousands of years. All religious systems include anthropomorphisms. Ancient societies projected motives and emotions onto natural phenomena—angry winds, vengeful storms—and animals and natural events were often named and ascribed personalities. Later, even physics was to be influenced by an anthropomorphic teleology. Aristotle described a rock's downward tumble not as the result of a force between bodies, but as the rock acting to achieve the desired end of being on the ground. Both ancient and modern literature as well as folk psychology are replete with anthropomorphic language. The characterizations of Aesop—the happy dog, the persistent tortoise, the industrious ant—resonate and endure to this day.

Reproach for anthropomorphisms has appeared for nearly as long as the anthropomorphisms themselves. Xenophanes (sixth century BC) was the first to give voice to the negative tone of anthropomorphism; soon, the term was appropriated to mean the blasphemous descriptions of gods as having human forms. Modern critiques date to the 17th-century philosophers Francis Bacon and Benedictus de Spinoza. In fact, the rise of modern science is matched by the diminishment and increasing censure of anthropomorphic descriptions of natural phenomena. Many of our practices toward animals, and the traditional view of humans as the acme of the animal kingdom, would be difficult to maintain in the face of a collapse in the division between man and animals.

In its current usage, anthropomorphism is tinged with the bad flavor that the anecdotalism of late 19th-century scientists like Charles Darwin and George Romanes left in science's mouth. While on the one hand epitomizing modern science, Darwin also embraced a classically anthropomorphic attitude toward animals, ascribing everything from emotions to insight to animals with abandon—and the future sciences of zoology, biology, and ethology developed in reaction against this. A comparison of the languages of description makes the distinction clear: Darwin spoke of "ants chasing and pretending to bite each other, like so many puppies" (1871, p. 448). A century later, a more typical description of the study of ants (taken from the Web site of the Polish Nencki Institute's ethology research group) investigates the "neurochemical mechanisms underlying the phenomena of social reward and social cohesion in ant colonies" and "the role of social context in the control of expression/suppression of various elements of ant behavior." Similarly, while Darwin noted that dogs could be variously magnanimous and sensible, shameful and modest, sensible and proud, these words are notably absent from contemporary ethological descriptions of dogs.

Explanations for Anthropomorphism

Why do we anthropomorphize? Anthropomorphism's endurance marks it as likely useful—or at least not irreparably harmful—in explaining and predicting animal behavior. Just as the developing child uses animism—the attribution of life to the inanimate—to make sense of the sensory chaos of his environment, anthropomorphism may have arisen as a strategy to make familiar an uncertain world. In normally developing humans, our characteristic propensity to attribute agency to others can become a theory of mind, and will find use in social interaction. In the development of the human species, anthropomorphism may have provided a means by which to anticipate and understand the behavior of other animals. With themselves as models, our human forebears could ascribe motivation, desire, and understanding to animals to determine with which ones they might want to cooperate or from which ones they should flee—as well as which ones they want to eat.

If there *is* an evolutionary explanation, we might expect other animals to engage in some version of the behavior. In fact, many animals do appear to attribute animal characteristics to inanimate objects or occurrences—what anthropologist Stuart Guthrie has called zoomorphism. In *The Descent of Man*, Darwin described his own dog growling and barking at an open parasol moving in a breeze, as though in the presence of "some strange living agent" (1871, p. 67). Primatologist Jane Goodall observed chimps making threats toward thunderclouds. Other ethologists have noted animals shying from, stalking, or attempting to treat as prey or playmate a variety of natural objects. Nonhuman animals seem to be subject to a similar version of animistic perception as humans.

However, we do not anthropomorphize all animals: gorillas and dogs are regularly anthropomorphized, but worms and manta rays rarely are. Frogs' lack of anthropomorphizable characteristics may have led to their dismal fate at the dissecting table when dissection was a mainstay of biology classes. What are the behaviors and physical features of animals which prompt us to anthropomorphize them?

The answer no doubt has much to do with the ease with which the animal can be mapped to the human, in terms of isomorphisms of features and similarities of movement. Physically, phylogenetic relatedness accounts for some anthropomorphizing (for example, of great apes and monkeys); simple ease of matching of parts may account for other differential treatments (an eel's lack of limbs, the facelessness of a limpet). In particular, discernable and flexuous facial features, the ability to form a mouth into a smile, and the ability to move the head expressively and reactively are reliable prompts to certain kinds of anthropomorphisms. Paleontologist Stephen Jay Gould and ethologist Konrad Lorenz both noted that animals with neotenized features, for example large heads and big eyes, may prompt affiliation and selection because these are features of human juveniles.

Arguments against and for Anthropomorphism

The primary complaint against anthropomorphizing extends the reaction to the anecdotalism of Darwin and others: anthropomorphism is not based in science.

There is no objective theory formation or testing, no careful consideration of evidence; there is merely unreflective application of human descriptions to nonhumans. Some argue that anthropomorphism is a category error, that is, the treatment of an entity (an animal) as a member of a class (things with minds and emotions) to which it does not belong, or the comparison of that entity to one (such as a human) belonging in a different category. Describing a dog as feeling guilt, they claim, is like saying that ideas are green. Those who assert that there are distinctively human traits might so argue: if the trait is, by definition, what separates humans from animals, then to treat an animal as possessing the trait is a logical error. If consciousness is a defining characteristic of humans, for instance, then to claim consciousness in nonhumans is a category mistake.

Indeed, some anthropomorphisms are clearly wrong for just these reasons. Happiness is commonly attributed to an animal on the basis of an upturn of the corners of its mouth; that which appears to be a smile, however, may be a fixed physiological feature (as with dolphins) or a sign of fear or submission (as with chimpanzees), not happiness. Similarly, an animal's yawn is likely not a sign of boredom, as might be assumed by extrapolation from our own behaviors; instead, it denotes stress.

Still, the implied suggestion that any mental ability exhibited by human beings is necessarily exclusive to humans is itself premature. A number of researchers are increasingly proposing a careful application of anthropomorphic terms to explain and predict animal behavior. Interestingly, it is the professional observers of animals who often become, with exposure and despite their training, more likely to anthropomorphize. These advocates suggest that anthropomorphisms are not necessarily incorrect. On the contrary, they say, anthropomorphisms are used in reliable ways and are useful. The comparative psychologist Donald Hebb discovered, for instance, that taking pains to eliminate anthropomorphic descriptions resulted in a diminished understanding of the behavior of his chimpanzees. Anthropomorphisms, carefully applied, may be coherent guides to predicting the future behaviors of animals. The psychologist and biologist Gordon Burghardt proposed using a critical anthropomorphism in science which accepts the inevitability of the tendency to see animals in this way, yet uses informed anthropomorphisms to develop hypotheses that can be empirically tested.

The Future of Anthropomorphism

The claims of anthropomorphism are, often, scientifically unproven—simply extrapolations from our own condition. The onus of science is to find the means to confirm or refute these assertions. Hence the future treatment of anthropomorphism by science should include empirical testing of specific attributions. In the case of attributions of mental states, the process should include a deconstruction of the concepts attributed—from love and guilt to happiness and depression—and a determination of any behavioral correlates, as well as what would count as confirming or disproving evidence of the presence of the attributional state.

The status of anthropomorphism, and the content of its attributions, is highly relevant in the ongoing discussion of the role of animals in our society: as pets, as food and entertainment, and in medical and behavioral research. They can be used to effect change in public perception or even policy. Ascribing personalities to animals is demonstrably more effective than raw statistics in getting the public to consider an animal's or species' plight. An analysis of the content—the work of cognitive ethology—will be relevant to animal law and animal rights movements. If, for instance, attributions of human-like emotional experiences and cognitive abilities to chimpanzees turn out to be correct, the question of the rights we should grant that animal is raised.

Historically, anthropomorphisms have been used to attempt to uncloak, demystify, or get traction in domains unknown (and perhaps unknowable) to humans, such as the subjective experience of an animal. They might be best thought of as attributions of human qualities to nonhumans not proven to bear these qualities. The science of cognitive ethology may provide such proofs. Anthropomorphism will likely continue regardless.

See also Critical Anthropomorphism

Further Reading
Burghardt, G. M. (1985). Animal awareness: Current perceptions and historical perspective. *American Psychologist, 40,* 905–919.
Crist, E. (1999). *Images of animals: Anthropomorphism and animal mind.* Philadelphia: Temple University Press.
Darwin, C. (1981). *The descent of man; and selection in relation to sex.* Princeton: Princeton University Press. (Original work published 1871.)
Datson, L., & Mitman, G. (2005). *Thinking with animals: New perspectives on anthropomorphism.* New York: Oxford University Press.
Guthrie, S. E. (1997). Anthropomorphism: A definition and a theory. In *Anthropomorphism, anecdotes, and animals,* ed. R. W. Mitchell, N. S. Thompson, and H. L. Miles, 50–58. Albany, NY: State University New York Press.
Hebb, D. O. (1946). Emotion in man and animal: An analysis of the intuitive process

of recognition. *Psychological Review, 53,* 88–106.

Horowitz, A. C., & Bekoff, M. (2007). Naturalizing anthropomorphism: Behavioral prompts to our humanizing of animals. *Anthrozöos 20:* 23–35.

Kennedy, J. S. (1992). *The new anthropomorphism.* New York: Cambridge University Press.

Mitchell, R. W., Thompson, N. S., & Miles, H. L. (Eds.). (1997). *Anthropomorphism, anecdotes, and animals.* Albany, NY: State University of New York Press.

"Neurochemical mechanisms underlying the phenomena . . ." Taken from http://www.nencki.gov.pl/en/working_groups/neurophysiology/lab_03.html. September 28, 2008.

Alexandra C. Horowitz

ANTHROPOMORPHISM: CRITICAL ANTHROPOMORPHISM

Anthropomorphism can be useful in studying and interpreting animal behavior if applied critically. This means anchoring anthropomorphic statements and inferences in our knowledge of the species' natural history, perceptual and learning capabilities, physiology, nervous system, and previous individual history. That is, if we ask what we as humans would do in the animal's position, or how we would feel if treated like the animal, we must apply all the information we know about the animal as well as our own experience. For example, given what we know about dogs, it would be safe to infer that a kicked dog that is writhing and squealing is feeling pain. Putting ourselves in the dog's place is acceptable in this situation, since dogs are mammals with a physiological organization similar to ours. We would not be safe in concluding that the dog is feeling pain in exactly the same

way we do, however. We are on less solid ground, from a critical anthropomorphic perspective, in concluding that an earthworm on a fishing hook is feeling pain in any way comparable to our pain when stuck. This is because we know far less about the earthworm nervous system. We could, though, conclude that the experience is aversive to the worm, since it avoids or tries to remove itself from such situations. Worms squirm to avoid predation, so such behavior is adaptive.

An important use of critical anthropomorphism is to help pose and formulate questions and hypotheses about animal behavior. Although we can never directly experience what another animal, including another human being, thinks or feels, we can make predictions as to what the animal or person would do using anthropomorphic methods. Insofar as we ground them on real similarities across individuals, our predictions may be very accurate and replicable. Enough research may even allow us to claim that the subjective mechanisms are comparable as well as the behavioral responses. Many of the greatest comparative psychologists and ethologists have acknowledged their use of anthropomorphic insights in formulating ideas and generating experiments in animal behavior. However, this is rarely stated in scientific reports, especially in this century. As the scientific culture has shifted, there needs to be more encouragement of the process of critical anthropomorphism in all areas of animal care, agriculture, and research.

Why is critical anthropomorphism necessary? In numerous instances an insistence on avoiding anthropomorphism in the sense used here has impeded research progress. Certain behaviors such as vigilance, greeting, aggression, fear, indecision, and dominance can only be recognized once we know the normal behavioral repertoire. Thus, courtship and fighting have been confused and mislabeled in species. Mating behavior, which involves neck biting in many mammals, may be anthropomorphically mislabeled aggression or fighting. In contrast, dominance wrestling in rattlesnakes was considered mating because observers did not know the sexes of the participants. When it was discovered that two males were involved, scientists stuck to their biases and said these must be homosexual snakes, or that snakes were too dumb to tell the genders apart! Why was the behavior considered sexual? Well, the entwining of the snakes certainly appeared to be so anthropomorphically and, besides, the snakes never bit or tried to injure each other as, the scientists assumed, seriously fighting animals should try to do. Now we know that rattlesnakes are not immune from their own venom and biting would quickly kill both antagonists. The wrestling allows the strongest male to obtain access to female snakes without either animal being killed.

Further Reading

Burghardt, G. M. 1985. Animal awareness: Current perceptions and historical perspective. *Amer. Psychol.*, 40:905–919.

Hart, L., ed.. 1998. *Responsible conduct of research in animal behavior*. Oxford: Oxford University Press.

Lockwood, R. 1985. Anthropomorphism is not a four letter word. In *Advances in Animal Welfare Science* (Ed. by M. W. Fox & L. D. Mickley), pp. 185–199. Washington, DC: Humane Society of America.

Mitchell, R. W., N. S. Thompson, & H. L. Miles, eds. 1996. *Anthropomorphism, anecdotes and animals: The emperor's new clothes*. New York: SUNY Press.

Ristau, C., ed.. 1991. *Cognitive ethology: The minds of other animals*. San Francisco: Erlbaum.

Rivas, J. & G. M. Burghardt. 2002. Crotalomorphism: A metaphor for understanding anthropomorphism by omission. In *The cognitive animal: Empirical and theoretical perspectives on animal cognition*. (Ed. By M. Bekoff, C. Allen, & G. M. Burghardt), pp. 9–17. Cambridge, MA: MIT Press.

Gordon M. Burghardt

ANTHROZOOLOGY

See Animal Studies

ANTIVIVISECTIONISM

Antivivisectionism is a widely accepted label for uncompromising opposition to the use of live animals in scientific research. No area of human activity affecting members of other species is more controversial than animal experimentation, or more likely to trigger reactions from advocates of animal rights and animal welfare. Vivisection literally means the cutting up of living organisms for the purpose of study or research. Historically, this is an accurate description of the way in which experiments upon, generally, unanesthetized animals were performed. Antivivisectionism became a very strong movement in 19th century Victorian England, where increasing attention was being paid to animal pain and suffering, leading ultimately to passage of the Cruelty to Animals Act 1876, the world's first law specifically regulating animal research. By comparison with earlier centuries, relatively little of today's experimentation upon animals is of a highly invasive sort. But the word vivisection has persisted in the vocabulary of protest, taken on a wider meaning over time, and now denotes all procedures of scientific research that result in the injury and/or death of animals.

Antivivisectionists are by definition abolitionists, demanding a total end to animal experimentation, whether accomplished immediately or gradually, but they may also have more limited and pragmatic goals, such as ending certain kinds of experiments on nonhumans that are deemed morally unacceptable (e.g., consumer product safety testing, burn experiments, or pain experiments performed without anesthesia or analgesia). By contrast, animal welfarists, although they oppose cruelty, generally accept the use of animals as subjects of research, but campaign for more humane treatment and for reduction, refinement, and replacement in relation to animal usage.

Animal experimentation has been opposed by antivivisectionists on a number of grounds: (a) inapplicability or limited applicability to humans of the data gathered owing to cross-species differences and artificial laboratory settings; (b) methodological unsoundness (embodying poor scientific procedures); (c) dangerously misleading and harmful results; (d) wastefulness, inefficiency, and unreasonable expense; (e) triviality; (f) redundancy; (g) motivation by mere curiosity; (h) cruelty; (i) availability of alternatives, and (j) desensitization of researchers and their coworkers. Scientists who are animal users regularly argue that great advances in medicine and human (and animal) health would not have occurred without animal experimentation.

Antivivisectionists regard this as a dubious counterfactual assertion, claiming in return, however, that most of the important breakthroughs (e.g., increased human longevity, control of infectious diseases) would have occurred, or even did occur, without animal experimentation. Along these lines it has been argued that, from a historical perspective, personal hygiene, improved nutrition, physical fitness, and public works sewage systems have done more to improve health and longevity than any other measures. It is also claimed that animal experimentation has in many cases retarded rather than advanced progress. For example, the lifesaving antibiotic penicillin showed negative results in lab animals, while thalidomide (a drug sold to pregnant women in a number of countries during the 1950s and 60s as an antiemetic and sedative) appeared safe based on initial animal testing. Some antivivisectionists acknowledge that medical science has benefited from animal research, but still put the case that the future need not resemble the past in terms of how health research is to be conducted.

In recent decades, much greater attention has been paid to the ethics of animal experimentation. Virtually every scientist using live animals for research today works under some sort of ethical regulation and scrutiny and within some legal framework, however loose. Codes of conduct take many forms, and compliance with whatever system is in place may be either mandatory or voluntary, and may be subject to scrutiny by ethics review panels comprising peers or peers plus nonspecialists (often including one or more members of the public). Activities taking place under the auspices of granting agencies, professional organizations, research institutions, and journals that report the results of research typically must conform to ethical standards assigned by these entities. At the same time, many professional philosophers and others have focused on the issues surrounding animals' moral status, with important implications for the ethics of animal research. Animal rights and animal liberation theories draw strict limits as to what is morally permissible in this field, and not infrequently forbid animal experimentation altogether. Several radical action groups, a few of which practice guerrilla tactics (e.g., making clandestine raids on laboratories to free animals or photograph experimental procedures, picketing the homes of researchers, or even threatening their lives), have secured a prominent place in the public protest arena. These influences have in one way or another generated controversy, inspired some, and alienated others, with constructive debate and change sometimes resulting. A move among scientists toward increased accountability and openness can be discerned, but at the same time some have adopted a siege mentality in regard to defending their work and workplaces.

Two philosophical issues in the larger debate over experimenting on nonhumans concern cost-benefit analysis and what may be called the central ethical dilemma. Generally, attempts to justify animal experimentation from an ethical standpoint appeal to a cost-benefit analysis. That is, they weigh the costs to animals (in terms of harm, suffering, and death) against the benefits to humans of the research in question. In the ethics of research using live human subjects, however, two other conditions must be met: (a) subjects must give their voluntary, informed consent, and (b) costs and benefits must be calculated with reference to the individual subjects concerned or else, with their consent, at least with reference to other humans who may benefit. In point of fact (b) follows from the principle that it is never ethically acceptable (because of justice considerations) to make some worse off in order by that same act to make others better off, when no benefits compensate for the losses suffered by those who end

up being disadvantaged. But in the domain of animal experimentation considerations, (a) and (b) are deemed inapplicable. Critics claim that this move is prejudicial to animals and may be challenged as inconsistent, ethically wrong, and in violation of ordinary feelings of compassion. The central ethical dilemma is that the more we learn from the biological and behavioral sciences, the greater the range of similarities we see between human and other animal species, and hence the greater is our motivation for continuing to do animal research in order to understand ourselves better, but by the same token closer perceived similarity creates a heightened sense of moral responsibility toward nonhumans. It is increasingly difficult to argue, on the one hand, that animals are very like us and, on the other, to deny that they should be treated very much as we would wish to be treated.

How ever these issues are to be sorted out by individuals and society, certain things remain clear. Knowledge is not an end in itself. If it were, horrible research in the name of science, carried out routinely on hapless animals or humans, could be ethically justified. Therefore the burden of moral responsibility and justification always lies with those who would experiment on animals (or humans).

Further Reading

Baird, R. M., ed. 1991. *Animal experimentation: The moral issues*. Amherst, NY: Prometheus Books.

Day, N. 2000. *Animal experimentation: Cruelty or science?* Berkeley Heights, NJ: Enslow.

Dolan, K. 1999. *Ethics, animals and science*. Oxford: Blackwell Science.

Gluck, J. P., T. Dipasquale, & F.B. Orlans, eds. 2002. *Applied ethics in animal research*. West Lafayette, IN: Purdue University Press.

Greek, C. R., & J.S. Greek. 2002. *Sacred cows and golden geese: The human cost of experiments on animals*. New York: Continuum International.

Groves, J. M. 1997. *Hearts and minds: The controversy over laboratory animals*. Philadelphia: Temple University Press.

Guerrini, A. 2003. *Experimenting with humans and animals: From Galen to animal rights*. Baltimore: Johns Hopkins University Press.

Haughen, D. M., ed. 2006. *Animal experimentation: Opposing viewpoints*. Farmington Hills, MI: Greenhaven Press.

LaFollette, H., & N. Shanks. 1997. *Brute science: Dilemmas of animal experimentation*. New York: Routledge.

Michael Allen Fox

ARGUMENT FROM MARGINAL CASES

See Marginal Cases

ART, ANIMALS, AND ETHICS

In recent years, animals have increasingly become serious subject matter for artists, as evidenced by the number of exhibitions taking animals and/or human-animal relationships as the key curatorial theme (see sidebar). However, despite this growing popularity of the animal theme, currently relatively few artists present the animals themselves as specific individuals, and even fewer overtly address the ethics surrounding human-animal relationships and/or the use of animals in art, either in the artwork itself or in statements made about the artwork. Instead the majority of artists tend to use animals as metaphors or symbols for the human condition, or as generic signifiers for the natural world. As discussed by Steve Baker (2001) the way in which animals are represented

is important because it affects the way we think about, and hence treat animals. Consequently, the use of animals in art to stand in for something or someone else is problematic because it can result in the animals becoming marginalized, which allows the artist to avoid addressing the broader ethical issues surrounding the way humans interact with animals. Artist and social activist Sue Coe is known to "... object strongly to the idea of using animals as symbols, because by using an animal or its (image) as a symbol of or for something else, that animal is effectively robbed of its identity, and its interests will thus almost inevitably be overlooked." (Baker, 2006, p. 78) This disregard for the animal's interest is of particular concern where animals have been caused to suffer or even be killed in the name of art.

In a 1976 performance work titled *Rat Piece,* American artist Kim Jones burned three rats alive, pouring lighter fluid on them as they ran around a cage screaming in pain and terror. Jones' performance was a response to his experiences during the Vietnam War when he and his fellow Marines were plagued by rats which they would capture, place in cages, and burn to death. It might seem reasonable to assume that Jones' *Rat Piece* was of its time and that causing animals to suffer this way in the name of art would not be seen as acceptable in the 21st century. However, in recent years a number of artists have produced art that has involved the death of an animal or animals, even if not always in such a prolonged and torturous manner as was the case with Jones' *Rat Piece.*

The death of animals for the sake of art can take several forms. British artist Damien Hirst is renowned for his works that preserve animals such as cows, pigs, sheep and sharks in tanks of formaldehyde, sometimes whole, at other times cut into pieces. While Hirst has no contact with the animals he uses until after he orders them to be delivered to him dead, Belgian artist Wim Delvoye has a more complex relationship with the pigs he uses as part of his Art Farm project. Delvoye began by working with the skins of dead pigs, but has since bought a farm in China specifically to house and raise the pigs for his artworks. The pigs are placed under a general anesthetic and are tattooed with various designs before eventually being slaughtered and skinned. The skins themselves become the final artwork, either pinned flat to walls or sometimes made into a three-dimensional form of the pig. Delvoye argues that his pigs are allowed to grow old (i.e., they are slaughtered later than they would have been for commercial production) and that the pigs benefit from being valued as artworks rather than just as meat (O'Reilly, 2004, p. 26).

In other cases, artists kill the animal(s) themselves, or are in some way directly involved with an animal's death, with the death at times being an integral part of the artwork. Austrian Actionist artist Hermann Nitsch is notorious for his *Orgien Mysterien Theater* (orgies-mysteries theatre) which he has been organizing since the late 1960s. These ritualized events often last several days and involve the slaughter of a number of animals such as sheep, goats, and cattle. The animals' entrails are at times trampled upon and the performance participants are covered in the animals' visceral remains.

More recent examples of animal death in the name of art include a work from 2000 by Marco Evaristti, titled *Helena,* which comprised 10 blenders, each containing a live goldfish. Visitors to the gallery had the option of turning the blenders on, and several people chose to kill the fish, resulting in the gallery director being charged with animal cruelty after a complaint was made by an animal protection organization. In 2003, an exhibition by Nathalia Edenmont was also the target of protests from animal rights groups. Edenmont's exhibition showed photographs of dead animals such as rabbits, cats and mice, often decapitated and wearing Elizabethan style decorative collars. What caused such a fuss was the fact that Edenmont killed the animals herself specifically for the artworks. In 2008, an exhibition by Adel Abdessemed was closed down just a week after it opened at a gallery in San Francisco, after intense lobbying by groups such as In Defense of Animals and People for the Ethical Treatment of Animals. The center of the controversy was a video loop showing six animals—a horse, a sheep, a deer, a cow, a pig, and a goat—being bludgeoned to death with a sledgehammer. While Abdessemed apparently did not kill the animals himself (he is said to have filmed the normal practice of killing animals on a farm in Mexico), the apparently gratuitous presentation of their violent deaths prompted controversy. While the aforementioned artists have all attracted the wrath of animal protection organizations and the general public alike, a work by Guillermo "Habacuc" Vargas, touched a particular nerve. In 2007, Vargas tied up a sick and emaciated street dog as part of a work titled *Exposición No.1* at a gallery in Nicaragua. Not long afterward a petition calling for a "Boycott to the presence of Guillermo Vargas 'Habacuc' at the Bienal Centroamericana Honduras 2008" began to be widely circulated via email, as Vargas apparently planned to remake the work for the Honduran Biennial. Photographs which accompanied many of the emails showed a starving dog, tied by a piece of rope to a wire across a corner of the gallery. On an adjacent wall the words

"Eres lo que lees" ("You are what you read") were spelled out in dry dog food. The international outrage was sparked by reports that Vargas had allowed the dog to die, refusing to give it food or water. While there is no dispute over the fact that Vargas tied up a severely emaciated dog in the gallery as part of his artwork, whether or not the dog died is difficult to substantiate, as the information available is contradictory.

The artworks discussed above inevitably engage with the ethics surrounding the use of animals in art due to the animals' suffering and/or death. Considering that our relationship with animals is currently so firmly intertwined with causing their deaths, either for food, as pests, for sport or simply because they are unwanted, it is perhaps not surprising that animal death and/or suffering for the sake of art is seen as valid by some artists. In those cases where the artwork has required the death of farm or other food animals, it can be argued that the animals were destined for slaughter anyway. Damien Hirst's preserved shark artwork *The Physical Impossibility of Death in the Mind of Someone Living* is particularly interesting in this respect, as not only was the tiger shark ordered to be caught and killed specifically for the artwork, but due to poor preservation techniques the original animal needed to be replaced with another tiger shark, again killed especially for this purpose. From an animal rights/welfare perspective, causing an animal to suffer or die in the name of art is always unjustifiable, regardless of the artist's intentions, and because of this all the aforementioned artists have attracted the attention of animal advocates. As Steve Baker points out,

> Contemporary art, along with literature and non-documentary film,

is a field in which the killing of animals can undoubtedly figure as a subject, but where it is not necessarily clear how the field can usefully contribute either to knowledge of the other-than-human or more-than-human-world, or to what might broadly be called the cause of animal advocacy (Baker, 2006, p. 70).

However, Baker has also argued that artists' creative freedom in using animals should not be too heavily restricted, because in using animals this way artists can prompt debate over the ethical issues surrounding human-animal relationships (Baker & Gigliotti 2006, 2–3).

The use of animals in art is not only controversial because of violent acts against the animal. In the case of Eduardo Kac, the controversy is over the fact that he commissioned a scientific laboratory to produce a genetically altered rabbit for the project *GFP Bunny* (2000). The rabbit, which Kac named Alba, had a green fluorescent protein sourced from a jellyfish gene inserted into her genome so that she would glow under ultraviolet light. The genetic alteration of an animal in the name of art opens up a range of ethical questions and has been the subject of much debate. Kac himself has stated clearly that he had the utmost concern and sense of responsibility for Alba's welfare and wanted to care for her in his home (although ultimately the laboratory refused to relinquish her). However, as Baker has pointed out, the technology used to produce Alba is implicated in the deaths of huge numbers of laboratory animals (Baker 2003, 35–36).

There are some artists, however, whose artwork is strongly informed by an animal rights ideology and who use their work to engage the viewer with the ethical issues surrounding human-animal relationships. Perhaps the best known of these is Sue Coe, whose politically charged paintings, drawings, and prints depict the suffering of animals for meat production as well as in laboratories. Coe has produced several illustrated books on these subjects, such as *Dead Meat* (1995) and *Sheep of Fools* (2005). American artist Mary Britton Clouse not only makes art about animals, but also founded a chicken rescue society and is a founding member of the Justice for Animals Art Guild, which has as its purpose "to oppose art that harms or exploits animals, and explore ways to support artists whose ethics and philosophies value the rights of animals" (Justice for Animals Art Guild http://www.brittonclouse.com/jaag.htm). British artist Britta Jaschinski makes photographs which are based on her concerns about the plight of zoo animals, while New Zealand artist Angela Singer makes work using recycled taxidermy such as trophy heads of deer to highlight the cruelty of hunting. Taxidermy animals are also used by a number of other artists, including Mark Dion, Jordan Baseman and Thomas Grünfeld. However, where Dion has a written a manifesto covering the responsible use of living plants and animals, other artists are not so forthright about what they feel their ethical responsibility is toward the taxidermy animals that are used in their art, prompting questions about how the animals are sourced and presented. The questioning of artists' intentions and ethical stance when they use animals in their work is important, because artists not only reflect how society regards animals, they can also help shape our ideas about animals and how we should treat them.

See also Museums and Representation of Animals

SELECTION OF RECENT EXHIBITIONS WITH AN ANIMAL THEME

- The Animal Gaze, various venues, London, 2008.
- Fierce or Friendly, Tasmanian Museum and Art Gallery, Hobart (Australia), 2007/2008.
- Fierce Friends: Artists and Animals 1750–1900, the Van Gogh Museum, Amsterdam and Carnegie Museum of Art in Pittsburgh, PA, 2006.
- Unsettled Boundaries, visual arts component of the 2006 Melbourne International Festival of the Arts, including the major exhibition, The Idea of the Animal, RMIT Gallery, Melbourne (Australia), 2006.
- Becoming Animal, the Massachusetts Museum of Contemporary Art, North Adams, MA, 2005.
- Animal Nature, Regina Gouger Miller Gallery, Carnegie Mellon University, Pittsburg, PA, 2005.
- Animals and Us: The Animal in Contemporary Art, Galerie St. Etienne, New York, NY, 2004.
- Animals, Haunch of Venison Gallery, London, 2004.
- The Human Zoo and A Painted Menagerie: the animal in art 1600–1930, Hatton Gallery, Newcastle University, 2003.

Further Reading

Baker, S. 2000. *The postmodern animal*. London: Reaktion.

Baker, S. 2001. *Picturing the beast*. 2nd ed. Champaign: University of Illinois Press.

Baker, S. 2003. *The eighth day: The transgenic art of Eduardo Kac*. Ed. S. Britton. and D. Collins. Tempe: Institute for Studies in the Arts, Arizona State University.

Baker, S. 2006. "'You kill things to look at them': Animal death in contemporary art." In *Killing animals (The animal studies group)*, 69–95. Urbana & Chicago: University of Illinois Press.

Baker, S., & C. Gigliotti. 2006. *We have always been transgenic*. AI & Society, 20.1, http://www.ecuad.ca/~gigliotti/gtanimal/BAKERGIG.htm.

Berger, J. 1980. "Why look at animals?" In *About Looking*. Ed. J. Berger. New York: Pantheon.

Coe, S., & J. Brody. 2005. *Sheep of fools*. Seattle: Fantagraphic Books.

Coe, S., & A. Cockburn. 1995. *Dead meat*. New York: Four Walls Eight Windows.

Gigliotti, C. 2006. *Leonardo's choice: The ethics of artists working with genetic technologies*. AI & Society, 20.1, http://www.ecuad.ca/~gigliotti/gtanimal/CGIGLIOTTI.htm.

Justice for Animals Arts Guild. http://www.britonclouse.com/jaag.htm.

O'Reilly, S. 2004. "Wim Delvoye." *Contemporary*, No.59 (May), 26.

Thomson, Nato, ed. 2005. *Becoming animal: Contemporary art in the animal kingdom*. Massachusetts: Massachusetts Museum of Contemporary Art.

Wolfe, C. 2006. "From dead meat to glow in the dark bunnies: Seeing 'the animal question'." in *Contemporary Art*, Parallax 12.1.

Yvette Watt

ASSOCIATION OF VETERINARIANS FOR ANIMAL RIGHTS (AVAR)

The Association of Veterinarians for Animal Rights (AVAR) was founded in 1981 by Nedim C. Buyukmihci and Neil C. Wolff. The term rights, as opposed to welfare, was chosen for the title of the organization because it exemplified the different philosophy of this approach. Although veterinarians are already involved in animal welfare, this is clearly inadequate to protect nonhuman animals' interests.

In veterinary medicine, the standard of caring for nonhuman animals is usually based on what is deemed adequate veterinary care. Nonhuman animals are treated as the property of their owners. Although there usually is a sincere attempt to relieve suffering and improve the quality of life for these animals, there are no meaningful limits to what may be done with them. When one examines the issues without prejudice and with humility, there do not appear to be any morally relevant differences between humans and other animals that justify denying other animals similar rights, consideration, or respect, based upon their interests or upon whether what we propose to do matters to the individual.

Further Reading

Buyukmihci, Nedim C. 2006. Consistency in treatment and moral concern. *Journal of the American Veterinary Medical Association* 206(4): 477–480.

Mason, Jim, and Peter Singer. 1990. *Animal factories,* 2nd ed. New York: Harmony Books.

Pluhar, Evelyn B. 1988. When is it morally acceptable to kill animals? *Journal of Agricultural Ethics* 1(3): 211–224.

Regan, Tom. 1983. *The case for animal rights*. Berkeley: University of California Press.

Singer, Peter. 1990. *Animal liberation*. New York: New York Review of Books.

Nedim C. Buyukmihci

AUTONOMY OF ANIMALS

The original meaning of autonomy, as applied to ancient Greek city-states, is self-rule. More recently, the term has been applied to individuals, actions, and desires. To answer the question "Are any animals autonomous beings who are capable of performing autonomous actions?" requires not only carefully studying animals, but also determining what sorts of actions qualify as autonomous.

Autonomous actions must at least be intentional actions. Every intentional action involves a desire and a belief that help to explain why the action was performed. Tom Regan argues that beings capable of intentional action are capable of one kind of autonomy, what he calls preference autonomy (preference being another word for desire). From this analysis, assuming that a dog can (1) desire a bone and (2) believe, as she trots into the backyard, that she can find a bone there, then the dog is capable of acting autonomously. However, one can be capable of acting autonomously but fail to do so for any of several reasons. For example, physical constraints such as locked doors can prevent a dog from going into the backyard. Force can prevent intentional actions from being autonomous. If you intentionally give money to someone, but only because he threatened you with a gun, your action is coerced, not free or autonomous. Moreover, sometimes we act intentionally, and even freely, but without sufficient understanding of what we are doing for our action to be autonomous. If a hospital patient intentionally and freely signs a form that states agreement to participate in psychiatric research, but the patient believes that the form simply entitles her to therapy following hospitalization, the patient has not autonomously agreed to participate in research.

Autonomous action clearly involves more than simply intentional action. One analysis, favored by Tom Beauchamp, is that actions are autonomous if they are performed (1) intentionally, (2) with understanding, and (3) without controlling influences (e.g., force) that determine the action. But certain other writers, such as Gerald Dworkin and David DeGrazia, would argue that these conditions are not sufficient for autonomous action. Apparently, based on the present analysis, a bird feeding her young would, under normal circumstances, count as acting autonomously (assuming that birds can act intentionally). Because autonomous beings are beings capable of acting autonomously, one's answer to the question "Are any animals autonomous beings?" will depend, in part, upon one's view of autonomous action. Those with relatively undemanding requirements are likely to conclude that many animals are autonomous. The view that anyone capable of intentional action is autonomous implies that all animals capable of having the appropriate sorts of desires and beliefs qualify. Which animals have such desires and beliefs is an extremely complex question, involving difficult conceptual issues in the philosophy of mind and various kinds of scientific evidence regarding animals. Tom Regan somewhat cautiously argues that normal mammals beyond the age of one year are capable of intentional action. David DeGrazia contends that most or all vertebrates and perhaps some invertebrates can act intentionally.

From a multitier perspective, animals are autonomous beings only if they can critically evaluate the preferences that move them to act and sometimes modify them on the basis of higher-order preferences and values. This is a high standard, requiring considerable capacity for abstraction and an advanced form of self-awareness. Perhaps such abstraction and self-awareness require language. There is a strong case that some apes have achieved language comprehension and production, and that some dolphins have achieved language comprehension. The most suggestive evidence from the language studies of the possibility of animal autonomy may be evidence that apes apologized for such actions as biting a trainer and relieving themselves indoors. Typically, apologies express regret for one's actions, but one might also regret the motivations that moved one to act. At present it seems unclear, from the multitier view, (1) whether autonomy might be possible for those animals which lack language, and (2) whether any animals are, in fact, autonomous beings.

See also Consciousness, Animal

Further Reading

Beauchamp, Tom L. 1992. The moral standing of animals in medical research. *Law, Medicine, and Health Care* 20(1–2): 7–16.

Christman, John, ed. 1989, *The inner citadel: Essays on individual autonomy*. New York: Oxford University Press.

DeGrazia, David D. 1996. *Taking animals seriously: Mental life and moral status*. Cambridge: Cambridge University Press.

Dworkin, Gerald. 1988. *The theory and practice of autonomy*. Cambridge: Cambridge University Press.

Regan, Tom. 1983. *The case for animal rights*. Berkeley: University of California Press.

David D. DeGrazia

B

BEAK TRIMMING

See Chickens

BESTIALITY

Until approximately the 16th century, the term bestiality referred either to a broad notion of earthy and often distasteful otherness or to sexual relations between humans and nonhuman animals.

The earliest and most influential condemnations of bestiality are the Mosaic commandments contained in Deuteronomy, Exodus, and Leviticus. Deuteronomy, for example, declared, "Cursed be he that lieth with any manner of beast" (27:21), and Exodus ruled that "whosoever lieth with a beast shall surely be put to death" (22:19). Besides mandating death for humans, Leviticus dictated that the offending animal must also be put to death—probably because, it was thought, the animal had been polluted by the Devil. It is hard to know the precise intentions of those who originally condemned bestiality, but in Judeo-Christianity, there have been three principal beliefs about the origins of its wrongfulness: (1) it is a rupture of the natural, God-given order of the universe; (2) it violates the procreative intent required of all sexual relations between Christians, and (3) it produces monstrous offspring that are the work of the Devil.

In some societies, such as in Puritan New England from the 1620s until the mid-19th century, bestiality was regarded with such alarm that even the very mention of it was condemned. It was therefore also referred to as "that unmentionable vice" or "a sin too fearful to be named" or "among Christians a crime not to be named." Nowadays, bestiality is variously described as "zoöphilia", "zoöerasty", "sodomy," and "buggery," and its meaning is almost always confined to human-animal sexual relations.

Since the end of World War II, especially, bestiality has been one among several categories of nonreproductive sexual practices toward which society in general has tended to exercise a growing tolerance. Indeed, in the last 50 or so years, those offenders whose sexual activities with animals have been reported to legal or medical authorities have faced considerably lesser charges than they had historically, such as breach of the peace or offending against public order. Instead of criminal prosecution, offenders have typically been sent either for counseling or for psychiatric treatment or, with probably greatest deterrent effect, they have been subject to public ridicule in their local communities.

In the past 10 years, however, there has been a great reversal of how bestiality has been viewed in the Unites States. In 27 or so states, bestiality has been recriminalized and defined as a form of cruelty.

Among these states there is considerable variation in the level of punishment that is attached to sexual relations with animals. In some states the maximum penalty is a fine and imprisonment of one year, and in others the maximum incarceration is five years.

Information about the incidence and prevalence of bestiality is quite unreliable, especially given its private nature and the social stigma attached to it. Bestiality can occur in a wide variety of social contexts. These include adolescent sexual experimentation, typically by young males in rural areas; eroticism (sometimes termed "zoöphilia," practiced by "zoos"), a rare event where animals are the preferred sexual partner of humans; aggravated cruelty, especially by young males or in cases of partner abuse; and commercial exploitation, as in pornographic films or in live shows of women copulating with animals in bars or sex clubs.

The prevalence of bestiality probably depends on such factors as the level of official and popular tolerance, opportunity, proximity to animals, and the availability of alternative sexual outlets. Some sexologists have claimed, with the use of interviews and questionnaires, that eight percent of the male population has some sexual experience with animals, but that a minimum of 40 to 50 percent of all young rural males experience some form of sexual contact with animals, as do 5.1 percent of American females. But because of the poor sampling techniques of such studies, these figures should be treated with great caution.

Further Reading

Beetz, Andrea M. and Anthony L. Podberscek (Eds.). *Bestiality and zoophilia*. West Lafayette, IN: Purdue University Press.

Dekkers, M. (1994). *Dearest pet*. (P. Vincent, Trans.). London: Verso.

Kinsey, A. C., Pomeroy, W. B., & Martin, C. E. (1948). *Sexual behavior in the human male*. Philadelphia: W. B. Saunders.

Kinsey, A. C., Pomeroy, W. B., Martin, C. E., & Gebhard, P. H. (1953). *Sexual behavior in the human female*. Philadelphia: W. B. Saunders.

Liliequist, J. (1991). Peasants against nature: Crossing the boundaries between man and animal in seventeenth-and eighteenth-century Sweden. *Journal of the History of Sexuality, 1*(3), 393–423.

Miletski, H. (2002). *Understanding bestiality and zoophilia*. Germantown, MD: Imatek.

Rydström, J. (2003). *Sinners and citizens: Bestiality and homosexuality in Sweden, 1880–1950*. Chicago: University of Chicago Press.

Piers Beirne

Further Reading

Adams, C. J. (1995a). Bestiality: The unmentioned abuse. *The Animals' Agenda, 15*(6), 29–31.

Adams, C. J. (1995b). Woman-battering and harm to animals. In J. Donovan & C. J. Adams (Eds.), *Animals and women: Feminist theoretical explorations* (55–84). Durham: Duke University Press.

Beirne, P. (2002). On the sexual assault of animals: A sociological view. In A.N.H. Creager & W. C. Jordan (Eds.), *The animal/ human boundary: Historical perspectives* (193–227). Rochester, NY: Rochester University Press and Davis Center, Princeton University.

Singer, P. (2001, March/April). Heavy petting. *Nerve*.

Piers Beirne

BESTIALITY: HISTORY OF ATTITUDES

Bestiality refers first to people acting like animals, in a bestial way. However, its second meaning, sexual contact

SEXUAL ASSAULT OF ANIMALS

Historically, sexual relations involving humans and animals have tended to be condemned and investigated—or, in the interests of tolerance, ignored—exclusively from an anthropocentric perspective. Yet sexual relations with humans often cause animals to suffer great pain and even death, especially in the case of smaller creatures such as rabbits and hens.

Today, both the feminist movement and the animal rights movement have started to rethink the moral and ethical status of bestiality. Sexual relations between humans and nonhuman animals are beginning to be seen as wrong for the same reasons we see sexual assault by one human against another human as wrong—because it involves coercion, because it produces pain and suffering, and because it violates the rights of another being.

It is impossible to know whether animals can ever consent to sexual relations with humans, so it is best to treat all such cases as forced sex. Sexual relations involving humans and animals are therefore more appropriately termed animal sexual assault.

between humans and animals, is the most frequent current use of the word. Attitudes about bestiality have changed over time, and these attitudes are revealing of people's general perception of animals.

The early Christian medieval world inherited both texts and traditions that described human/animal intercourse. In the classical Greco/Roman texts, gods in the form of animals had intercourse with humans, and tales drawn from folklore also preserved anecdotes of such sexual contact. Pagan Germanic tradition also preserved tales of bestiality, whether between human and animal, or between humans one of whom took the shape of an animal.

The Christian tradition did not accept bestial intercourse, but there was a change over time in the perception of the severity of the sin. During the earliest prohibitions, bestiality was regarded as no more serious than masturbation. By the 13th century, however, Thomas

Aquinas ranked bestiality as the worst of the sexual sins, and the law codes recommended harsh penalties for the practice.

There seem to be two primary reasons for this change. The first is that by the late Middle Ages churchmen became more concerned with the presence of demons interacting with humans. As part of this preoccupation, tales of bestiality increasingly referred to intercourse with demons, the succubi and incubi that seemed ubiquitous. The increased concern with bestial intercourse seems also to reflect a growing uncertainty about the separation of humans and animals. Preoccupation with and legislation against bestial intercourse expressed an attempt to secure the separation of species when it seemed endangered.

As church laws were taken over in the late Middle Ages by kings who wanted to exert more authority over their kingdoms, what had once been identified as sinful then became identified as illegal. It is in this form that laws against bestiality persisted into the modern world.

Further Reading

Aelian. 1959. *On the characteristics of animals* Cambridge. Boston: Harvard University Press.

Brundage, James. 1987. *Law, sex and Christian society.* Chicago: University of Chicago Press.

Dekkers, M. 1994. *Dearest pet.* London: Verso.

Payer, Pierre. 1984. *Sex and the Penitentials.* Toronto: University of Toronto Press.

Salisbury, J. E. 1994. *The beast within.* New York: Routledge.

Joyce E. Salisbury

BLESSING OF THE ANIMALS RITUALS

It is unclear when Blessing of the Animals rituals first occurred in the Christian tradition, though most likely they reflect a conflated Christian-pagan practice. Certainly as the roles of animals in human culture shift, so do the purposes of animal blessings. By the early 21st century, Blessing of Animals rituals in Western Christianity focused on domestic, companion species (dogs and cats in particular), whereas earlier blessings incorporated work and agricultural animals, such as mules, oxen, and horses. The earliest evidence is visual, including images of Saint Anthony Abbot (a fourth-century Christian holy man) blessing animals along with poor or afflicted humans. St. Anthony, whose feast day is on January 17th, is the patron saint of animals. Documentary evidence shows that this mid-January blessing ritual, in recognition of his feast day, occurred into the early 20th century in cities such as Rome. Reports indicate that humans brought a wide range of animals to the steps in front of Catholic churches throughout the city for the blessing. Written reports, along with images from as early as the 15th century, indicate a

tradition of animal blessings connected to the saint. It is also possible that Catholic Rogation Days, which included a blessing of farm fields, also incorporated the blessing of farm animals.

In the 20th century, Blessing of Animal rituals became increasingly prevalent in Western Christianity, from the United States to Canada to Australia and, to a lesser extent, in Europe. These rituals follow a standard pattern. Often geared to attract families with children, they tend to have a human-focused impetus. They are usually held outside, in front of the church building or in a park close to the religious institution, though occasionally they are held in sanctuaries. Many of the large and influential Christian denominations developed these blessings: Roman Catholic, United Methodist, Presbyterian (USA), Disciples of Christ, and the Episcopal Church. However, they also tend, more than many other religious rituals, to be ecumenical or interfaith in nature—even secular in sponsorship at times.

As the position of pets shifts in Western cultures, so does the incorporation of these companion animals into the religious life of the humans who live with them. In other words, as pets become more central to the lives of some humans, these humans seek ways to incorporate their companion animals into all facets of their lives. Thus, Blessings of the Animals/Pets is growing rapidly.

While there is no standard ritual, it is helpful to discuss one in particular, since it might be the catalyst for the growth of these blessings. A large and influential Episcopal Church in New York City, the Cathedral of Saint John the Divine, holds arguably the largest and most impressive Blessing of Animals. It set the stage and provided the model for subsequent rituals. Beginning in the 1980s, the Cathedral held an annual Blessing of Animals on the Sunday closest to the Feast of Saint Francis (October 4th). As of the beginning of the 21st century hundreds, if not thousands, of these blessings occur annually, mostly in connection with St. Francis's feast day.

The blessing at Saint John the Divine provides a helpful template for understanding the phenomenon. Officially the ritual is titled "The Holy Eucharist & Procession of Animals." Many years the sanctuary is filled to capacity, with over 3,000 humans and as many as 1,500 animals present. Congregants wait in line with their companion animals for hours in order to find a place in the sanctuary. After a formal Eucharistic service, including music and dancing, the central doors of the sanctuary are opened and the procession takes place. It should be noted that these doors are only opened three times each year: Christmas, Easter, and the Blessing of Animals. Myriad animals with differing cultural positions process: camels (exoticized), cattle (usually food in the United States), bees, fish, hedgehogs, and hawks, for examples. Following the Eucharistic liturgy, humans along with their companion animals move outside and are offered the opportunity for an individual blessing for each animal. The entire event takes several hours. In addition to the ritual, a fair is held. Representatives from various animal protection organizations, such as dog rescue groups and farm animal awareness services, come to share information.

Father Rand Frew, left, of St. John the Divine, and Vince Sharp, of the Turtle Back Zoo in West Orange, New Jersey, carry a boa constrictor outside the Cathedral of St. John the Divine during the Feast of St. Francis of Assisi in New York City. Individuals attending the ceremony were invited to bring their pets who were blessed following the service. (AP Photo/Jennifer Szymaszek)

Other Christian—ecumenical and interfaith—as well as secular Blessings take place throughout the year. For example, not-for-profit or municipal entities such as local animal shelters sometimes sponsor Blessings. Often a local clergyperson or group of interfaith leaders presides. Cats, dogs, guinea pigs, ferrets, parrots, turtles, hermit crabs, and snakes are among the ritual participants at these quasi-religious events. In addition to the blessings, many animals up for adoption are brought there, as groups try to find them good permanent homes. This connection to animal welfare issues is becoming increasingly important as part of the annual blessings.

It is difficult to determine the core purpose of the Blessings or to conclude with any certainty why they spread so rapidly in the late 20th and early 21st century. This phenomenon probably accompanies the growth of the pet industry, the ownership of pets, and other companion animal-related issues of the same time period. However, it should be recognized that the rituals can be problematic, in particular for some of the animals. While ritual, spectacle, and performance are certainly connected, the spectacle and forced performance of companion animals is ethically questionable. It is possible that Blessings of Animals serve no purpose for the animals, but only provide humans with a circus-like atmosphere and a sense that they are expanding their ethical horizons or, of even less value for the lives of animals, these Blessings are simply a way to bring new humans into various religious communities.

Blessings of Animals also fit within the larger environmental or green movement within some forms of Christianity, so animals become symbolic of a commitment to God's creation or to the human stewardship component of creation stories. This is indeed a focus at the Cathedral of Saint John the Divine. In the ideal world—the world of the peaceable kingdom that is prophesied variously in Western religious traditions—other-than-human animals are often included. A number of animal welfare organizations, such as the Humane Society of the United States, suggest that a focus should be on animals in confined animal feeding operations. Public Blessings of these animals could draw awareness to this mass production system that relies on inhumane systems of confinement, among other issues.

However, it could also be argued that Blessing of the Animals rituals suggest a shifting attitude toward animals, specifically in the early 21st century. While they were excluded from sanctuaries for generations, now animals are being invited to return (once a year, anyway). They have sacred significance and are worthy of blessing. This is indeed an expansion of the religious sensibility that dominated the Western world in the post-Enlightenment era. Interpreting the cultural impact will take decades. In the meantime, the numbers and variety of Blessings continue to expand. As the roles of animals shift, so do the roles of Blessings, from those that acknowledge animals' usefulness to humanity to those that also recognize their role as humans' companions and, in some cases, to Blessings that recognize their own intrinsic value.

See also entries beginning with "Religion and Animals"

Further Reading
Hobgood-Oster, L. (2008). *Holy dogs and asses: Animals in the Christian tradition.* Urbana, IL: University of Illinois Press.
McMurrough, C. (1939). Blessing of animals: Roman rite. *Orate fraters, 14*(2), 83–86.

Laura Hobgood-Oster

BLOOD SPORTS

Definitions

Blood sports are organized activities in which animals are placed at great risk of injury or death for human entertainment. Among the most common current blood sports are bullfighting, dogfighting, cockfighting, and non-subsistence hunting and fishing. Non-subsistence or sport hunting includes hunting with weapons (rifles, shotguns, bows, etc.) as well as with other animals, such as falcons or dogs. Bullfighting is an example of animal baiting, in which an animal is goaded into aggression through pain, taunting, cornering, and so on. Dogfighting and cockfighting are typically not considered baiting because the animals are usually not goaded to fight at the time of the event, but rather are trained and bred to fight beforehand.

The animals subjected to blood sports routinely suffer terribly both during the events and during training for fights. Losing animals are often killed. In baiting and fighting, the wounds animals suffer are often intentionally grisly and painful. Fighting animals are chosen for the damage they can do to each other, and they are bred and trained to be relentlessly savage. When using weapons to hunt, sport hunters usually try not to inflict intentionally ghastly wounds on the animals they kill. In most communities, making prey suffer is frowned upon. However, the deaths of animals killed during coursing (using predators to hunt prey) can be more gruesome.

Sport Hunting

Hunting is one of the longest-standing ways in which humans interact with other animals. Much prehistoric art depicts commonly-hunted animals and sometimes even hunting scenes. Hunting in general has greatly affected animal populations and the environment. Hunters themselves have long been quite active on both sides of land and animal management programs: Avid hunters were among the pioneers of land management (and many are still among the most active), and yet poaching (and sometimes other forms of hunting) continues to push animals to extinction in many parts of the world.

Although humans and our ancestors have hunted since at least the Paleolithic era, hunting simply for entertainment probably established itself with the advent of domestication and agriculture. Early sport hunting was largely restricted to the upper class, who had the time and resources for it. These privileged classes typically approached a hunt as they would a battle, and hunting was probably regarded as practice for war. Destroying powerful animals made the royalty appear powerful to their subjects and probably even to themselves.

In medieval Europe and in the East, hunting became associated with land ownership. Although falcons, bows, spears and even swords continued to be used in hunts into the 1600s, hunting with dogs became the most common way to hunt. Coursing dogs were followed as they chased their prey. The hunters usually did not kill the animals themselves; instead they watched as their dogs tore the animal apart. Most types of coursing were illegalized in England in 2004.

As weapons became more effective, complex hunting codes were used to make sport hunting more difficult, in order to ensure that a wounded animal was killed. Nonetheless, the casual cruelty to animals that pervaded many aspects of human life affected sport hunting. Animals were sometimes herded into confined areas and shot wholesale.

Whereas subsistence hunters try to kill an animal as efficiently as possible, sport hunters may not. A sizable minority of sport hunters in the United States, for example, prefer the challenge of hunting with bows or black powder rifles instead of more effective, modern weapons. Sometimes sport hunters use modern weapons, but in ways that make clean kills difficult, forcing wounded animals to die slowly. Bison, for example, were shot en masse from moving trains; in Alaska wolves are currently shot from airplanes.

Historically, hunting, both for food and for entertainment, has taken a heavy toll on animals. Hunting in general led to the extinction of dodos and passenger pigeons. It is also commonly believed that hunting strongly contributed to the extinctions of mammoths, mastodons, Caribbean monk seals, *baiji* (river dolphin), aurochs, steppe bison, Steller's sea cows, giant kangaroos, giant antelopes, great auks, moas, Carolina parakeets, etc. Hunting in general is also helping to push a great many more animals to the brink of extinction: tigers, whales, European minks, Asiatic rhinoceroses, dugongs, some seals, and many fish, to name a few.

At least as early as medieval Europe, animals were imported into areas depleted by hunting and land loss. Later, as animal populations continued to decline overall, hunters contributed to conservation efforts. Indeed, concerned that both their sport and natural treasures were at risk, some sport hunters were among the creators and strongest proponents of land and animal management programs. India created a reserve in 1861. Various African and North American nations began programs shortly thereafter; other countries have since followed this lead. An increasing number of countries regulate both where and what animals can be hunted as well as how the animals can be killed. Although poaching remains a very serious problem in many areas, sport hunters in many regions voluntarily abide by hunting rules.

Animal Fighting

Around the world, many types of animals are forced to fight. Dogs, roosters, horses, kangaroos, camels, beta fish, and various types of insects are some of the current participants. Animal fighting is strongly ingrained in some cultures, often despite laws to protect both people and animals.

It is believed that cockfighting was first practiced in Southeast Asia thousands of years ago. It spread westward via Persia, Greece, and Rome. It also may have begun on its own elsewhere. Roosters are made to fight until one of them is too severely mauled to continue fighting; the loser often dies. Fighting roosters may have their wattles docked (cut off) to prevent them from ripping and bleeding during fights. Sometimes the birds' spurs are covered with longer spikes or blades to make the fight bloodier and quicker. Naked-heel fights conducted without spurs can last for hours—too long for the attention span of most of those who watch this sort of event.

Dogs were commonly used in war by 700 BCE. They may have been forced to fight each other as early as this as well. Indeed, dogfighting was common in Roman Europe, if not elsewhere.

The Romans were not alone in pitting various animals against each other. Dogfighting was practiced in Japan by the Kamakura period (1185–1333 CE). It was promoted among the samurai, many of whom felt dogfighting kept their own ferocity sharp during times of peace. Infamously, the *daimyo* of the Tosa province (present-day Kochi) and Akita prefecture were strong proponents of dogfighting; the fighting dogs bred in these areas are now well known.

The first documentation of cricket fighting comes from China's Song Dynasty (1213–1275 CE). Cricket fighting in China became much less popular after the Communist revolution because of its association with the bourgeoisie. It is illegal in Hong Kong and the Macao Province, but its popularity is unfortunately increasing in other parts of China, as well as in other countries.

Bearbaiting was popular in Tudor England; Henry VII constructed a large bear garden at Whitehall. Elizabeth I was especially fond of bearbaiting; she was said to giggle like a schoolgirl at the suffering of the animals. Her interest in baiting helped increase its popularity. The fighting pits of her reign also began to stage the brutal deaths of a wider range of animals. Bulls, boars, rats, badgers, and even more exotic animals all died in the pits. To turn the animals into the nasty beasts needed for an entertaining spectacle, they were subject to all sorts of abuse and cruelty. There are reports of bears being whipped, beaten, stoned, starved, and forced to sleep on beds of thorns. Before a fight, bulls might have their noses stuffed with cayenne pepper.

Various groups attempted to outlaw baiting, but it was not until the social revolutions of the Victorian era that efforts to ban animal cruelty and many blood sports started to succeed. An especially grand milestone was the Cruelty to Animals Act of 1835 in England. Strongly lobbied for by the Society for the Prevention of Cruelty to Animals (the first humane organization, now called the Royal Society for the Prevention of Cruelty to Animals), this act amended a previous act (the 1822 Act to Prevent Cruel and Improper Treatment of Cattle). It illegalized cruelty—including fighting and baiting—toward bulls, bears, dogs, and sheep. The ban spread to England's possessions around the world. In 1836, Massachusetts was the first state to ban cockfighting; Louisiana was the last, not banning it until 2008. The American Society for the Prevention

Engraving of a Henry Alken painting depicting a tethered bull being baited with dogs and sticks, ca. 1810. Such cruel blood sports as bull- and bearbaiting were popular in Europe for centuries but were banned in most countries by the 19th or 20th centuries. (Getty Images)

of Cruelty to Animals advanced the protection of animals (including from blood sports) with the passage of an anticruelty law in 1866.

After the Humane Act of 1835, English owners of bullbaiting dogs—bulldogs—focused more on fights between dogs than on fights with bulls. Bulldogs were originally bred to help farmers herd and manage bulls, not kill them. The heavy build and strength that was useful against bulls was not such an asset against other dogs, so it is believed that the bulldogs' owners began crossing them with the swift and equally tenacious terriers, creating bull terriers. Staffordshire bull terriers, pit bull terriers, and American Staffordshire terriers all trace their lineages to these putative bulldog-terrier mixes.

Various forms of animal fighting are still legal in parts of the world. Dogfighting, for example, is still legal in Russian and Japan. Even where it is illegal, it can be popular. Dogfighting and cockfighting are arguably the predominant blood sports today, next to hunting. Cockfighting is popular in parts of the United States, Latin America, Africa, Southeast Asia, the Philippines, Indonesia, and the Near East. The adoption of game-bred dogs by the American pop culture has surely exacerbated its spread (while also letting others learn how loving these animals can be). Dogfighting is no longer a poor, rural problem. The conviction of pro football quarterback Michael Vick in 2007 attests to how far this crime has spread into small towns, cities, and even suburbs.

When the culture and authorities allow, fighting events in many places take on the appearance of a fair, with whole families—young children included—watching. At these events, traditional and professional dogfights and cockfights follow strict rules. These rules do not protect the animals; they simply ensure fair fights. Matches are regulated, and animals are highly trained. There are variations in the ways fights are managed. According to Cajun rules, dogfights are held in pits that are 15 to 20 feet square with 2- to 3-foot walls. Diagonal scratch lines are made in opposite corners 12 feet apart. Before a match, the dogs are weighed and washed. Washing prevents owners from covering their dogs with poison or substances that could make it harder for the other dog to maintain a hold. At the start of a match, the dogs are placed facing each other behind the scratch lines. The referee commands the players to release their dogs. The matches are hauntingly quiet, as the dogs grab and relentlessly rip open each other's mouths, faces, throats, and legs. If a dog moves so that his head and shoulders are not facing his opponent, a turn is called and the dogs are separated and repositioned behind the scratch lines. The dog that turned is held by his owner. The other dog is released and allowed to attack the held dog. If the released dog attacks the held dog, the held dog is released and the match continues. If the released dog does not attack the held dog, the match is over. The match is also over any time a dog stops fighting, dies, jumps out of the ring, or is pulled out by the owner. Losing dogs are often severely beaten, drowned, electrocuted, or hanged to death.

In contrast to regulated dogfights are the casual fights that occur on streets, and in parks and neighborhoods. Whereas traditional matches are organized in advance, fights can now happen between strangers with a simple "Wanna fight your dog?" These dogs are usually not well trained. They may simply be forced to be more aggressive—toward other dogs, other animals, and people. Because they are less structured, these fights are more chaotic and dangerous to bystanders.

Owners of fighting animals gain prestige among their peers because they are associated with a terrible animal, and are more or less able to control it. It comes as no surprise that people for whom all other avenues of empowerment have been cut off may turn to building their image with a powerful dog who loyally obeys its owner to the brutal end.

Gambling has long been strongly associated with many blood sports, and dog and cockfighting are no exceptions. In addition, when raids and seizures are made of dog or cockfighting rings, authorities very often find illegal weapons as well as drugs. There is also reason to believe that animal fighting is run by organized crime in some areas.

Breeding

Like the wolves from which they evolved, dogs do not naturally fight each other to the death. Normally, dogs will only posture and feint, but not resort to actual fighting. When fights do occur, they are swift and very rarely lethal. This tendency to break off an attack when they see submission cues has been bred out of game-bred dogs. Throughout history, different types of dogs have been forced to fight; one era's fighting dog is another era's loving companion. In addition, fighting dogs are bred not only for gameness: As any owner of a pit bull will attest, these dogs are very loyal. Lineage is important to committed dogfighters. As much money as owners can make gambling, they can make more by breeding their dogs.

Training and Conditioning

Breeding is important to prepare a dog to fight, but it is not enough. In order to fight, dogs must be trained to do so. A fighting dog's training usually starts at a very young age. Separating young pups from their mother and littermates can make them more violent toward other dogs when they grow up. Every opportunity is taken to enhance the aggressiveness and tenacity of these dogs. To build their endurance, dogs are made to run on treadmills or chase a small bait animal that is kept just out of their reach. Bait animals can be squirrels, rabbits, or even stolen pet dogs and cats (as was found in 2004 to have been happening for years in parts of Arizona). After a training session, the dog is usually allowed to rip apart the live bait animal both as a reward and a way to develop the dog's taste for blood. So-called trainers increase a dog's strength by having him wear heavy chains or weights for long periods. The dogs are also made to jump up and hang from ropes by their mouths to develop their lunge and bite. Throughout all of the training, the dog's tolerance for pain is pushed to the limit; it is not so much the stronger dog who wins, but the one who can withstand more pain and damage.

Roosters are trained much as dogs are. They run on treadmills to develop their stamina. They may wear gloves on their feet or have their spurs covered to let them practice fighting without undue injury. They are also set to taunt other roosters to increase their aggressiveness.

The Future

Laws banning blood sports can help save countless animals from horrendous deaths—when they are enforced. Much progress has been made to prevent these horrific events, but the problem is still extensive. The Humane Society of the United States estimates that there are at least 40,000 professional dogfighters in the United States—a country in which dogfighting is illegal. Law enforcement officers and governments do not always see the importance of these crimes or their association with other crimes. Grassroots advocacy to enforce these laws and pass new ones can only help.

Many of those who turn to animal fighting seem to do so because they have few other ways to create an impressive self-image. In addition to offering economic and educational assistance, communities are finding nonviolent ways for dogs to strut their stuff. Owners can gain satisfaction and pride when their dog's strength wins at a weight pull instead of a deadly fight.

See also Bullfighting; Cockfighting; Dogfighting; Hunting, History of Ideas

Further Reading

Fleig, D. (1996). *History of fighting dogs.* (W. Charlton, Trans.). Neptune, NJ: T.F.H. Publications.

Geertz, C. (2000). Deep play: Notes on the Balinese cockfight. In *The interpretation of cultures.* New York: Basic Books.

Homan, M. (2000). *A complete history of fighting dogs.* New York: Howell Books House.

Pushkina, D., & Raia, P. (2008). Human influence on distribution and extinctions of the late Pleistocene Eurasian megafauna. *Journal of Human Evolution, 54*(6), 769–782.

Shirlyn, H., & Lyons, T. (1999). Field crickets. *Insecta Inspecta World.* Retrieved March 12, 2006, from http://www.insecta-inspecta.com/crickets/field/index.html

William Ellery Samuels, Lieve Meers, Debbie Coultis, and Simona Normando

BULLFIGHTING

Bullfighting, or *corridas* in Spanish, is considered a form of art and of cultural heritage by its supporters and a severe form of animal cruelty by a growing number of people all over the world. This has led to passionate debates. A popular motto among the participants in anti-bullfighting demonstrations is "Torture, neither art not culture," expressing their conviction that intentionally inflicting pain on an animal for the purpose of entertainment can never be acknowledged as art. They argue that art and culture should imply the promotion of knowledge and excellence in order to enrich us to become wiser, more humane, and compassionate. Neither art nor entertainment should be based on abusing or making fun of the weaker—either humans or animals. Circuses used to exhibit people with deformities or peculiar physical features. Not long ago, there was a type of *corridas* designed for children (*charlotadas*), in which dwarves dressed as clowns or in other funny costumes, hit and jumped over the bull. Some have pointed out that if the same actions performed in bullfighting were done on a domestic animal, it would be considered a felony according to the Spanish Penal Code (Art. 337). The official statistics compiled by the Spanish Government reveal that in 2007 (*Estadisticas Taurinas* 2007) there were 2,622 bullfighting events that used 12,167 animals. A related controversial issue, which has become the target of a tax resistance campaign, is that, according to the advocacy group, Platform Stop Our Shame (SOS), bullfighting is subsidized with more than 560 million Euros of public money annually (*Fundación Altarriba "Dinero público."*). Platform SOS is asking for that money to be invested in social aid, education, or public health instead.

Types of Corridas

Bullfighting exists in various forms in several countries all over the world: Spain, Portugal, France, México, Colombia, Venezuela, Perú, Guatemala and, more recently, the United States. Although there are three main styles (Spanish, Portuguese and French), all of them, even the ones that are called bloodless, are based on exhausting and injuring the bull by using spikes, spears, swords, and daggers to cause immense pain (*I Jornada sobre ganado de lidia*, 1999) and blood loss in order to weaken the bull.

In the Spanish style, the *corrida* is divided into three parts called *tercios* (thirds). In the first part, the bull (already stressed by the transport) enters the ring while bullfighters (*toreros*) wave capes (*capotes*) to try to make the bull charge, and then the *picador* (a horse rider) sticks a *pic* (lance) into his back. The great pain, blood loss and stress inflicted on the animal makes him lower his head, which exposes the neck to the *banderilleros,* who will plant *banderillas* (barbed sticks with harpoon-like ends) on the withers of the bull. They finally run the bull in circles until it is dizzy and stops chasing. In the third part (called quite eloquently "the third of death"), the matador plays with the bull holding the *muleta* (a red cape) and a sword in order to dominate and exhaust him. It is then, when the tortured and completely worn out animal stands with his feet together and his head low that the matador thrusts his sword between the shoulders trying to reach the heart. The bull does not always die immediately, so a dagger is driven into the base of the skull to paralyze him (*descabello*).

But sometimes this measure does not work, and the bull remains fully conscious while his ears and tail are cut off as trophies.

Another Spanish style is called *Rejoneo*, in which the bullfighter inflicts the same tortures on the bull, but does so riding a blindfolded horse. Even though the horse has been trained to avoid the bull and it wears padding (a measure taken because the sight of injured horses with their intestines hanging out was too unpleasant for the audience), every year horses are severely injured and eventually die due to the injuries inflicted by the horns of the bull (Vicent, 2001).

The Portuguese style (*corrida de touros* or *tourada*) includes three types: The *Cavaleiro,* where a horse-rider dressed in traditional 18th-century costume tries to stab three or four *bandarilhas* (like the Spanish *banderillas*) into the back of the bull. The second type is called the *Bandarilheiros*— similar to the Spanish *matadores,* who simply play with the bull with a red cape. Finally, there is the *Forcados,* a group of eight men who provoke the bull without any protection or weapon. The front man tries to grab the bull's head, aided by the others, in order to achieve the *pega de touros* (bull catch). The Portuguese style is often viewed as cruelty-free because the bull is not normally killed in front of the audience, but the killing takes place out of the sight of the public. Nevertheless, even though the bull is not killed, the stress that he undergoes in order for the audience to have fun should be taken into account.

The French styles include the *Course Camarguaise* (in the *Camargue* region of Provence) and the *Course Landese* (in the region of Landes, on the French South-Atlantic coast). In the *Course Camarguaise*, the first part consists of a running of the bulls (*encierro*) and the second part, the course itself, takes place in a portable arena in which the participants (*raseteurs*) snatch rosettes or tassels off the bulls. Once the course is over, the bulls are herded back to their pen. The *Course Landese* is a competition between teams (*cuadrillas*) using cows instead of bulls. Cows have a rope attached to their horns controlled by one man (*Teneur de corde*) while the *entraîneur* positions the cow to face and attack the player. The *écarteurs* will try to dodge around the cow, holding their ground until the last moment, and the *sauteur* will leap over it. The cow is not killed but, again, is being abused, exhausted,

and stressed by a hostile environment. Not all bullfighting in France is French-style; the Spanish style is becoming more and more popular, with bulls often killed in public.

Other Bloodsports

Bulls are not only abused in bull rings but also in other *fiestas,* where they are harassed by the public, stressing them and often leading to a painful death. Among the myriad of blood *fiestas* there are some that stand out for their cruelty.

***Toro de la Vega* (Bull of La Vega)**   This tournament takes place in Tordesillas, in

An assistant bullfighter stabs a dying bull to death during a Novillero bullfight at the San Isidro festival at the Las Ventas bullring in Madrid, Spain. (AP Photo/Paul White)

Castilla León (Spain) every September, in honor of the Virgin of the Peña. A bull is harassed with spears by the villagers and forced to cross a bridge where they start hurling lances at him. The bull suffers from severe injuries caused by the lances, a terrible agony that can last up to some hours, ending when the eventual winner of the tournament throws the fatal blow. The winner has the right to cut off the bull's testicles and exhibit them at the end of the lance. The intrinsic cruelty of the tournament and the fact that the government acknowledges it as an event of national tourist interest have placed it in the center of anti-bullfighting campaigns. In past years, activists travel every September to ask for mercy for the bulls of *Tordesillas,* where they are confronted by the villagers.

Bull of Coria This event takes place in Coria, Cáceres (Spain) on the 23rd of June to honor Saint John and, together with the Bull of La Vega, represents those considered the most violent of the thousands of blood fiestas all over Spain. The bull is released from the barnyard and the villagers run him to the bull ring. Once in the arena, he is attacked by the public with darts from blowpipes. Spectators try to hit him in the eyes and testicles for several hours until he is finally shot (Fundación Altarriba, http://www.altarriba.org/2/verguenza/caceres-coria-english.htm. FAACE Web site, 2008; Bull of Coria, http://www.faace.co.uk/Coria.htm).

***Bous embolat* or "Fire Bulls"** *Bous embolats* take place in the Comunidad Valenciana region of central and southeastern Spain (including the provinces of Alicante, Valencia, and Castellón) and in the *Terres de l'Ebre* region, though these events are forbidden in many other regions. In this fiesta, several teams compete to see which is the fastest to place and light two balls of fire on the tips of the horns of a restrained bull. Once the balls are lit, the bull is let loose and the public harass it. The bull inevitably suffers, due to the fear of the fire and the burns caused, especially in the eyes. At times, the bull has died from being burned alive.

Attitude Changes to Bullfighting

In recent years there has been a remarkable change in Spanish attitudes toward bullfighting. According to a 2006 Gallup poll, 72 percent of Spaniards have no interest in bullfighting (*Investiga,* 2006: "Interest in Bull Fights") and only eight percent of Spaniards consider themselves supporters. In 1989, a campaign to declare cities as opposed to bullfighting started in Catalonia, and so far 47 cities have joined, achieving a major success in 2004 when Barcelona took a crucial step by agreeing to become an anti-bullfight city.

In Catalonia, animal-protection law prohibits the construction of new bull rings and, in fact, at the time that this essay is being written, it is in the spotlight due to a campaign to officially ask the Catalan Parliament to debate the ban on bullfighting. Additionally, the growing rejection of the mistreatment of animals has even reached the Spanish Parliament, where a group of MPs has created the Parliamentary Association for the Defense of Animals, lobbying against bullfighting and also supporting the ban on cat and dog fur, as well as the ban on seal-derived products within the EU. More evidence that winds of change are blowing comes from the city of Paterna in the province of Valencia, where the continuation of festivals featuring bulls was rejected in a historic public ballot in a region that was formerly especially fond of bull festivals.

The use of animals in feasts, either as a questionable pastime or as a symbolic combat between the supposed rational and the beast, is definitely facing the beginning of the end; a growing number of people demand a more compassionate society in which animals are no longer the victim or entertainment to alleviate humankind's miseries. Those who support the banning of bullfighting believe that until the bulls graze peacefully in the meadows far away from the suffering that they have undergone over the years, Spain cannot be called a civilized country.

Web Sites with Bullfighting Information

The following Web sites advocate against bullfighting:

Animanaturalis: www.animanaturalis.org

FAACE, Bull of Coria. http://www.faace.co.uk/Coria.htm

Fundación Altarriba, Shame on Coria. http://www.altarriba.org/2/verguenza/caceres-coria-english.htm

Fundación Altarriba, The bull of La Vega. http://www.altarriba.org/2/verguenza/valladolid-tordesillas-english.htm

Investiga, Interest in Bull Fights (*Interes en las corridas del toros*). http://www.ig-investiga.es/encu/toros06/intro.asp

League Against Cruel Sports: http://www.league.org.uk

STOP OUR SHAME: www.stopourshame.com

WSPA: http://www.wspa-international.org

Further Reading

Manuel, Vicent. 2001. *Antitauromaquia*. Aguilar.

Purroy Unanua, Antonio, & Agrónomo, Ingeniero. 1999. *I Jornada sobre ganado de lidia (Ponencias)*. Escuela Técnica Superior de Ingenieros Agrónomos de Madrid. Pamplona: Ediciones Mundi-Prensa.

Núria Querol i Viñas

C

CAGING

See Chickens

CAPTIVE BREEDING ETHICS

Most of our planet suffers from some amount of environmental degradation, and trends suggest that the situation will worsen before it improves, if it improves at all. Consequently, conservationists increasingly focus on restoration efforts, and restoration ecology is a rapidly growing field. Reintroducing animals from captivity into areas where they no longer persist represents one tool in the restorationists' toolkit. This entry focuses on reintroduction, rather than on releasing animals to augment existing populations (restocking) or introductions of animals to areas outside their historical range. The latter two are generally inadvisable, although they can be useful under special circumstances. Reintroduction involves difficult ethical questions that many scientists have raised. These are examined here.

At one time, zoos and aquariums argued that breeding animals in captivity for eventual reintroduction to the wild would grow to become the defining rationale for their continued social relevance and future existence (Reading & Miller, 2001).

Today, however, zoos and aquariums recognize that, while important, captive breeding for reintroduction represents a relatively insignificant part of what they do (Reading & Miller, 2001; Hutchins et al., 2003). Snyder and colleagues (1996) caution against relying too heavily on captive breeding and reintroduction for conservation, and instead suggest that conservationists should employ this tool only when other options are unavailable. Still, zoos, aquariums, government wildlife agencies, and other groups likely will increase the amount of captive breeding they undertake as a part of restoration programs. With this increase in captive propagation for reintroduction, it is important to consider the ethical concerns of this approach to conservation.

The ethics of even engaging in captive propagation for reintroduction at all should be considered. Frederic Wagner (1995) asks "Just because we can breed animals in captivity for reintroduction, does that mean we should?" Is reintroduction just a human endeavor to "redecorate nature," as Marc Bekoff (2000, 2006) suggests? Alternatively, Robert Loftin (1995) asks if we have a moral obligation to prevent human-caused extinction; and, if so, is captive breeding and reintroduction justified? After all, humans have already "redecorated nature" extensively through global and local species extinctions and introductions. Do we have any responsibility

101

to try to prevent extinction and restore nature, at least to some degree, even if doing so in some way mentally or physically "harms" individual animals?

The larger ethical consideration of whether or not to engage in captive breeding for reintroduction often relates to broader worldviews and core (or more strongly held, central) values. In this situation, the main ethical consideration is how we balance the welfare and rights of individual animals against the value of captive breeding to reintroduction programs and our obligations to sustain populations, species, and ecological communities and processes (Norton, 1995). Michael Hutchins and colleagues (2003, p. 964) describe this as " . . . issues of individual animal welfare versus overall species and ecosystem conservation." This is an important consideration, because sometimes actions designed to benefit populations will conflict with the interests of individual animals held in captivity (Wuichet & Norton, 1995).

Tom Regan (1995) suggests that there are three basic worldviews with respect to holding animals in captivity (in particular, he was discussing zoos, not breeding facilities for reintroduction *per se*). These are utilitarianism, animal rights, and environmental holism. Briefly, the utilitarian doctrine, as championed by Peter Singer (1980), argues that we should afford rights to sentient species—those able to experience suffering and pleasure—or we risk engaging in what he calls speciesism (favoring some species, most notably human, over other species). Singer argues that we should engage in actions that result in the greatest good for all sentient organisms. Thus, we must take into account all the costs and benefits of our actions. Tom Regan (1983) argues from a strong animal rights stance which values the individual rights of all animals. He suggests that we should minimize depriving individuals (of all sentient species) of their basic rights. Is subjecting animals to our wants nothing more than environmental fascism (Regan, 1983)? Finally, environmental holism grew out of Aldo Leopold's essay "Land Ethic," which argues that, "A thing is right when it tends to preserve the integrity, stability, and beauty of the biotic community. It is wrong when it tends otherwise" (1968L, pp. 224–225). Thus, followers of this worldview believe that the interests of the entire biotic community trump the rights of the individual. Conservationists, for example, often argue from this perspective to justify holding animals in captivity for "the good of the species" (c.f. Hutchins et al., 2003).

What do these different worldviews suggest with respect to captive propagation for reintroduction? Regan (1983) argues that any type of captivity or manipulation of a sentient animal represents a form of "environmental fascism." Are other animals sufficiently different from humans to warrant different treatment? ask Joy Mench and Michael Kreger (1996). As we learn more about other animals, we find fewer distinctions, yet no one, not even Peter Singer or Tom Regan, suggests that we should treat all animals equally. So should different species be afforded different rights? What about species that are not sentient or feel no pain (Bostock, 1993)? One of the great difficulties in evaluating different ethical stances is our ability to assess the impacts of captivity on individuals of other species. Would an individual animal trade greater freedom for the greater security and amenities (such as adequate food, water, and shelter) of captivity? Many humans agree to these tradeoffs, albeit usually on a different scale (consider the post 9/11 societal changes in the United States and other countries, and the fact that many poor people in former Marxist countries look back with nostalgia at a time when the state ensured their basic needs).

It is important to note that most re-introductions fail, and that other approaches to conservation usually hold greater promise (Griffith et al., 1989; Beck, 1995; Reading & Miller, 2001). In addition, just because we can (or do) breed a species in captivity and reintroduce it does not necessarily mean we should. How do we reconcile the low rates of reintroduction success with issues of animal welfare and rights? Is it humane to reintroduce animals given the fact that most of the animals released will die? Dale Jamieson (1995a) argues that since captive breeding and reintroduction play only a marginal role in conservation, we should instead focus our limited resources on protecting habitat.

If we agree on the importance of captive breeding for reintroduction, additional ethical considerations arise. Is a commitment to the ethical treatment of animals in captivity sufficient if those animals contribute to ecological restoration via reintroduction? Joy Mench and Michael Kreger (1996) argue that most people are concerned that animals be spared pain and suffering to the greatest extent possible, that they have a good quality of life, and are not used for "trivial" purposes. But is simply addressing the concerns that "most people" have sufficient? Michael Hutchins and colleagues (2003) ask, "How far do we need to go in addressing the welfare of animals held in captivity, short of fully replicating nature?" Defining an animal's physical and especially psychological well-being is a

very difficult task. Since we can never fully understand other species, John Wuichet and Bryan Norton (1995) suggest that we necessarily fall back on anthropomorphically biased opinions about what the well-being of an individual animal really means.

Wuichet and Norton (1995) believe that our treatment of animals in captivity should strive to achieve a level of physical and psychological well-being comparable to or better than that of life in the wild. In other words, the captive environment should be as authentic as possible. Do we have a moral obligation to maximize survival prospects for individual animals no matter the cost, as Loftin (1995) states? The reality is that resource constraints will always enter into the equation, precluding most, if not all, programs from going as far as they would like in attempting to replace nature on a smaller scale (Snyder et al., 1996). So how far should or must we go?

Research on successful reintroduction suggests that increasing the "naturalness" of a captive environment would also maximize reintroduction success rates. Indeed, reintroductions that use animals from other wild populations (i.e., translocations) usually succeed far better than programs that use animals bred in captivity. Using captive-bred animals for reintroduction requires addressing a host of biological considerations that, in turn, have ethical implications. These include maintaining genetic diversity (and therefore aggressively managing who mates with whom), acclimatizing animals to their release environments, and providing environmental stimuli for adequate development of the full array of important behavioral skills, as well as avoiding habituation to humans (Snyder et al., 1996; Miller et al., 1999; Reading

104

Captive Breeding Ethics

et al., 2004). But, are captive environments that expose animals to predation and other survival risks morally justified even if they increase the survival of released animals (Beck, 1995)? This brings up the tricky question of how we balance issues of animal welfare with the welfare of species and considerations of different techniques (Wagner, 1995).

The ethical considerations of breeding animals in captivity for eventual reintroduction to the wild are complex. Divergent worldviews argue from different ethical standpoints as to whether or not such activities should even occur. Many people espousing strong animal rights and animal welfare ethics suggest that captive breeding and reintroduction are always morally wrong. Others, arguing from an environmental holism or land ethics perspective, embrace a strong ethical obligation to restore populations extirpated by people, and therefore believe that the interests of the entire biotic community trump the rights of individuals. Yet even those who support using captive breeding and reintroduction in general must judge whether or not such an approach is appropriate, given the circumstances surrounding each individual case. To the extent that captive breeding programs do exist, difficult ethical questions still remain with respect to how far we must go in replicating nature in the captive environment, as well as our obligations to individuals held in captivity and those destined for release back into the wild. Most people agree that we should go as far as resources allow in providing the most realistic captive environment possible. Such an approach would also increase reintroduction success rates. We will likely never fully resolve the difficult ethical questions surrounding captive breeding and reintroduction on our increasingly altered planet.

Further Reading
Beck, B. (1995). Reintroduction, zoos, conservation, and animal welfare. In *Ethics on the ark: Zoos, animal welfare, and wildlife conservation*. (Ed. by B. G. Norton, M. Hutchins, E. F. Stevens, & T. L. Maple), 155–163. Washington, DC: Smithsonian Institution Press.
Bekoff, M. (2000). Redecorating nature: Reflections on science, holism, humility, community, reconciliation, spirit, compassion, and love. *Human Ecology Review* 7: 59–67.
Bekoff, M. (2006). *Animal passions and beastly virtues: Reflections on redecorating nature*. Philadelphia: Temple University Press.
Griffith, B., Scott, J. M., Carpenter, J. W., & Reed, C. (1989). Translocation as a species conservation tool: Status and strategy. *Science* 245: 477–480.
Hutchins, M., Smith, B., & Allard, R. (2003). In defense of zoos and aquariums: the ethical basis for keeping wild animals in captivity. *Journal of the American Veterinary Medical Association* 223(7): 958–966.
IUCN. (1987). Translocation of living organisms: introductions, reintroductions, and restocking. IUCN Position Statement. Gland, Switzerland: IUCN.
Jamieson, D. (1995a). Zoos revisited. In: *Ethics on the ark: Zoos, animal welfare, and wildlife conservation*. (Ed. by B. G. Norton, M. Hutchins, E. F. Stevens, & T. L. Maple), 52–66. Washington, DC: Smithsonian Institution Press.
Jamieson, D. (1995b). Wildlife conservation and individual animal welfare. In: *Ethics on the ark: Zoos, animal welfare, and wildlife conservation*. (Ed. by B. G. Norton, M. Hutchins, E. F. Stevens, & T. L. Maple), 69–73. Washington, DC: Smithsonian Institution Press.
Leopold, A. (1968). *A Sand County almanac and sketches here and there*. New York: Oxford University Press.
Loftin, R. (1995). Captive breeding of endangered species. In *Ethics on the ark: Zoos, animal welfare, and wildlife conservation*. (Ed. by B. G. Norton, M. Hutchins, E. F. Stevens, & T. L. Maple), 164–180. Washington, DC: Smithsonian Institution Press.

Mench, J. A., & Kreger, M. D. (1995). Animal welfare and public perceptions associated with keeping wild mammals in captivity. In 1995 AZA Annual Conference Proceedings. Association of Zoos and Aquariums, Bethesda, MD, 376–383.

Mench, J. A., & Kreger, M. D. (1996). Ethical and welfare issues associated with keeping wild mammals in captivity. In *Ethics on the ark: Zoos, animal welfare, and wildlife conservation*. (Ed. by B. G. Norton, M. Hutchins, E. F. Stevens, & T. L. Maple), 5–15. Washington, DC: Smithsonian Institution Press.

Miller, B., Ralls, K., Reading, R. P., Scott, J. M., & Estes, J. (1999). Biological and technical considerations of carnivore translocation: A review. *Animal Conservation* 2(1):59–68.

Norton, B. (1995). Caring for nature: A broader look at animal stewardship. In *Ethics on the ark: Zoos, animal welfare, and wildlife conservation*. (Ed. by B. G. Norton, M. Hutchins, E. F. Stevens, & T. L. Maple), 102–121. Washington, DC: Smithsonian Institution Press.

Reading, R. P., & Miller, B. J. (2001). Release and reintroduction of species. In *Encyclopedia of the world's zoos*. (Ed. by C. Bell), 1053–1057. Chicago: Fitzroy Dearborn Publishers.

Reading, R. P., Kleiman, D. G., & Miller, B. J. (2004). Conservation and behavior: Species reintroductions. In *Encyclopedia of animal behavior*, Vol. 1: A-C (Ed. by M. Bekoff), 426–435. Westport, CT: Greenwood Press.

Regan, T. (1983). *The case for animal rights*. Berkeley: University of California Press.

Regan, T. (1995). Are zoos morally defensible? In *Ethics on the ark: Zoos, animal welfare, and wildlife conservation*. (Ed. by B. G. Norton, M. Hutchins, E. F. Stevens, & T. L. Maple), 38–51. Washington, DC: Smithsonian Institution Press.

Singer, P. (1990). *Animal liberation*, 2nd Ed. New York: Random House.

Snyder, N. F.R., Derrickson, S. R., Beissinger, S. R., Wiley, J. W., Smith, T. B., Toone, W. D. et al. (1996). Limitations of captive breeding in endangered species recovery. *Conservation Biology* 10: 338–348.

Wuichet, J., & Norton, B. (1995). Differing conceptions of animal welfare. In *Ethics on the ark: Zoos, animal welfare, and wildlife conservation*. (Ed. by B. G. Norton, M. Hutchins, E. F. Stevens, & T. L. Maple), 235–250. Washington, DC: Smithsonian Institution Press.

Wagner, F. (1995). The should or should not of captive breeding. In *Ethics on the ark: Zoos, animal welfare, and wildlife conservation*. (Ed. by B. G. Norton, M. Hutchins, E. F. Stevens, & T. L. Maple), 209–214. Washington, DC: Smithsonian Institution Press.

Richard P. Reading and Brian J. Miller

CATS

The domestic cat is the most popular companion animal in the United States today, with more than 80 million living in American households. With regard to the welfare of cats in our society, there are three issues of primary concern: the use of cats in biomedical research, the problem of unowned, free-roaming cats, and the high euthanasia rate of cats in animal shelters.

In 1881, British zoologist St. George Mivart published a textbook called *The Cat: An Introduction to the Study of Backboned Animals, Especially Mammals* in which he described the cat as "a convenient and readily accessible object for reference" in studying mammals, including humans. Since the publication of Mivart's book, cats have been used in research primarily to learn about the specific functions of nerve cells and about how the brain processes visual information. Research with cats has contributed to advances in treating various disorders of the eye, including "lazy eye," glaucoma, and cataracts, as well as recovery from damage to the brain and spinal cord from injuries and strokes. Cats also have been used to study particular medical problems they have in common with humans, such as hearing disorders, diabetes, and acquired immune deficiency syndrome (AIDS).

Domestic felines come in many shapes and sizes, including the nearly hairless Sphynx. They are used in a variety of laboratory experiments. (Photos.com)

Research in these areas is contributing to both feline and human health.

Compared to other nonhuman animals, the numbers of cats used for biomedical research is small and continues to decline. In 1995, fewer than 30,000 cats were used for research purposes in the United States, representing only two percent of all research animals used that year, excluding rats and mice. At present, cats reportedly comprise less than one percent of all animals used in research. Furthermore, the institutions conducting research with cats in the United States, Great Britain, and many other countries must comply with strict regulations for animal care and use specified by their respective animal welfare laws.

An issue of even greater concern is the ongoing problem of cat overpopulation, particularly the problem of free-roaming, unowned, feral cats. Although the number of such cats is difficult to determine, estimates of their numbers is as high as 70 million across the United States. Several factors may account for the existence of so many homeless cats. First, many people believe that cats can survive easily on their own and choose to abandon their pets when it is inconvenient to keep them. Also, pet cats with access to the outdoors sometimes stray from home. Those who are not identified with a tag, microchip, or tattoo and do not return home on their own may become permanently lost. In addition, unneutered pet cats allowed outdoors may mate with stray cats whose litters may be born outside, further contributing to the homeless cat population.

The question of what to do about these free-roaming or feral cats has been hotly debated among the humane community, wildlife agencies, and cat advocacy groups. Two primary management philosophies exist. Some believe that it is better to trap and humanely kill these animals. Those who advocate this policy argue that, even with help from human caretakers, these animals suffer and die a miserable death. They also are concerned about the spread of disease, both within the cat population and to humans, and the impact of these animals on wildlife populations, especially birds and small mammals.

On the other hand, many groups support TNR (trap, neuter, return) as long as there are people willing to feed and provide veterinary care for outdoor cat colonies. The arguments in favor of this method are that neutering the animals will eventually reduce the size of the colony and eliminate problem behaviors such as spraying, howling, and fighting that cause problems in residential areas. TNR advocates also argue that this is a moral issue and, as domestic animals, these cats deserve our assistance. Furthermore, even if a colony is removed, other cats will move into the area.

In addition to the problem of unowned homeless cats, issues regarding cats in animal shelters continue to be of great concern. Of all cats entering United States animal shelters each year, 70 percent are euthanized and only two percent are

returned to their owners. In an effort to reduce euthanasia numbers and increase the chances of returning lost cats to their homes, humane organizations advocate spaying and neutering, identifying cats with a tag or microchip, and keeping pet cats indoors.

Further Reading

American Society for the Prevention of Cruelty to Animals (n.d.). *Position statement on feral cat management.* Retrieved October 7, 2008 from http://www.aspca.org/site/PageServer?pagename=pp_feralcat.

American Veterinary Medical Association. (2007). *Market research statistics.* Retrieved October 7, 2008 from http://www.avma.org/reference/marketstats/ownership.asp.

AVMA Animal Welfare Forum. (1995). Veterinary perspectives on the use of animals in research. *Journal of the American Medical Association* 206(4).

Berkeley, E. P. (1982). *Maverick cats: Encounters with feral cats.* New York: Walker.

Clifton, M. (Ed.). (November, 1992). Seeking the truth about feral cats and the people who help them. *Animal People.*

Fitzgerald, B. M., and Turner, D. (2000). Diet of domestic cats and their impact on prey populations. In D.C. Turner and P. Bateson (Eds.), *The domestic cat: The biology of its behaviour,* 123–144. New York: Cambridge University Press.

Humane Society of the United States. (2008). *The Humane Society of the United States urges U.S. Fish and Wildlife Service not to shoot feral cats on San Nicolas Island.* Retrieved October 7, 2008 from http://www.hsus.org/press_and_publications/press_releases/hsus_urges_usfws_not_shoot_feral_cats_san_nicolas_island_061808.html.

Johnson, P. D. (2006). *The cat in biomedical research.* Retrieved October 7, 2008 from www.uac.arizona.edu/VSC443/catmodel/catmodel07.html.

Mivart, S. G. (1881). *The cat: An introduction to the study of backboned animals, especially mammals.* London: John Murray.

Pet Food Institute. (2006). *Pet population data.* Retrieved October 7, 2008 from http://www.petfoodinstitute.org/reference_pet_data.cfm.

U.S. Department of Agriculture (1996). *Animal welfare enforcement: Fiscal year 1995.* APHIS Publication No. 41-35-042. Washington, D.C.: U.S. Department of Agriculture.

R. Lee Zasloff

CHICKENS

The domesticated chicken is derived from the wild jungle fowl of Southeast Asia, and was originally domesticated over 10,000 years ago. The world is now populated with over 16 billion broiler chickens and 5 billion laying hens, with the highest numbers being found in China, Brazil and the United States.

Broilers have been selected for their prodigious appetite, rapid growth, and massive development of the pectoral muscles that provide breast meat. They are usually kept in large mixed-sex flocks in litter-floored housing, and harvested for meat at around six weeks of age. Laying hens, selected to produce over 300 eggs per year, are thinner and more agile than broilers, and have little value as a source of meat. They are kept in flocks of adult females, often in small groups in cages. Chicks of broiler and layer strains are supplied by hatcheries that incubate fertile eggs obtained from breeding flocks.

Close association with humans has made the chicken the most abundant of all bird species, but success at the species level comes at a cost to individual chickens. For the majority of consumers, low cost is the primary determinant when selecting poultry products. To minimize the cost of production, most chickens are given little space or behavioral freedom. Producers defend their housing and management practices on the basis that modern chickens are not well adapted for life in nature, and would not be productive under intensive farming conditions if they weren't healthy and content. Nevertheless, an ability to respond to instinctual urges and learned preferences is undoubtedly desirable from the chicken's perspective.

Cage Housing of Laying Hens

Major controversy surrounds the housing of laying hens in cages. Producers provide the minimum cage space needed to maintain high egg production. This space allocation is determined by the ability of hens to access food and water, and to avoid overheating in hot weather. Genetic selection for group living at close quarters has produced hens that are tolerant of one another, sharing space rather than attempting to defend access to food and water through aggression. Maintaining hens at close quarters in cages is also possible because feces drop through the wire floor of the cage, reducing disease risk from intestinal parasites. Conveying feces away on a manure belt makes it easy to avoid problems with high ammonia concentrations and, because the cage floor slopes so that eggs roll out as soon as they are laid, dirty and cracked eggs are minimized. Furthermore, cages are stacked in multiple tiers so that the vertical space of the chicken house is used efficiently.

Nonetheless, the behavioral restriction of hens in cages has prompted calls for more roomy conditions. Providing additional space allows for greater ease in performing comfort behaviors such as preening and stretching, as well as locomotory behaviors such as walking, running and jumping. Greater activity strengthens bones, making them somewhat less susceptible to fractures when hens are removed from cages and killed

Chickens held in cages at Whiting Farms in Colorado. This farm has about 85,000 chickens who are harvested for feathers that will be used for making fly fishing flies. (AP Photo/John Marshall)

at the end of their productive lifespan. It is unclear how much additional empty space is desirable from a hen's perspective. What is in the space may have greater salience. Thus, the European Union has mandated that, from 2012, hens may no longer be kept in plain cages. Cages must be furnished with a perch, nest and litter material to facilitate expression of perching, nesting, foraging and dust bathing behavior.

Cage-Free and Free-Range Housing

Whether furnished cages provide sufficient behavioral freedom is a matter for debate, and some people favor banning cages outright. In affluent countries, the market for cage-free and free-range eggs is expanding, leading egg producers to replace a proportion of their cage housing with housing comprising a combination of slatted floors and littered areas, or aviaries with multiple wire-floored tiers, with or without access to the outdoors. Litter, nest boxes and perches are provided, although the ideal quantity and layout of these resources has not been well established.

Despite popular opinion, many welfare problems have been encountered in these facilities. Compared to cage housing, these include increased risks of cannibalism, feather pecking, bone fractures, smothering, bacterial diseases, and parasitism. Predation is added to the list for hens given access to free range. The extent of these risks depends on specific details of housing design, genetic strain of chickens, their rearing conditions, and the producer's experience with this type of housing. In particular, strong genetic selection needs to be applied to develop strains of chickens that are better adapted for living in these facilities. Rearing chickens with access to perches from an early age also mitigates some of these problems. For free-range hens, the welfare implications of unexpectedly being denied access to the outdoors due to inclement weather or disease threats from wildlife (e.g., avian influenza) have not been determined.

To reduce contact with feces, only about one-third of the floor space in cage-free hen housing is covered with litter. In contrast, broilers must be kept on all-litter floors to cushion their heavy bodies and prevent breast blisters. In either case, litter must be kept dry to minimize the release of ammonia from feces, especially in warm weather. Ammonia irritates the eyes and respiratory passages, and can create lesions on the feet and hocks of heavy, inactive broilers. Controlling ammonia depends as much on proper ventilation and management of the drinkers as it does on the space allowance per chicken. If the litter is too dry, ammonia is replaced by problems with dust.

Rapid Growth of Broilers

Broiler chickens have large appetites and grow rapidly, which places them at risk of developing cardiovascular and skeletal disorders. These risks have been reduced to some extent by genetic selection and manipulation of day length to constrain early growth but stimulate rapid growth later on. However, the limited mobility of modern strains of broilers, and the potential for pain from leg and joint disorders, has prompted calls for the use of slower-growing, less productively efficient broilers that display more active behavior, including use of perches.

Broiler breeders would become unhealthy if allowed to eat like broilers for prolonged periods. Therefore, their feed intake is strongly restricted to control growth and promote reproductive fitness in adulthood. The resultant hunger can lead to the development of unwanted behaviors such as spot pecking. Feeding a high-fiber diet partially alleviates this problem.

Induced Molting

Laying hens molt after they have been laying eggs at a high rate for about one year. Until recently, molting was induced by complete feed withdrawal for up to two weeks, prompting loss of abdominal fat and leading to improved survival rates, egg production, and eggshell quality during a second laying cycle. Although this increased the longevity of survivors, it put the lightest hens in the flock at the risk of anorexia and death. Consequently, prolonged feed withdrawal has been outlawed in the European Union and abandoned in the United States, and molting is now induced by feeding a low-nutrient diet.

Beak Trimming

Beak trimming (or debeaking) involves amputating up to two-thirds of the upper beak and less of the lower beak. It is effective in reducing damage from feather pecking and cannibalism, which are serious welfare problems in laying hens and broiler breeder hens kept cage-free in large flocks. By making manipulation of feed more difficult, beak trimming reduces feed wastage, although it may also reduce the ability of hens to remove parasitic mites through preening. Unfortunately, beak trimming causes pain, fuelling bans in Sweden, Norway, Finland, and Switzerland, and a UK ban scheduled for 2011. Pain from the procedure can be lessened by use of analgesics and limiting beak trimming to the first 10 days of life. However, genetic selection against feather pecking and cannibalism is the most promising long-term solution.

Slaughter

Due to their value for meat, broiler chickens are usually transported only short distances to slaughter. In contrast, end-of-lay hens have little value for meat, compounded by food safety concerns about bone fragments in meat resulting from bone fractures. As such, transportation distances can be great to reach the few slaughterhouses willing to accept these hens, and hens face an elevated risk of dying in transit. Difficulties in marketing mean that hens are increasingly killed on the farm using carbon dioxide gas. Because the hens are killed almost immediately following catching, the duration of suffering is brief.

Controversy surrounds the most humane method of rendering broilers unconscious prior to slaughter. With developments in technology, it is likely that the current practice of hanging chickens upside down on shackles and stunning them electrically will be replaced by stunning using a mixture of carbon dioxide and inert gas.

Future Trends

To promote chicken welfare, there is a growing trend toward introduction of science-based welfare assurance and labeling schemes, either through legislation or under the auspices of various animal welfare organizations, supermarkets, and poultry industry groups. These standards vary in the extent to which they emphasize natural behavior, chicken feelings

and preferences, and physical health and productivity, and their implementation depends upon the quality of audits and their appeal to the public. Some consumers have shown willingness to pay more for cage-free and free-range products. However, allowing for greater behavioral freedom has introduced other well-being problems. A holistic approach is needed to that enhances overall well-being, and also takes into consideration impacts on human health and safety, wildlife, and the environment.

Further Reading

Appleby, M.C., J.A. Mench, & B.O. Hughes. 2004. *Poultry behaviour and welfare*. Wallingford, UK: CABI Publishing.

Council of the European Union. 1999. Council Directive 1999/74/EC of 19 July 1999 laying down minimum standards for the protection of laying hens. Official *Journal of the European Communities*. L 203/53–57. Retrieved from http://eur-lex.europa.eu/LexUriServ/LexUriServ.do?uri=OJ:L:1999:203:0053:0057:EN:PDF.

Council of the European Union. 2007. Council Directive 2007/43/EC of 28 June 2007 laying down minimum rules for the protection of chickens kept for meat production. Official *Journal of the European Union*. L 182/19–28. Retrieved from http://eur-lex.europa.eu/LexUriServ/LexUriServ.do?uri=OJ:L:2007:182:0019:0028:EN:PDF.

National Chicken Council. 2005. Animal Welfare Guidelines and Audit Checklist. Washington DC: National Chicken Council. Retrieved from http://www.nationalchickencouncil.com/files/AnimalWelfare2005.pdf.

Perry, G.C. (ed.) 2004. *Welfare of the laying hen*. Wallingford, UK: CABI Publishing.

RSPCA. 2008. RSPCA Welfare Standards for Chickens, February 2008. Southwater, UK: Royal Society for the Prevention of Cruelty to Animals. Retrieved from http://www.rspca.org.uk/servlet/Satellite?blobcol=urlblob&blobheader=application%2Fpdf&blobkey=id&blobtable=RSPCABlob&blobwhere=1158755026986&ssbinary=true.

RSPCA. 2008. RSPCA Welfare Standards for Laying Hens and Pullets, March 2008.

Southwater, UK: Royal Society for the Prevention of Cruelty to Animals. Retrieved from http://www.rspca.org.uk/servlet/Satellite?blobcol=urlblob&blobheader=application%2Fpdf&blobkey=id&blobtable=RSPCABlob&blobwhere=998045492811&ssbinary=true.

The LayWel Project. 2006. Welfare implications of changes in production systems for laying hens. European Commission, 6th Framework Programme, contract No. SSPE-CT-2004-502315. Retrieved from: http://www.laywel.eu/.

United Egg Producers. 2008. Animal Husbandry Guidelines for U.S. Egg Laying Flocks, 2008 Edition. Alpharetta, GA: United Egg Producers. Retrieved from http://www.uepcertified.com/docs/UEP-Animal-Welfare-Guidelines-2007–2008.pdf.

Weeks, C., & Butterworth, A. 2004. *Measuring and Auditing Broiler Welfare*. Wallingford, UK: CABI Publishing.

Ruth C. Newberry

CHIMPANZEES IN CAPTIVITY

There are an estimated 2,400 chimpanzees living in captivity in the United States. Approximately 940–980 live in biomedical research laboratories, 270 live in accredited zoos, 625 live in sanctuaries, and an estimated 550 chimpanzees are living in various conditions in the entertainment industry, in roadside attractions, and as people's "pets." There are approximately 370 captive chimpanzees living in Japan, approximately 980 chimpanzees living in zoos in Europe, and about 50 in Australia and New Zealand. Although it is impossible to know the exact number of chimpanzees in captivity worldwide, it is safe to say that the numbers have been decreasing gradually, as importing chimpanzees from Africa is illegal and breeding is very tightly controlled. However,

the number of captive chimpanzees in Africa, where chimpanzees live naturally, is on the rise. An increasing number of rescued baby chimpanzees, orphaned as a result of the illegal bushmeat trade, are being protected in seminatural sanctuaries across the continent. Many hope that with efforts to protect habitat, and through educational campaigns to protect native animals, these wild born chimpanzees may be freed from captive existence someday, but that is not possible for the thousands of captive chimpanzees living in the rest of the world.

History of Captivity

Originally chimpanzees were brought into captivity by curiosity seekers and collectors, and the chimpanzee captives did not live long. There are reports of a few young chimpanzees living in captivity in private European collections and used as entertainers prior to the 20th century, but it was not until the early years of the 1900s that more systematic efforts to study chimpanzees in captivity began. Psychologist Wolfgang Kohler, who in 1913 became the director of the Anthropoid Station of the Prussian Academy of Science in Tenerife in the Canary Islands, was the first to study captive chimpanzee insight and problem-solving abilities. At the same time, a Russian comparative psychologist, Nadya Ladygina-Kohts, was documenting the emotional development of an infant chimpanzee named Joni. Both studies were short-lived. A decade later, Robert Mearns Yerkes began what was ultimately to become a very successful effort to create and sustain captive chimpanzees in the United States, but his initial efforts also ended with the early and tragic deaths of Chim and Panzee, both of whom died on separate visits to the primate collection of Madame Rosalia Abreu in Cuba. By the 1930s, Yerkes was successfully breeding chimpanzees in captivity. The colony that he began with four chimpanzees in New Haven, Connecticut, moved to Orange Park, Florida, and ultimately to Emory University in 1965 with 66 chimpanzees. It now exists as the Yerkes National Primate Research Center in Atlanta, Georgia, and has produced five generations of captive chimpanzees.

Early studies of captive chimpanzees were designed to provide basic physiological and behavioral information that would aid in the maintenance of captive populations. Researchers sought to understand the nutritional needs of chimpanzees and their reproductive habits, as well as to learn about their development, their intelligence, and their distinctive personalities. Yerkes was very clear that while it was important to investigate chimpanzees in order to understand them better, that understanding was ultimately in the service of bettering "man"—in his words "to contribute to the solution of our intensely practical, medical, social, and psychological problems." (Yerkes, 1916, p. 233) To that end, chimpanzees in the early years were used in a variety of experiments including lobotomy research, infectious disease research, radiation exposure, organ transplantation studies, and drug and alcohol addition studies. Infant chimpanzees were also used in deprivation studies that involved removing them from their mothers and depriving them of human contact, contact with other chimpanzees, and natural stimuli including light, sound and, in at least one case, all tactile stimulation.

In the 1950s, chimpanzees were being used in military experiments which involved crash tests, exposure to extreme G-forces, decompression, and radiation. Before sending humans into space, NASA and the Russian Federal Space Agency began sending animals into space, and chimpanzees were among the early space explorers. In 1953, the Holloman Aeromedical Field Laboratory's Space Biology Branch in Alamogordo, New Mexico imported more than 60 chimpanzees from Africa to use in biodynamic and aeronautical research. The chimpanzees were similar enough to humans that it seemed reasonable to use them to reveal the suspected effects of space travel, and they were smart enough to be trained in complicated tasks similar to those that astronauts would need to perform in flight. Because of their similarity to humans, chimpanzees were shot into space on test runs before humans went. On January 31, 1961, Ham, a trained three-and-a-half-year-old chimpanzee, was the first chimp-o-naut. Only after Ham returned did Alan Shepard become the first American to travel in space. The second chimp-o-naut, Enos, a five-and-a-half-year-old, was sent up on November 29, 1961 and, following her success, John Glenn orbited the earth three times in 1962.

As biomedical research on chimpanzees was rapidly increasing in the 1960s, so too was our understanding of chimpanzees as smart, sensitive, and highly social animals. Jane Goodall began her groundbreaking study of chimpanzees in the wild, and behavioral researchers in the United States began teaching chimpanzees to use human language and other symbolic communication techniques to reveal their intelligence. Having seen an early film of researchers from the Yerkes colony attempting to teach a young chimpanzee, Viki, how to speak, Allen and Beatrix Gardner, psychologists at the

University of Nevada in Reno, embarked on a project to teach chimpanzees sign language. In 1966, the Gardners acquired one of the Air Force chimpanzees, raised her as they would a child, and trained her in American Sign Language. That chimpanzee, Washoe, was the first chimpanzee to acquire human language. She is reported to have learned and used over 200 signs. Other language projects also got underway at this time. In 1967, another couple of psychologists, David and Ann Premack, first reported their success in teaching a wild born chimpanzee, Sarah, to use plastic tokens to represent words. These tokens varied in shape, size, texture, and color, and Sarah formed sentences by placing the tokens in a vertical line. Sarah used nouns, verbs, adjectives, pronouns, and quantifiers; over the years she was also taught concepts such as same/different and negation and she also learned how to distinguish "greater than" and "less than." Sarah is probably most well known as the chimpanzee used in the experiments that started the subfield in comparative and developmental psychology called the "theory of mind research," in which nonhumans and non-linguistic humans were studied to determine whether they understood that others have mental states and what those mental states might be. Project Lana was another chimpanzee language experiment begun at Georgia State University by Duane Rumaugh in 1970. Lana was a chimpanzee born at Yerkes, and she was taught a language system of lexigrams called Yerkish. Lana used an electronic keyboard, and when she pressed a key with a lexigram on it, the key would light up and the lexigram would appear on a projector. Lana learned to create sentences such as "Please machine give juice." When the wrong "word" appeared, Lana would erase it and replace it with the correct word.

In 1970, Washoe moved to the Institute of Primate Studies (IPS) at the University of Oklahoma, where other types of behavioral studies on chimpanzees were being performed. Psychologist Roger Fouts, who did his graduate work with Washoe in Reno, moved with her and began teaching sign language to more chimpanzees. Researchers across the country were becoming more intrigued by these language studies, which at once seemed to provide insights into the minds of our closest living relatives and at the same time threatened to undercut human uniqueness. In 1973, Herbert Terrace of Columbia University decided that he was going to attempt to teach a chimpanzee sign language under highly controlled conditions. Nim Chimpsky, born at IPS, was that chimpanzee. Like Washoe, Nim was raised by humans and pampered by a series of doting human caregivers in New York City. However, each day Nim had to go to the Columbia University lab, sit at a desk, and learn signs. Though Nim appeared to learn about 125 signs, Terrace concluded that Nim did not really understand their meaning and was unable to put them together in any way that resembled grammatical sentences. In 1977, Terrace ended the project and sent Nim back to Oklahoma. Then in 1982, as the Institute for Primate Studies was unable to get funding to continue, Nim and 20 other sign language-using chimpanzees were sent to New York University's Laboratory for Experimental Medicine and Surgery in Primates (LEMSIP), possibly to be used in hepatitis research. Before the demise of IPS, Roger Fouts had taken Washoe to Central Washington State, where she lived until her death in October 2007. Cleveland Amory and the Fund for Animals ultimately secured Nim's release and retired him to Black Beauty Ranch, where he lived until his death in March 2000, but almost all of the other chimpanzees from Oklahoma ended up in biomedical research laboratories.

From the beginning of the practice of keeping chimpanzees in captivity in the United States, their sale and movement from one laboratory to another was fairly common. Early on, laboratories and zoos would swap chimpanzees back and forth as well. Keeping chimpanzees in captivity is expensive, and many laboratories could not sustain the expense, which was the case in Oklahoma. Sometimes when the funding runs out, a chimpanzee is placed in a laboratory in which similar experiments are being performed. For example, Sarah was moved from the University of Pennsylvania psychology lab to the Ohio State University Chimpanzee Cognition Center. Back in 1965, the Air Force facility in Alamogordo, New Mexico stopped using chimpanzees and was taken over by Fred Coulston, who was then affiliated with the Albany Medical College, which was sold to New Mexico State University in 1980 and, in 1993, with over 335 chimpanzees living at the facility, it was sold again to another Alamogordo facility controlled by Coulston, called White Sands, where an additional 200 chimpanzees lived. The Holliman facility and White Sands combined and became known as the Coulston Foundation, and specialized in toxicology and immunology studies. A few years later, with the closure of New York University's Laboratory for Experimental Medicine and Surgery in Primates (LEMSIP), where many of Nim Chimpsky's cohort had been sent, the Coulston Foundation acquired over 100 additional chimpanzees that were to be subjected to additional types of infectious disease

research. With over 650 chimpanzees, the Coulston Foundation controlled the largest number of captive chimpanzees in the world until animal welfare violations were exposed. In 2002, on the verge of financial collapse, Coulston sold its facility and donated its remaining 266 chimpanzees to a sanctuary called Save the Chimps. Some of the chimpanzees that were born in Oklahoma and taught to communicate with sign language, then moved to LEMSIP in New York and exposed to hepatitis, then moved to Coulston in Alamogordo where they continued to be experimented upon, now live on grassy islands in Florida in social groups with other chimpanzees who endured similar experiences.

Though chimpanzees are quite expensive to keep in captivity (current estimates are between $300,000-$500,000 for lifetime care) and finding adequate facilities was never easy, that did not stop laboratories from breeding chimpanzees. In the mid-1970s, chimpanzees were classified as endangered in the wild and as threatened in captivity. What this meant was that importing them from Africa was prohibited, but keeping them in captivity was not. So, laboratories increased their breeding programs to ensure the continued availability of chimpanzees as experimental models. In 1986, the NIH launched a large breeding program to increase the number of chimpanzees available for AIDS research. Ironically, that same year the British government banned the use of chimpanzees in research on ethical grounds, arguing that, given how close chimpanzees were to humans, to treat them as expendable was immoral. By the 1990s, it became clear that chimpanzees were not an appropriate model for HIV research because they did not develop AIDS. But now the government was faced with a challenge. There was a decrease in the use of chimpanzees for experimentation, but a surplus of chimpanzees that required long-term care. Estimates placed the total number of chimpanzees in laboratories at that time at around 1,800. Since chimpanzees can live 50–60 years in captivity, and simply killing the chimpanzees would have generated a large public outcry; euthanasia was prohibited as a method of population control. The NIH imposed a temporary breeding moratorium and convened a working group to study the captive chimpanzee problem. In 1997, the Chimpanzee Management Program (ChiMP) made a series of recommendations that included extending the breeding moratorium and continued monitoring of the surplus captive chimpanzee problem. The breeding moratorium became permanent in 2007.

The United States remains the only country in the world, except possibly for Gabon, that uses chimpanzees in invasive biomedical research. The last research facility in Europe using chimpanzees stopped in 2004, when biomedical research with chimpanzees became illegal in the Netherlands. Japan ended biomedical experimentation on chimpanzees in 2006.

The Welfare of Captive Chimpanzees

The Animal Welfare Act enacted in 1966 regulated the care and use of animals in laboratory research, in facilities that exhibit animals, in transportation, and by dealers, and that included chimpanzees. Early guidelines outlined minimum space requirements that allowed chimpanzees to be individually housed in single cages that measured 5×5 feet and were only 7 feet high. Chimpanzees are strong, active, highly social animals. Isolation housing, exposure to unavoidable psy-

chological distress, repeated anesthetizations, and exposure to painful procedures, while legal, were contrary to the well-being of the chimpanzees. In 1985, as the result of pressure brought by animal protection groups as well as primatologists, the Act was amended to include providing space for normal exercise and "a physical environment adequate to promote the psychological well-being of primates." But what that meant was subject to debate.

In 1988, the Jane Goodall Institute published a detailed set of recommendations to help provide some specific ways to resolve the debate and promote the psychological well-being of chimpanzees. Some of these recommendations included:

- Chimpanzees should always be housed with at least one other conspecific, unless ill and thus in need of special care.

- Under no circumstances should a chimpanzee be housed without visual and auditory contact on at least two sides.

- All enclosures should have windows to the outdoors, and any newly constructed facility should include an outdoor enclosure.

- Caregivers or scientists working with chimpanzees must be given extensive training in the nature of chimpanzee life and behaviors.

- Caregivers should be selected for their compassion and dedication to the wellbeing of the chimpanzees.

- All handling procedures should be performed in a way that reduces the stress and pain experienced by the chimpanzee.

- No experiment should be initiated on a chimpanzee without the prior acquisition of funds to provide for the retirement and care of the individual chimpanzee post-experimentation.

- Any chimpanzee no longer in use for biomedical research should be allowed to retire.

While instructive, the full JGI recommendations were not ultimately adopted. Instead, the regulations required each facility that housed chimpanzees to design their own "environmental enhancement plan." In 1999, the vagaries of the regulations were addressed in a new set of guidelines that were a bit more specific, but nonetheless allowed flexibility for the facilities in devising their plans, as long as the plan specifically addressed the social needs of individuals, provided enrichment to prevent self-injurious behavior, and considered the special social needs of infant primates and others with particular physical characteristics. In addition, facilities were to provide sufficient space to allow chimpanzees to engage in species-typical behavior, and enclosures had to contain complexities, objects that could be manipulated, and varying feeding mechanisms to provide environmental enrichment.

Because of the growing awareness of the cognitive and emotional sophistication of chimpanzees and the distress captivity was likely to cause them, discussions of the ethical issues involved in the proper care of chimpanzees in captivity became more pressing. Public commentary on the USDA regulations was extensive. New discussions about care in zoos also become more sophisticated. The American Zoo and Aquarium association established the Chimpanzee Species Survival Plan in order to improve the care of chimpanzees in zoos and to carefully monitor

their reproduction, and in the 1990s the international Great Ape Project (GAP) was launched. Scientists, primatologists, and ethicists involved in GAP advocated the expansion of the "community of equals" to include chimpanzees and to grant all great apes the right to life, the protection of individual liberty, and a prohibition on torture. In the summer of 2008, Spain adopted a resolution that would extend these basic rights to great apes and would outlaw using them in experiments, circuses, TV commercials, or films.

In the United States in 2000, concerns about the ethical use of chimpanzees in research and the continual problem of "surplus" chimpanzees led the US government to pass the Chimpanzee Health Improvement, Maintenance, and Protection (CHIMP) Act, which provides lifetime sanctuary for chimpanzees owned by the federal government, and some others that are no longer needed for research. In September 2002, after a competitive selection process, Chimp Haven, a 200-acre state-of-the-art, naturalistic sanctuary in Caddo Parrish, Louisiana, was selected to become the National Chimpanzee Sanctuary System. The first chimpanzee residents arrived in April 2005. When the CHIMP Act was passed in 2000, there was a last minute rider added that allowed for the possibility that chimpanzees could be recalled from retirement if they were the only chimpanzee that could satisfy a specific research need and their removal would not disrupt their social group. In December 2007, the option to remove chimpanzees from the sanctuary was eliminated with the passage of the Chimp Haven Is Home Act. To date over 150 chimpanzees have been retired to Chimp Haven.

In 2008, the Great Ape Protection Act was introduced in Congress. As of this

World-renowned conservationist Jane Goodall gives a little kiss to Tess, a female chimpanzee at the Sweetwaters Chimpanzee Sanctuary. (AP Photo/Jean-Marc Bouju)

writing, it is still pending. This act, supported by animal welfare organizations and some primatologists, would prohibit invasive research on great apes, prohibit funding for such research, prohibit the transportation of great apes for such research, and require that all great apes be retired to sanctuary. Whether or not these efforts are successful, important work remains to be done for all currently captive chimpanzees to ensure that they receive the highest level of care. To promote their psychological well-being, all captive adult chimpanzees must be provided with the opportunity to develop stable social relationships with other chimpanzees. Captive chimpanzees need to live in a chimpanzee community where they can communicate with others of their kind and learn to exhibit species-typical behaviors. At a minimum, providing them with access to the outdoors, space to develop social relations and to avoid conflict, materials to nest, fresh fruits and vegetables, and enrichment to keep their active minds stimulated, are essential for the well-being of captive chimpanzees. Keeping chimpanzees in captivity denies them their freedom, but their wildness remains within them. It is possible, with diligence and care, to respect their wild dignity, and this is what every one of them deserves.

See also Sanctuaries; Sanctuaries, Chimpanzees in

Further Reading

Brent, L. 2001. *The care and management of captive chimpanzees.* San Antonio: American Society of Primatologists.

Call, J., & Tomasello, M. 2008. Does the chimpanzee have a theory of mind? 30 years later. *Trends in Cognitive Science*, 12, 187–192.

Fouts, R., 1997. *Next of kin: What chimpanzees have taught me about who we are.* New York: William Morrow & Co.

Goodall, J. 1971. *In the shadow of man*. New York: Houghton Mifflin Co.

Jane Goodall Institute. 1988. Recommendations to USDA on improving the psychological well-being for captive chimpanzees. *Journal of Medical Primatology* 17, 116–121.

Hess, E. 2008. *Nim Chimpsky: The chimp who would be human*. New York: Bantam.

Hughes, P., & Cassidy, D. 2005. *One small step: America's first primates in space*. New York: Chamberlain Bros.

Kohler, W. 1957. *The mentality of apes*. New York: Penguin.

Ladygina-Kohts, N. N. 2002. *Infant chimpanzee and human child: A classic 1935 comparative study of ape emotions and intelligence*. Oxford: Oxford University Press.

Linden, E. 1986. *Silent partners*. New York: Crown.

Premack, D., & Woodruff, G. 1978. Does the chimpanzee have a theory of mind? *Behav. Brain Sci.* 1, 515–526

The First 100 Chimpanzees: http://first100chimps.wesleyan.edu/

Yerkes, R. M. 1916. Provision for the study of monkeys and apes. *Science* 43, 231–234.

Yerkes, R.M. 1943. *Chimpanzees: A laboratory colony*. New Haven: Yale University Press.

Lori Gruen

CHINA: ANIMAL RIGHTS AND ANIMAL WELFARE

China's reforming post-socialist state has produced breathtaking economic growth in the last 30 years. While the world applauds the Chinese economic miracle, it has great concern over the sustainability of its long-term development. In addition to the widely recognized problem of environmental degradation, cruelty against animals in China has reached an unprecedented level. In recent years, a vivid and quite confrontational discussion on animal rights and animal welfare, topics rejected as unworthy of serious academic discussion only a generation ago, has erupted in China.

Progressive foreign ideas on animal protection came to China in the early 1990s. Chinese philosophers spearheaded the academic exploration. Why philosophers? In the early 1990s, the issue of animal rights was basically an academic research interest rather than a topic of policy implications. Yang Tongjin, a philosopher at the Chinese Academy of Social Sciences (CASS), an official think tank, published one of the first articles on the Western concept of animal rights (Yang, 1993). However, Dr. Yang's article did not spark further interest. It was not until the mid-1990s that animal rights began to attract attention in China again.

Chinese Proponents of Animal Rights

In 2002, Qi Renzong, a Chinese philosopher at CASS, published a seminal article espousing the ideas of animal protection. Qiu apparently tailored his article to address the several questions that would be evoked by his arguments.

An attitude among most Chinese was that China was not ready for tackling issues of cruelty against animals, because there should be more concern about the many people who were living a hard life. Qiu did not question the importance of human rights advancement, yet he believed recognition and protection of animal rights would only help promote human rights (Qiu, 2004). In response to the view that China was not materially or philosophically ready for discussing or protecting animal rights, Qiu offered the following comments in the article:

In my opinion, we have the conditions now to discuss animal rights.

These conditions are rising awareness of the public for environmental and animal protection, media exposure of the maltreatment of and cruelty to animals, experience of animal protection work, rising rights consciousness, and the achievement of a growing prosperous society (Zhao, 2004).

As far as Qiu was concerned, the subject of animal rights could no longer be neglected. He introduced the three key elements of the rights claim: the subject of rights, indirect objects of rights, and direct objects of rights. He highlighted three basic positions related to human-animal relations. These are, first, that humans having no obligation toward animals; second, that humans have an indirect obligation toward animals; and third, that humans have a direct obligation. He then presented the theological, philosophical, Confucian, and ethical arguments for the three positions. In terms of animal liberation, Qiu introduced the concept of speciesism and its various manifestations in animal abuse.

In the end, Qiu brought readers' attention to the three tactics for animal liberation. He rejected the status-quo position, believing it was pessimistic and obstructive. However, he also questioned the abolitionist arguments that, to him, were unrealistic and counterproductive. He called instead for actions to improve animal welfare at the present time. He believed a gradualist approach would better serve the goal of animal liberation in the future.

The Opposing Voices

Zhao Nanyuan, a professor at China's prestigious Tsinghua University, launched a frontal attack on Qiu's article. In his 2004 essay "The essence of animal rights arguments is anti-humanity," Zhao saw ulterior motives behind Qiu's arguments. To Zhao, what Qiu introduced was nothing but a full shipload of "foreign trash" (Zhao, 2004b). Zhao wanted readers to be vigilant, because animal rights advocates were, in his opinion, determined to convert their ideas into policies and actions.

Animal rights, according to Zhao, incorporate misguided ethical arguments. "Ethics allows the talking of nonsense and it, as a result, often makes people astray and acting ridiculously contrary to their original intentions." Ethics limit and inhibit freedom, Zhao charged. Therefore, like famine, plague and wars, moralists who propagate ethical standards are creators of human disasters. This is, according to Zhao, why the intentions of moralists like Professor Qiu were suspicious (Hu, 2004). Zhao rejected the view that nonhuman animals are sentient beings and that they are entitled to rights.

Zhao's most provocative position in this essay was his sweeping attack against rights proponents, animal lovers, and Western animal rights ideology as a whole. To him, the influx of Western animal rights ideas was no accident. Chinese advocates of animal rights, in Zhao's opinion, are treacherous and are serving the West's neo-imperialist objectives in China. They enjoy defaming their own country, Zhao claimed, and helping the West to demonize non-Western civilizations. Zhao calls China's animal advocates a bunch of psychopaths with emotional and personality flaws. In his view, these Chinese, like all other animal rights advocates, are "anti-humanity" elements. He warned these people to stop acting as members of the "fifth column" of the neo-imperialists. He suggested that they should learn from the South Koreans, who stood firm against Western protests of Korean dog-eating culture (Jie, 2005).

The Intensification of the Debate

Zhao's provocative rebuttal reflects the frustration of these ideologically charged opponents. Zu Shuxian, a rights advocate, made a powerful response by citing scientific evidence and the work of Charles Darwin (Li, 2004). All mammals have emotions, he argued. Not only can many of them imitate others, and make and use tools, they also have memories. To Zu, the mental similarities between humans and nonhuman animals cannot be overemphasized. Arguing that existence is independent of scientific inquiry, Zu rejected Zhao's assertion that science has failed to prove that animals are sentient beings. Refuting Zhao's claim that animal protection advocates are psychopaths, Zu reminded the readers that great scientists like Albert Einstein once called on future generations of scientists "to free ourselves by widening our circle of compassion to embrace all living creatures . . ." With regard to the various charges and accusations made by Zhao, Zu likened them to the character assassination common in pre-reform China between 1949–1978.

Zhao's provocative article was also criticized by Zheng Yi, an overseas writer who published *China's Ecological Winter* (2002). Zheng believes the science of ecology fully demonstrates the fundamental contributions of biodiversity to human survival. As members of this diverse ecological system, Zheng argues, nonhuman life forms deserve human moral consideration, as the latter owes its survival and prosperity to the former.

Animal Welfare Concepts

Animal welfare is also a foreign idea introduced into China in the reform era. China's increasing openness to the outside world, rising animal protection awareness, and increasing media exposure of cases of cruelty against animals in the country underlie the Chinese discussion of animal welfare.

The lot of nonhuman animals in China also divides the interested parties. To the opponents of animal rights, there is no animal welfare crisis in China. Qiao Xingsheng, a college teacher, argues that animal suffering under conditions of mass production is a necessary evil. He denies that the animal welfare problem is developing into a crisis (Li, Xiaoxi, 2004). Qiao Xingsheng praises China's fine treatment of animals under current regulations.

Zhao Nanyuan flatly denies that there is cruelty against animals in China. To him, cruelty against animals is a fabrication by hostile Westerners and Chinese lunatics who, in Zhao's view, love animals more than their fellow human beings. Reported cruel acts, according to Zhao, are sensational stories whipped up by the media or by evil-minded animal lovers (Mang, 2004). To Zhao, Chinese culture is above reproach in its treatment of nonhuman animals. Both Zhao and Qiao conclude that China should do nothing at present regarding animal welfare (Mao, 2003).

Anti-Cruelty in Laws and Regulations

Anti-cruelty is a new subject of policy-making. Today, some 70 laws and regulations include articles related to animal welfare. Yet most only touch on the issue in vague terms and are not enforceable. In China, there is not yet any comprehensive anti-cruelty legislation.

Why is China so behind the rest of the world in animal welfare legislation? One study found four main obstacles (Mo & Zhou, 2005). First, treatment of animals in general is not a concern for policymakers (Mo & Zhou, 2005). Second, existing Chinese laws are discriminative in coverage. Except for endangered species, most animals fall outside legal protections. Third, existing laws and ordinances are no deterrence to cruelty against animals. Law enforcement is a major challenge. Fourth, articles in the existing laws are mostly overarching principles that have low enforceability.

Animal Welfare Legislation

To its proponents, animal welfare legislation is long overdue. Academics Song Wei and Wang Guoyan agree that there is a void in this policy area in China. China's sustainable development calls for animal welfare legislation to stop, for example, wildlife devastation. China is also more likely to export animal products if it improves farming conditions (Qiao, 2004).

Other proponents expressed similar views on the development and social importance of anti-cruelty legislation. Mao Lei, a *People's Daily* reporter, states: "For the sake of development, our legislative action on animal welfare ultimately serves the interest of us humans in the long run." Legal restrictions placed on humans are worthwhile and necessary (Qiao, 2004). In her legislative proposal to the National People's Congress, Li Xiaoxi called on the national legislature to outlaw cruel hunting and livestock-raising practices.

She referred to SARS and bird flu to emphasize the need for legal construction in animal welfare (Qiu, 2004).

Mang Ping's article "Animal welfare challenges human morality: animals should be free from fear and trepidation" touches on both the practical and philosophical aspects of animal treatment. Pragmatically, poor animal welfare causes economic losses. Philosophically, the author argues, as sentient beings, animals should be given moral consideration on farms, in transport, and when their lives end. Rejecting the opposition's arguments that animals cannot fulfill obligations, Mang asks if there is better obligation fulfillment than sacrificing one's own life in return for humane treatment. Mang also rejects arguments that animal welfare legislation does not fit China's conditions. She argues that China has a tradition of kindness to animals (Song & Wang, 2004).

The opposing views are also clearcut. Qiao Xingsheng sees no ground for animal welfare legislation at the present time. Culturally, he points out, people in China do not see animals as equals. Legislatively, anti-cruelty law is a Western concept and therefore does not suit China. Adopting such laws in China is practically unenforceable (*Song & Wang, 2004).). Liang Yuxia, a researcher at the CASS, agrees that anti-cruelty legislation is too progressive and unenforceable at present (Zhao, 2004b).

In his article "The strange tales and absurd arguments of the animal welfare proponents," Zhao Nanyuan rejects the view that animal welfare impacts human health. He argues that SARS and bird flu have nothing to do with poor animal welfare. Factory farming, he claims, better controls diseases. Animal welfare legislation, Zhao argues, could lead to meat price hikes, thus affecting people's right to eat meat. Therefore, "advocacy of animal welfare violates human rights." This is, he alleges, an action to be resolutely resisted because it is antihuman. He asks why China would adopt laws that are anti-humanity (Zhao, 2004a).

Nevertheless, actions for animal welfare legislation in China have gathered momentum. In August 2003, a proposal on animal welfare legislation was submitted to the National People's Congress. The proposal called for expanding the list of state-protected species. It suggested that four other types of animals (farm, lab, entertainment, and working animals) should also be protected.

Opposition to the proposal was swift. Chinese author Jie Geng launched a point-by-point critique of the proposal. The fact that anti-cruelty laws exist in the West, he argued, does not mean that China should also have them. He implied that cruelty against animals is not as serious in China as it is in the West. He argued that the outside world can take no actions against China for its lack of anti-cruelty laws. He reminded his audience that Korea has not been excommunicated from the WTO for its dog-eating culture (Zu, 2004).

Conclusions

Chinese exploration of the subjects of animal rights and animal welfare is a new development in this rapidly changing society. The animal rights and welfare debate is a public discussion initiated by independent-minded scholars and activists. Such autonomous societal initiatives were not possible in the pre-reform era.

No intellectual pursuit is value-free. In China, intellectual fervor has always carried normative concerns. The evolving debate on rights and welfare for

animals is no exception. Those who have called for attention to animal rights and welfare are calling for policy change in animal-related policy areas. As we have shown, the opponents who reject these calls aim to maintain the policy status quo.

Importantly, the debate is politically significant. Animal advocacy groups will continue to push for policy change. Together with other domestic NGOs, they contribute to the rise of civil society. Their activism, agenda setting initiatives, and success in facilitating policy change will eventually redefine state-society relations on the Chinese mainland.

Further Reading

Hu, Jun. 2004. Gai bu gai wei dongwu lifa (Should we legislate animal welfare?). Accessible via http://www.ycwb.com/gb/content/2004-05/18/content_693002.htm; downloaded July 20, 2004.

Jie, Geng. 2005. Dian ping dongwu fulifa: zhongguo falujie yingdang guanzhu de huati. Accessible via http://www.xys.org/xys/ebooks/others/science/report/mao3.txt; downloaded January 2, 2005.

Li, Jingyue. 2004. Dongwu fuli, ni chancheng haishi fandui? (Are you for or against animal welfare?), *zhonghua dushu bao* (*The Chinese Readers' News*), April 28, 2004. Accessible via http://arts.tom.com/1004/2004/4/28 63573.html; downloaded April 28, 2004.

Li, Xiaoxi. 2004. Guanui dongwu fuli lifa de jianyi (A proposal on animal welfare legislation). Submitted to the National People's Congress, January 2004 (unpublished article and copy, courtesy of Li Xiaoxi).

Mang, Ping. 2004. Dongwu fuli kaoyan renlei diode: shengxu ye ying mianyu kongju (Animal welfare challenges human morality: animals should be free from fear and trepidation). *China Youth Daily*. Accessible via http://news.xinhuanet.com/news center/2002-11/13/content_627869.htm; downloaded February 20, 2004.

Mao, Lei. 2003. Dongwu fuli lifa keburonghuan (Animal welfare legislation in China cannot be postponed any more).

People's Daily, January 14, 2003. Accessible via http://www.people.com.cn/GB/news/6056/20030114/907578.html; downloaded December 1, 2003.

Mo Jinghua and Zhou Xianchong. Wo guo dongwu fuli xianzhuang jiqi falu baohu chutan (China's current animal welfare conditions and a preliminary exploration of the question of legal protection) accessible via http://www.riel.whu.edu.cn/show.asp?ID=1772, downloaded February 7, 2005.

Qiao, Xingsheng. 2004. Dongwu fuli lifa buneng tuoli zhongguo guoqing (Animal welfare legislation cannot be divorced from China's national conditions). Accessible via http://www.people.com.cn/GB/guan dian/1036/2515143.html; downloaded October 11, 2004.

Qiu, Renzhong. 2004. It is time to discuss animal rights in China. Accessible via http://shc.jdjd.cn/article5/dongwu.htm; downloaded April 20, 2004.

Song, Wei, and Wang, Guoyan. 2004. Dongwu fuli de hexing shi shengme (What is the essence of animal welfare?). *People's Daily,* January 14, 2003. Accessible via http://www.people.com.cn/GB/news/6056/20030114/907573.html; downloaded on September 28, 2004.

Yang Tongjing. 1993. Dongwu quanli lun yi shengwu zhongxing lun (Animal rights theory and ecocentric arguments), *The Journal of Studies in Dialectics of Nature,* no. 8. Accessible via http://www.cass.net.cn/chinese/s14_zxs/facu/yangtongjin/key anchenggou/03.htm, downloaded September 15, 2004.

Zhao, Nanyuan. 2004a. Dongwu fuli de qitan kuailun (The strange tales and absurd arguments of the animal welfare proponents). Accessible via http://www.blogchina.com/new/display/31484.html; downloaded September 20, 2004.

Zhao, Nanyuan. 2004b. Dongwu quanli lung de yaohai jiushi fan relei (The essence of the animal rights arguments is antihumanity). Accessible via http://shc.jdjd.cn/article021007/dongwuql.htm, downloaded September 20, 2004.

Zu, Shuxian. 2004. Yaohai shi tichang chanren fandui lungli diode (The essence of Zhao Nanyuan's arguments is advocacy of cruelty and opposing human morality).

Accessible via http://www.fon.org.cn/index.php?id=3015; downloaded September 20, 2004.

Peter J. Li

CHINA: MOON BEARS AND THE BEAR BILE INDUSTRY

Bear bile has been used in traditional Chinese medicine (TCM) for over 3,000 years. The practice of caging endangered Asiatic black bears (known as "moon bears" because of the yellow crescents on their chests) and milking them daily for their bile started in Korea in the early 1980s, and soon spread to China.

It was suggested that bear farming would satisfy the local demand for bile, while reducing the number of bears taken from the wild. However, wild bears are still poached today for their whole gall bladders or as an illegal source of new stock for the farms. Bears are also bred on these farms.

Bears arrive at the Moon Bear Rescue Center in Chengdu, Sichuan Province in appalling physical and mental condition. Bears like Andrew, Freedom, Belton and Frodo have severed limbs as a result of being trapped in the wild. Crystal and Gail have had their canine teeth cut back, exposing pulp and nerves, and paw-tips sliced off to de-claw them, making them less dangerous to milk for their bile. These bears can spend up to 25 years in cages no bigger than their bodies.

Traditional Chinese Medicine

According to Chinese government figures, 7,002 bears remain trapped on

Animals Asia workers and volunteers give an emergency health check to a bear farmed for bile in China. Government wildlife officials defend China's raising of bears on farms to make bile for traditional medicine and have rejected a European appeal to shut down the industry. (AP Photo/Animals Asia)

factory farms, where they are milked daily for their bile.

Today, bile can be replaced with herbal and synthetic alternatives, which are plentiful, cheap and effective. Eminent experts such as Professor Zhu Zhenglin, who ran a TCM clinic in Chengdu, are dedicated to both the culture and usage of TCM, as well as to the end of bear farming. Professor Zhu writes:

> I believe that it is the time now and it is the responsibility of our new generation of traditional Chinese medicine practitioners that we further develop the TCM theories left by our ancestors, and not rely on the old beliefs of bear bile or tiger bone.

Similarly, TCM practitioner and academic Professor Liu Zhengcai, who is renowned throughout China, says:

> If people don't use bear bile, the industry will have no reason to exist. I always tell my patients and students that bear bile is not necessary, and is replaceable. Not using bear bile complies with the TCM theory of "harmony with nature."

Dr Zhu Guifang, a Chengdu businesswoman involved in the TCM tonic industry who recently visited the Animals Asia sanctuary, uses even stronger language:

> I have being selling TCM tonic food for 13 years, including bear bile and

bear bile wine . . . But after visiting your sanctuary, I am shocked and feel ashamed of having hurt these animals. Starting from today, I will never sell bear products again.

This support gives us hope that, like the Vietnamese government, the Chinese government will agree to outlaw the trade in bear bile. For the bears suffering and dying on the farms right now, freedom cannot come soon enough.

Day after day, their bile is drained through crude metal catheters implanted into their gall bladders, or via permanently open, infected holes in their abdomens. This latter method—the only method permitted in China—is known as the free-dripping technique, which the authorities claim is hygienic and humane.

However, veterinarians have found that bile extraction significantly increases the risk of disease in the bears and must, by the nature of the wound, cause pain. Anyone with a basic understanding of physiology knows that a permanent hole in the abdomen is a perfect vector for bacterial infection.

Some farmers have devised a fake free-drip technique whereby a Perspex catheter is hidden within the hole to prevent it from healing. This allows the farmers to deceive government inspectors by maintaining that the hole is naturally and permanently open, allowing bile to drip freely out as the regulations dictate.

Contaminated Bile

Animals Asia has urged Chinese authorities to look into the possible harmful side effects of contaminated bear bile. This organization has a growing dossier of evidence that the bears tapped for their bile are developing liver cancers at an alarming rate. Moon bears held in captivity rarely contract liver tumors unless they are very old, but almost half of the rescued bears that have died were euthanized because of liver cancer.

The bile is contaminated with pus, blood and even feces. A healthy bear's bile is as fluid as water and bright yellowy-orange to green in color. Veterinary surgeons have described bile leaking from the diseased gall bladders of the rescued bears as black sludge. They also consider it highly likely that cancer cells are present in the bile extracted from bears with liver tumors.

The prized ingredient in bear bile, ursodeoxycholic acid (UDCA), is used by TCM practitioners for a myriad of complaints, everything from hangovers to hemorrhoids. However, UDCA can be synthesized easily under laboratory conditions—the UDCA produced is pure, clean and reliable.

Dr Wang Sheng Xian, a Chengdu pathologist who analyzes the livers of bears that have died from liver cancer, has said: "The more I learn about the extraction of bile from bears, the more I would never recommend this kind of drug to my family and friends. This drug could be harmful to people." There are many effective and affordable synthetic alternatives as well as more than 50 herbal options.

"Although I respect TCM, what I have seen from the samples from caged bears makes me doubt that products like this work. I personally think we had better use alternative drugs and never extract bile from bears," Dr Wang said.

A Vietnamese pathologist has also expressed grave concerns for the health of both humans and bears after conducting clinical examinations of the damaged gall bladders of three moon bears rescued from Vietnamese bile farms by Animals Asia.

Dr. Dang van Duong, chief pathologist at the Bach Mai Hospital in Hanoi, said he was shocked by the condition of the bears and urged consumers to think twice before taking the bile from such diseased animals. He found a substantial thickening of the wall of the gall bladder, a consequence of the bile extraction process.

In Vietnam, bile is extracted with the assistance of an ultrasound machine, catheter and medicinal pump. The bears are drugged—usually with ketamine—and restrained with ropes, and have their abdomens repeatedly jabbed with four-inch needles until the gall bladder is found. The bile is then removed with a catheter and pump.

In 2007, Animals Asia's veterinary team released the report, "Compromised health and welfare of bears in China's bear bile farming industry, with special reference to the free-dripping bile extraction technique."

The report, which was distributed among conservation groups and Chinese health authorities, stated:

> AAF's veterinarians hypothesize that the etiology of the cancer [in farmed bears] is related to the chronic inflammation, infection and trauma caused by bile extraction. Research is under way to investigate this hypothesis. In another context, consideration must be given to the potential effects on humans of the consumption of bear bile that is so contaminated with pus and inflammatory material.

Two bears that arrived at the rescue center in March 2008 illustrate the state of the bear farming industry today.

Kiki and Chengdu Truth

Kiki was squeezed into a tiny cage, barely breathing, and one of the first priorities for an emergency health check. The staff at the Moon Bear Rescue Center tried to gently rehydrate and medicate him and offered him water and a fruit shake. He desperately wanted to drink, but as soon as he licked the delicious juice, he would frantically paw at his face as if in great pain.

There was terror in his eyes, and when he was anesthetized and staff looked closer, they saw that his right eye was rotten and full of pus, and his left eye was semi-rotten. Kiki was blind.

His teeth were smashed to pieces, and some teeth had also been torn away, together with sections of his rotten gum. He had ulcers on his lips and nose, and a broken jaw. He had a massive wound on his hind leg, with flesh necrotic and rotten to the bone.

In surgery, those caring for him saw that Kiki's abdomen was grossly distended with gas, which had been pressing dangerously and painfully on his heart and lungs. Just as they were deciding what to do first to help him, Kiki died on the surgery table. When he was opened up for autopsy, it was found that he was totally diseased with septicemia and liver cancer. It is impossible to imagine how he had withstood so much pain.

Another bear called Chengdu Truth was in a similarly poor condition. His footpads were hyperkeratotic—cracked and dry—showing he hadn't walked on solid ground for years. The bars of his cage were clearly embedded into his soles. He weighed just 65 kilos, when a healthy adult male should weigh 165 kilos.

In addition to a liver tumor weighing several kilos, he had infected puncture wounds over each shoulder. The farmer who owned him had known this bear was dying and had injected him time and time again with inappropriate antibiotics using unsterilized needles, causing the puncture wounds to fester and rot.

Chengdu Truth was so weak, so sick, and so thin that he could barely lift his head and, once the staff at the center knew he had liver cancer, there was no choice but euthanasia.

What Has Been Learned from the Bears

With 247 bears rescued in China by the end of 2008, The Moon Bear Rescue Center continues moving slowly, but surely, towards its end goal. The rescued bears are ambassadors for hope.

Initially displaying frightening aggression when they arrive at the sanctuary from the farms, the bears gradually discover their true natures. The transition in personality from an animal so violent and fearful of the human species to one that is trusting, inquisitive, and completely at ease with people, can only be described as remarkable.

What they teach those humans working with them and helping them is astonishing. Animals that have probably never before built nests are doing so today, as evidenced by skillfully created bamboo beds deep within their natural forest enclosure. Similarly, animals that have never before climbed trees are shinning up into branches several meters high with ease.

They remember and form close friendships with specific members of the group—often sleeping in twos or even threes in their hanging-basket beds. Some will gang up against another bear, or exclude a newcomer from their special circle of friends.

The bears remember the source and the cause of pain. They recognize individual people and will often huff in caution if a veterinarian steps into the room. They growl or explode with rage if an anesthetic jab stick appears.

Straw piles are collected and transported from one end of their grassy enclosure, up the stairs, and into their basket beds inside the den. One bear will often wait until another is distracted and then steal his or her toys or food. The staff at the Moon Bear Rescue Center knows that Crystal grieved when her best friend, Gail, died. She paced more, she ate less, she was sad, evident to all humans who saw her.

Frolicsome and happy, the bears love to play—either with the toys and enrichment items in their enclosures, or within the enclosed bamboo forests, or simply with each other.

The rescue project and China sanctuary benefit more than just the bears. Bear farmers are compensated with funds to start a new business, provided they close their premises and hand their licenses to us. The project employs over 140 local people, and sources local food and construction materials.

Open house days see the Moon Bear Rescue Center welcoming hundreds of visitors each month, including busloads of school children. Friends of Animals Asia support groups are springing up at universities throughout China. Together with their support, Animals Asia is educating a wider section of the general public about bear farming and the concept of animal welfare in China.

In Vietnam, where bile farming is now illegal—but still widely practiced—the center is preparing to welcome the first of 50 bears into a new sanctuary near Hanoi before the end of 2008. They will join the 24 bears already rescued.

Legislation to Help the Bears

In July 2000, Animals Asia signed a landmark agreement with the China Wildlife Conservation Association (CWCA) and the Sichuan Forestry Department to rescue 500 bears and work towards ending bear farming. Since October 2000, over 40 bear farms have been closed and licenses will no longer be issued.

In December 2005, members of the European Parliament passed a declaration calling on China to end bear farming. And political support is growing within China; members of the National People's Congress have visited the sanctuary in Chengdu and pledged to help end the industry.

Meanwhile, the staff at the Moon Bear Rescue Center believes that their promise to the rescued bears is that they will care for them for the rest of their lives. They will wake every day with the sun on their backs and without fear in their hearts.

Further Reading

Fan, Zhiyong. 1999. The People's Republic of China Endangered Species of Wild Fauna and Flora Import and Export Administrative Office CITES Management Authority of China (CNMA) and Song Yanling Institute of Zoology Academia Sinica. 3rd International Symposium on the Trade in Bear Parts. National Institute of Environmental Research, Seoul, Republic of Korea, 26–28 October 1999. Published by TRAFFIC East Asia. http://www.traffic.org

Loeffler, Kati, Robinson, Jill, Cochrane, Gail. 2006. Compromised health and welfare of bears subjected to bile extraction in China's bear bile farming industry with special reference to the free-dripping technique. Animals Asia Foundation, September 2006.

Animals Asia Foundation Web site: www.animalsasia.org.

Jill Robinson

COCKFIGHTING

While the question of how to define the proper relationship between humans and other animals can be fraught with controversy, there is one animal issue that draws near consensus: cockfighting. Across the spectrum, groups from The Humane Society of the United States to leading poultry trade groups like the National Chicken Council condemn cockfighting.

Cockfighting is a bloodsport in which two roosters (called gamecocks), selectively bred for aggression and pitted against one another, fight to the death. While blood flows and feathers fly, spectators stand ringside and gamble on the fight's outcome. A fighting rooster's gameness, or willingness to continue fighting in the face of exhaustion and mortal wounds, is considered a source of pride for cockfighters.

Origins and History

Cockfighting is thought to have originated in Asia. While no one knows when humans first captured wild roosters and forced them to fight for entertainment, cockfighting was likely practiced in some form as early as 2500 B.C.

As cockfighting spread across Asia and into Europe, it became popular in England in the 1700s. Early settlers from England and Ireland brought gamecocks with them into the United States, where the practice took root in all parts of the country. However, cockfighting was never without its critics—for both its animal cruelty and gambling aspects. In 1835, England became the first country to ban cockfighting, along with bull baiting, bear baiting, and other forms of staged animal fighting.

Tools of the Cockfighting Trade

Roosters naturally have a spur on the backs of their legs, which they use in self defense or during squabbles over territory. Cockfighters will usually saw off a gamecock's natural spur and tie a weapon to the bird's heel in its place.

Roosters are matched with others with the same weapons, and birds are within two ounces of the same weight in a match.

The most common cockfighting weapon in the U.S is the gaff. Similar in shape to a curved ice-pick, gaffs are tied to both legs for a fight. Another common cockfighting weapon is the long knife, which can reach three inches, and the short knife, which can range from one to one-and-a-half inches long. In a knife fight, the weapon is tied to the left leg.

In Puerto Rico, cockfighters use a weapon called a *postiza,* an artificial spike which was once made of turtle shell but is now made of hard, sharpened plastic.

Naked heel cockfighting pits the birds against each other without any weapons. However, the absence of sharp weapons does not necessarily reduce animal suffering—and may in fact enhance it. As one cockfighting sympathizer explained to a Congressional committee in May 2006, "The wounds inflicted with a gaff or another type of knife are cleaner wounds and the birds can recover better than with a naked heel."

Details of the Fight

Most cockfights are held in a derby format, with multiple fights throughout the event. Although there is no limit on the number of entrants, most derbies require between three and five roosters per entrant. While most entry fees stand at $100–200, some may be as large as $1,000. Entry fees are pooled and given to the derby's winner, with the money being split evenly if several entrants are tied at the derby's end.

During each individual fight, cockfighters will call out their bets by answering the gambler who has called out the sum he wants to bet. The amount of these bets is left to the discretion of the gambler.

The Opposition's Efforts

The Humane Society of the United States has been the national leader in efforts to stop cockfighting in the United States. Local and statewide organizations that have been very engaged against the cockfighting issue legislatively or otherwise include Animal Protection of New Mexico; the Louisiana SPCA; citizen groups in Arizona, Missouri and Oklahoma, and scores of local humane societies.

Cockfighting has always been criticized by animal protection organizations. After England banned the activity early on, over half the individual states in the United States followed suit in the 1800s, with much of the rest of the country banning cockfighting in the early 1900s. Animal protection advocates continued targeting cockfighting throughout the latter half of the 20th century, with a dramatic resurgence in legislative efforts to eradicate cockfighting beginning in 1998, spearheaded largely by The Humane Society of the United States.

The Final Nail

In 1998, cockfighting remained legal in just five states; Arizona, Louisiana, Missouri, New Mexico and Oklahoma. Elsewhere, it was a felony in 17 states

Cockfighting is a blood sport, now illegal in the United States and most of Europe. People bet on the outcome of fights and the birds greatly suffer and often die as a result of the staged fights. (AP Photo/John Gress)

and a misdemeanor in 28. That year, citizens in Arizona and Missouri gathered enough signatures to place cockfighting bans on their ballots, both of which passed overwhelmingly.

In 2000, animal protection advocates in Oklahoma gathered nearly 100,000 signatures to place a proposal to ban cockfighting on the ballot. Cockfighters went to court to prevent a vote on the issue, but their efforts only delayed a vote until 2002. On Election Day 2002, Oklahoma voters approved the ban, despite the expenditure of over half a million dollars by cockfighters opposing the ban.

Neither Louisiana nor New Mexico allow for citizen-inspired ballot initiatives, so animal advocates in those states focused their battle to ban cockfighting on their state legislatures. With neither state wanting to harbor the distinction as the last refuge of the bloodsport in the United States, newspapers throughout both states ran editorials calling for the legislature to ban cockfighting.

On March 12, 2007, New Mexico's Governor Bill Richardson signed into law a ban on cockfighting. When the Louisiana legislative session convened two months later, a cockfighting ban was high on the legislative agenda. The Louisiana Gamefowl Breeders Association, which had for years held off a cockfighting ban by hiring a top lobbying firm in Baton Rouge, saw the writing on the wall and agreed to a ban, with a phase-out over three years. Animal advocates countered and successfully whittled the phase-out period down to one year. On August 15, 2008, Louisiana became the 50th and final state to outlaw cockfighting.

During this time period, the U.S. Congress approved a ban on the interstate transport of any animal for an animal fighting venture. While the original language approved by Congress provided for a felony penalty, cockfighters hired former U.S. Senator Steve Symms to lobby on their behalf. Their efforts managed to reduce the penalties to a misdemeanor until 2007, when Congress upped the penalties with the passage of the Animal Fighting Prohibition Enforcement Act.

Looking Ahead

As of 2008, cockfighting is a felony in 37 states, and animal advocates continue efforts to make cockfighting a felony in all 50 states. The Humane Society of the United States has documented that cockfighting remains far more pervasive in the states with misdemeanor penalties. In Alabama, for example, where the maximum penalty is a $50 fine, cockfighters see the punishment as just the cost of doing business.

Despite the fact that cockfighting is prohibited nationwide, three monthly magazines openly serve the cockfighting underworld. *The Feathered Warrior, Grit & Steel,* and *The Gamecock* are available by subscription, at some cockfighting pits, and at small feed stores in remote areas.

As testament to the success of animal protection groups, circulation for each of these magazines has dropped by roughly half over the past 10 years, according to circulation reports that all periodicals must file with the U.S. Postal Service.

Shut Down by Scandal

With so much gambling money exchanging hands at cockfights, it is not surprising that payoffs have been made to some local officials. Cockfighting has been repeatedly linked to public corruption.

In 2004, South Carolina Agriculture Commissioner Charles Sharpe was indicted, and later convicted, of taking a $10,000 payment from a cockfighting group and then trying to pressure a local sheriff to leave a cockfighting pit alone.

In the following year, 2005, the FBI raided a large cockfighting pit in Cocke County, Tennessee. It was later revealed that the cockfighting raid was just one part of a larger investigation into the local sheriff's department. FBI agents believed many in the sheriff's department were taking bribes from cockfighters and other criminals. Ultimately, numerous deputies—including the number two man in the Cocke County Sheriff's Department—were prosecuted.

In 2006, the FBI indicted four members of the Honolulu Police Department, after evidence surfaced that officers had tipped cockfighters off about pending raids and had shared sensitive information with cockfight organizers.

In 2007, news reports began to circulate in Virginia that the sheriff of Page County was under investigation for taking bribes from a local cockfighting pit owner. After raiding the Page County cockfighting pit, the U.S. Attorney's Office for the Western District of Virginia initiated an investigation into the local sheriff's department. This led to the indictment of Sheriff Danny Presgraves, in October 2008, on a range of charges, including the allegation that he had taken bribes from cockfighters in exchange for not raiding a local cock-fighting pit.

Cockfighting remains both legal and popular in the Philippines, Mexico, and Puerto Rico, as well as some other places. The persistence of cockfight-ing in these countries helps keep the dying U.S. industry alive, by providing an overseas market for the breeders of gamecocks. However, pressure on fed-eral authorities to crack down on these illegal shipments promises to ensure that, in the United States, cockfighting is on its last legs.

Further Reading

Anti-cockfighting:
www.humanesociety.org/cockfighting
Cockfighting Hurts, a DVD produced by The Humane Society of the United States.
Pro-cockfighting:
Snow, Russell J. 2004. *Blood, sweat & feathers: The history and sport of cockfighting.* Beth-lehem, PA: Twiddling Pencil.
Sociological:
Dundes, Alan, e. 1994. *The Cockfight: A Casebook.* Madison: University of Wisconsin Press.

John Goodwin

COMPANION ANIMALS

Although often used as a synonym for pets, the term companion animals refers primarily to those animals kept for com-panionship. Pets is a broader category than companion animals, and includes animals kept for decorative purposes (for example, ornamental fish, birds), those kept for competitive or sporting activities (dog shows, obedience trials, racing), and those kept to satisfy the interests of hobbyists (specialist ani-mal collecting and breeding). In prac-tice, of course, any particular pet may overlap two or more of these different subcategories.

The practice of keeping animals pri-marily for companionship is certainly very ancient, and may have contributed to the process of animal domestication at least 12,000 years ago. Recent hunter-gatherers and incipient agriculturalists are well known for their habit of captur-ing and taming wild mammals and birds, and treating them with affection and concern for their well being. Among the native peoples of Amazonia, pet keep-ing is particularly widespread. The list of species commonly kept includes vari-ous monkey and rodent species; coati, opossum, deer, peccary and tapir; wild cats such as margay, ocelot and, occa-sionally, even jaguar, and a huge variety of birds, especially parrots and trumpet-ers. Strangely, although many of these animals belong to species that are hunted and killed routinely for food, as pets they are only rarely killed and eaten. In many of these cultures the art of animal taming and pet keeping is cultivated particularly by women, some of whom acquire a con-siderable local reputation for their skills in this area.

The existence of pet keeping in hunter-gatherer societies raises fascinating ques-tions about the function of this activity. Until recently, it was widely assumed that the keeping of pet animals for compan-ionship was a largely Western pastime as-sociated with unusually high levels of material affluence. Viewed from this perspective, pet keeping tended to be cat-egorized as an enjoyable but unnecessary luxury. During the last 30 years, how-ever, medical evidence has accumulated suggesting that companion animals contribute to their owners' mental and physical health. Close and supportive human relationships are known to exert a protective influence against many common life-threatening diseases, probably by buffering people from the negative health consequences of chronic life stress. It appears that companion animals may serve a similar function. In various studies it has been found that pet owners exhibit fewer physiological risk factors for heart disease than non-owners, as well as demonstrating improved survival with cardiovascular disease. In addition, pet owners tend to make less use of public health services, and display less deterioration in health in response to stressful life events. The presence of pets also induces short- and long-term reductions in heart-rate and blood pressure in people exposed to experimental stressors. These findings suggest that companion animals provide a means of augmenting the social support people receive from each other, and that this role may be just as important in hunter-gatherer societies as it is in our own.

Despite the apparent contribution of pets to human health and well being, the standard of care provided for these animals by their owners is often less than ideal. Unrealistic expectations combined with ignorance of animals' basic needs are the most common sources of companion animal welfare problems. Many pets are kept in unsuitable environmental conditions, and provided with inadequate diets and insufficient exercise and mental stimulation. Owners' efforts to control

A man and companion dog enjoy a scenic view. (Photos.com)

their pets' behavior can also result in the use of inappropriate and mistimed punishments that may cause the animals considerable distress. The global trade in exotic pets, especially wild birds, reptiles, amphibians and fish, has seriously depleted some wild populations, as well as causing inestimable suffering and death during capture, handling and transport. Since the middle of the 19th century, companion animal breeders have created a wide range of hereditary breed defects, especially in dogs, while pursuing their own arbitrary standards of beauty. Many of these defects condemn the animals to lifetimes of distress and discomfort, and some require corrective surgery. Painful cosmetic mutilations, such as tail-docking and ear-cropping, and elective surgical procedures such as declawing and debarking designed to eliminate behavior problems, are widely performed particularly in North America. The fate of unwanted pets is also a major cause for concern, although reliable figures on the numbers of animals involved are not available. Estimates range from 8 to 15 million dogs and cats lost, abandoned, or disowned by people each year in the United States, of which approximately 30–40 percent are reunited with their owners or adopted. The remaining 60–70 percent are killed humanely after a brief statutory holding period.

These darker aspects of pet keeping have prompted some animal advocates to argue that the entire phenomenon constitutes a violation of animals' rights and interests, and that pet keeping should be abolished alongside other forms of animal exploitation. This position ignores the fact that many human-companion animal relationships appear to be mutually beneficial and rewarding to both human and animal participants. It also tends to discount the potentially positive effect of these relationships on our perceptions of animals in general.

Further Reading
McNicholas, J., Gilbey, A., Rennie, A., Ahmedzai, S., Dono, J-A., & Ormerod, E. 2005. Pet ownership and human health: A brief review of evidence and issues. *British Medical Journal,* 331: 1252–1254.

Serpell, J. A. 1996. *In the company of animals,* 2nd Ed. Cambridge: Cambridge University Press.

Podberscek, A. L., Paul, E. S., & Serpell, J.A. 2000. *Companion animals and us.* Cambridge: Cambridge University Press.

National Council on Pet Population Study and Policy. 2001. *The shelter statistics survey, 1994–97.* http://www.petpopulation.org/statsurvey.html.

James Serpell

COMPANION ANIMALS, WELFARE, AND THE HUMAN-ANIMAL BOND

Animal behavior and animal welfare are linked in many ways. The manner in which humans interact with and change animals' behavior, as well as the behavioral outcomes quantifying animal welfare, are intertwined with each other.

Animals have been a part of human civilization for thousands of years, and have been used for a number of different reasons, ranging from food production, to helpers in the hunt, to human companions. While physical stature is often grounds for their use in our society, their behavior is a major reason that we own them for companionship. Their behavior can also be a reason for fracturing the human-animal bond, leading to abandonment and euthanasia.

There are numerous examples in society of humans domesticating animals for use. Cattle are domesticated for food production, as well as for beasts of burden. Horses are domesticated for beasts of burden, as well as for companionship. Dogs are domesticated for hunting, protection, and herding, as well as purely for companionship.

Not only is innate or natural behavior important, but learned behavior is also important. Currently there is no way to measure which has more influence on the outcome—it's the old nature vs. nurture—but we must understand that no animal lives in a vacuum. Certain breeds may be more predisposed to certain behaviors, but it is necessary to take the learned component into consideration.

Border collies, for example, have historically been bred for herding sheep; their behavior defining who they are. In addition to their selected behavioral ability to herd, they must also be responsive to their human handlers for guidance. Physical conformation plays a role in the dog's ability to perform its job, but without the herding behavior, they would be of little use to a shepherd. On the flipside, the herding behavior can become a problem for an owner who adopts a border collie for companionship in an urban environment.

Animal behavior, learning, and training have been studied for years. In contemporary times, the research of people such as B. F. Skinner and Ivan Pavlov has paved the way to study how animals and humans learn, via classical and operant conditioning.

As an overview, classical conditioning, the emotional aspect of learning and behavior, consists essentially of bringing internal reflexes under the control of a previously unconditioned stimulus. Most people are familiar with Pavlov's dogs (the bell is linked to food, and the dogs then respond to a bell by salivating). But classical conditioning happens with animals all the time, especially when fear responses are taken into consideration. For example, the average dog has not been conditioned to enjoy going to the veterinarian's office. No matter how kind or patient the veterinarian is, the dog is unable to cognitively understand that he or she is there to maintain good health or that the veterinarian has the dog's best interest in mind. The dog learns that the veterinarian causes discomfort with needles, ear cleanings, and anal gland expressions. The dog then becomes classically conditioned to associate the veterinarian's office or white lab coat with fear.

Operant conditioning is the process by which the likelihood of a specific behavior is increased or decreased through reinforcement or punishment each time the behavior is exhibited, so that the animal associates the pleasure (or displeasure) with the behavior. The seminal research was done by B. F. Skinner, with his development of the Skinner Box, where a rat learned that food was released when it pressed down on a lever. With this simple yet elegant study, Skinner was able to expand with more studies evaluating the shaping of behaviors, different models of pairing unconditioned stimuli with reinforcements, and different intervals of reinforcement.

A current example of humane operant conditioning in action is clicker training. The dog is conditioned (via classical conditioning) to understand that the sound of a click means that a piece of food is coming. When the owner observes a wanted behavior, such as when they are trying to teach the dog to sit, the owners clicks, marking the wanted behavior exactly (operant conditioning). The dog will then be more likely to sit. Eventually the owner pairs a command with the action, and now the dog has learned a new trick!

Even based on these proven methods of modulating animals' behavior, there are people who still heavily rely on more punishment-based methods. These methods can be used successfully in limited circumstances, such as with emotionally-stable military dogs and their well-trained handlers. This being said, the vast majority of average pet owners have a difficult time using these methods, let alone wanting to use these methods if more humane options are offered to them.

Unfortunately, neither the owners who select these trainers to work with their dogs nor the trainers themselves always understand the scientific principles behind operant and classical conditioning, including punishment. Owners may find it difficult to stop a trainer from performing a certain training technique, such as helicoptering (swinging the dog in a circle by its leash and collar), because they just paid a professional dog trainer to help them with their dogs' problem behavior. Also, a lot of these trainers come highly recommended, either by friends or by statements on their websites touting their training prowess. However, since trainers are currently not required by any state to obtain a dog training license (unlike hairdressers and plumbers), there is no governmental oversight of the correctness or humaneness of their training techniques, abilities, or credentials.

Even though these inhumane and scientifically incorrect methods of training are commonly used, there are very few instances of these trainers being charged with animal cruelty or abuse. Organizations that promote humane and scientifically correct training have spoken out against primarily punishment-based training methods (American Veterinary

Society of Animal Behavior, 2008). However, trainers have been sued for injuries to dogs in their facilities, as well as charged with animal cruelty.

Owners who choose to use these methods need to understand the principles of behavior modification. For the animal to be punished appropriately, certain rules must be followed: the punishment has to happen immediately (within a few seconds); it cannot not be too aversive (to avoid causing fear and anxiety); it should not directly relate to the owner (to avoid a classically conditioned aversion to the person); and it should work after a few tries. If an owner or trainer has to continually deliver punishment, such as collar corrections, it is not working.

So, with all of these methods in their back pocket, owners can, and do, affect an animal's good and not-so-good behavior in many ways, and this also affects the human-animal bond. For example, a family may reinforce a dog's begging from the table, such as by periodically feeding it from the table. Yet when the dog climbs up onto the chair for food, they get frustrated when the dog won't listen to them. In essence the dog has been listening very well! If the dog performs a certain behavior, whether it is sitting looking cute or jumping on the chair, it is reinforced with food that is much tastier than its dry kibble.

Another unwanted behavior that some cat owners reinforce is the cat waking them up in the middle of the night, either by meowing, pawing at the door, or pawing at their face. The owner may feel guilty. Perhaps they forgot to feed the cat that evening? Maybe they didn't give the cat enough food? So, understandably, they get up and feed the cat. Well, the cat just learned a valuable lesson! The owner continues with this night-feeding until they are sleep-deprived, even as they remember that they fed the cat a whole bowl of food that evening. The same methods by which these animals learned inappropriate behaviors can be used to train them to learn appropriate behaviors as well.

It has been shown that behavior problems are a primary reason for euthanasia or relinquishment to animal shelters. A lot of the problems can, and should, be addressed in a proactive manner, lest they break the human-animal bond. House soiling, unruliness, and aggression are common reasons for relinquishing dogs to shelters, and inappropriate elimination and aggression are common reasons for relinquishing cats. Not only are these reasons for relinquishment or euthanasia, but they are also reasons for a diminished human-animal bond. Perhaps the cat that urinated outside of its litterbox is now an outside cat, with an increased chance of being severely injured. While the potential primary reason for euthanizing a cat that was hit by a car is its injuries, if the veterinarian digs deeper, the real reason for euthanasia was that the cat was made to be an outside cat because it urinated outside of the litterbox.

But as veterinarians look to solving these problems, they use humane behavior modification methods of classical and operant conditioning. In more recent times there has been an effort to bring this information to owners, shelters, and other organizations, such as those training working dogs. The American College of Veterinary Behaviorists is a recognized specialty of the American Veterinary Medical Association, consisting of veterinarians fulfilling special postgraduate education and research responsibilities. Another organization is the American Veterinary Society of Animal Behavior,

a group of veterinarians who share an interest in understanding, teaching and treating behavior problems in animals. The Animal Behavior Society promotes the study of animal behavior, and also has a certifying arm for people who have reached a certain level of academic training. These organizations, among others, have produced position statements on the behavior and welfare of animals in regard to behavioral modification and training. There are other organizations related to dog and horse training, as well as captive and laboratory animal welfare and training. With these organizations, there are a good number of resources for owners to seek help with their pets.

With help from properly trained people, owners can help change the behavior of their pets. What owner wouldn't want a dog to sit when they came home, instead of jumping on them, or a bird who didn't scream when they left the room, or a cat that didn't scratch the furniture? If an owner provides opportunities for an animal to perform the correct behavior, rewards such behavior, and properly uses humane punishment techniques, the animal stops performing the inappropriate behavior and, subsequently, performs the appropriate behavior. The methods of classical and operant conditioning apply to everyday life, not just in the laboratory. We, as humans, are often focused on stopping a behavior, but fail to focus on what the animal does do correctly and reward that behavior.

In conclusion, there is obviously a close relationship between an animal's behavior, its welfare, and the human-animal bond. Owners need to understand their pets' behavior, how they influence it, and how they can change it for the better using humane techniques, in order to decrease relinquishment and euthanasia for problem behaviors. Humans need to appreciate the uniqueness and wonder of their animal companions.

Further Reading

American College of Veterinary Behaviorists, www.dacvb.org.

American Veterinary Society of Animal Behavior, www.avsabonline.org.

American Veterinary Society of Animal Behavior Position Statements. Accessed November 6, 2008. www.avsabonline.org.

Animal Behavior Society, www.animalbehavior.org.

Hart, B. L., and Hart, L. A. 1988. *Perfect puppy: How to choose your dog by its behavior.* New York: W.H. Freeman & Company.

Landsberg, G., Hunthausen, W., and Ackerman, L. 2003. *Handbook of behavior problems of the dog and cat.* New York: W.B. Saunders Company.

Lindsay S, R. 2000. *Handbook of applied dog behavior and training, Vol. 1: Adaptation and learning.* Ames: Iowa State Press.

Serpell, J. 1995. *The Domestic Dog: Its evolution, behaviour, and interactions with people.* Cambridge: Cambridge University Press.

Melissa Bain

CONSCIOUSNESS, ANIMAL

There would be no concern for animal welfare and no political movement towards animal rights unless people were convinced that some animals are conscious, sentient beings whose feelings and experiences have the positive and negative subjective qualities that make some experiences pleasurable and others unbearable. What convinces people that this is true? For many it is plain common sense—they believe they can see when an animal is happy or sad. But others have been trained to be skeptical of such appearances, and they seek scientific justification for claims about animal cognition and consciousness. Although consciousness is not the only morally significant property, others might include satisfaction of desires and goals, consciousness is perhaps the most significant for animal ethics. The 19th-century British philosopher Jeremy Bentham summed it up with the question "Can they suffer?" although, perhaps "Can they experience pleasure?" is just as relevant.

The past decade has seen the establishment of the field of animal cognition through the publication of textbooks, anthologies, and a dedicated journal. While much of the work on animal cognition is centered on primates, domestic dogs are rapidly becoming a model species for asking evolutionary questions about cognition, and there is fascinating work on birds, especially members of the crow family, and on insects, especially honeybees, and other species too numerous to mention here. However, most of the scientists contributing to this boom have explicitly bracketed questions about consciousness and have focused instead on cognition, roughly defined as the capacity of animals to flexibly and adaptively exploit the sources of information in their physical and social environments. Many, but not all, of the scientists doing this work share the common sense view about animal consciousness, but most of them believe that the topic is scientifically intractable. Nevertheless, as the range of flexibility and adaptiveness of animal cognition comes into better focus, it is hard to think that this doesn't tell us something about animal consciousness.

Scientific progress in understanding the molecular, neural, genetic, and hormonal mechanisms underlying animal behavior has been even more rapid over the past decade. Because of its clinical significance, pain has been studied intensively, but there has also been a boom in affective science—the scientific study of feelings and emotions—more generally. In affective neuroscience and behavioral genetics, where the scientific agenda is to discover underlying mechanisms for pain and stress responses, mice and rats are the most common animal models, although rhesus monkeys are also widely used by neuroscientists. Physiological, for example, hormonal, and behavioral approaches to affective science cover a wider range of species, from mink to trout. This diversity is partly due to the agenda of applied animal science for managing animals in agricultural and wildlife settings; hence, even fish have been studied for their responses to painful stimuli. As with animal cognition, most who work in the affective sciences have avoided addressing questions of consciousness directly. This sometimes puts these scientists in the difficult position of arguing that the animal models they study are good models for human emotions and feelings, while arguing that the lack of equivalence between humans and animals justifies the pain and distress that is caused by their experiments. However, by limiting themselves to objective behavioral and physiological measures, most scientists manage to sidestep the topic of consciousness in their professional publications.

There are, of course, many concerned scientists who share the common sense view about animals. Nonetheless, they take a more skeptical stance in their scientific work, evincing a "show me" attitude that refuses simply to go along with common sense. Scientific skepticism has very often trumped common sense (the earth does move!), and it must be taken seriously by those concerned with the ethical treatment of animals, because scientists' opinions are very important in forming laws and rules about how animals should be treated in research and agriculture. Sometimes special interests override consistent scientific treatment in the formulation of these rules. For example, based on the recommendations of a scientific panel, the British Animal Scientific Protection Act draws a line between vertebrates and invertebrates, but makes an exception to give the common octopus the same protection from harmful treatment as any mammal, bird, fish, reptile, or amphibian. Are octopi special among invertebrate animals in having conscious experiences? The United States Animal Welfare Act was amended in 2002 to exempt rats, mice, and birds. Despite the apparent arbitrariness of these lines, it nevertheless seems reasonable to draw a line somewhere, and scientific consensus may draw the line less inclusively than common sense would.

The central question here is the *distribution question:* Which animals are conscious and which ones aren't? Most people think that there's no black-and-white answer to this question. Perhaps earthworms, or goldfish have some degree of consciousness, just not the same as ours. But what would that mean? Could it mean that goldfish see, hear, smell, and taste things dimly (or in pale colors)? Or does it just mean that they are aware of fewer things than we are? We also know of many examples where animal senses are more acute than humans. Honeybees, for example, have five different color vision cones compared to our three, so they can differentiate between flowers that look the same to us. Do they have a higher visual consciousness than humans? Such questions are examples of the *phenomenal question:* What are the conscious experiences of other animals like? Before discussing the skepticism that many scientists have about such questions, it will be useful to say a bit more about what is meant by consciousness.

Meanings of Consciousness

The word consciousness can have several different meanings. When talking about animal consciousness, only some of these meanings are controversial. There are two senses of consciousness so ordinary that no one disputes their application to many animals. One is the sense in which animals can be awake rather than asleep or in a coma. Another connects to the ability of animals to sense and respond to features of their environment. It can be said that they are conscious or aware of those features. Consciousness in both these senses is identifiable in many different species of animal, and can be studied scientifically. Fish sleep, and earthworms are, in the relevant sense, awake and aware of several things in their environments.

Two other senses of consciousness are controversial when used to talk about nonhuman animals: conscious experience (also called phenomenal consciousness) and self-consciousness. These are also the senses most relevant for animal welfare and animal rights.

Conscious experience is difficult to define, but one way to get at it is to think about what happens when you see, smell, hear, taste, or feel things. Think about looking at a horse, for example. In the presence of a horse, you have more than the abstract knowledge that there's a horse in front of you; the horse *looks* a particular way, it *smells* a particular way, and if you were to lick it, it would *taste* a particular way, too. Because the

word consciousness is ambiguous, philosophers also use term *qualia* to describe the experiential qualities of phenomenal consciousness.

Self-consciousness refers to an organism's capacity to understand itself as an individual that is similar to but distinct from others. Of course, every animal normally has some way of discriminating itself from others; animals typically don't eat themselves, for example. But self-consciousness is generally characterized as involving some sort of self concept. The some sort of here is deliberately vague, because it is not at all obvious what that self concept should contain. One idea that has been very important is that a self concept should contain (and may even be derived from) a capacity for thinking about the thoughts (and other mental states) of others. A self concept, in this view, involves the fact that I have certain perceptions, experiences, thoughts, and desires, while you may perceive and think about things differently and have different desires. This is often called having a theory of mind, and because it involves thoughts that are themselves about thoughts, it is also sometimes called higher-order thought.

When people claim that it is impossible to objectively answer the question of whether nonhuman animals are conscious, they are usually indicating the difficulty of gathering good scientific evidence for either phenomenal consciousness or self-consciousness in animals. Some theorists think that these two things are related, that you have to have higher order thought or theory of mind to have phenomenal consciousness, but others think that phenomenal consciousness is a more primitive capacity that doesn't require the complex ability to think about oneself.

Testing Analogical Arguments for Consciousness

Direct evidence for phenomenal consciousness in animals is difficult to obtain. With people, we can ask them to describe their experiences, and although we must always be alert to the possibility of false reports, we can at least establish a degree of within and between subject reliability in their self-reports under various conditions, which gives us some confidence that they are describing something real. In the absence of such a rich means of communicating with animals, arguments for animal consciousness depend ultimately on analogies between humans and animals. Analogy arguments that are based purely on behavioral, anatomical, and physiological similarities have an inherent weakness: critics can always exploit some dissimilarity between animals and humans to argue that this is the relevant factor the animals are lacking. Stressing evolutionary continuity between humans and other animals may help a bit, but evolution occasionally produces novel traits, so there is no logical requirement that even our closest relatives have some trait just because humans have it. In the absence of more specific theoretical grounds for saying that animals are conscious, the combined argument is still vulnerable to objections based on specific dissimilarities. One way to get beyond the weaknesses in the similarity arguments is to try to give a theoretical basis for connecting what we observe about animals to phenomenal consciousness.

A theoretical connection would perhaps be possible if we could say what phenomenal consciousness is for by specifying its biological function(s). A good functional account would bring physiology, anatomy, and behavior together, showing how the mechanisms serve the functions by making specific behavior possible. In defending his higher-order thought theory of phenomenal consciousness, the philosopher Peter Carruthers argues that it evolved to represent the mental states of others, and he cites the absence of evidence that other animals, except perhaps chimpanzees, can do this to argue that they probably lack phenomenal consciousness. However, there are other possible functions of phenomenal consciousness in certain kinds of learning and flexible cognition, which would support a broader answer to the distribution question.

An article such as this perhaps raises more questions than it answers, but the topic would be of little philosophical interest if it were otherwise. And despite the fact that there have been centuries of argument about animal minds and consciousness, recent developments in the neurosciences and animal cognition that are yet to be fully integrated make this an exciting time for philosophers to be working in this area.

See also Affective Ethology; Animal Subjectivity; Whales and Dolphins, Sentience and Suffering in

Further Reading

Allen, C. 2004. Animal Pain. *Noûs* 38: 617–643.

Allen, C. and Bekoff, M. 2007. Animal minds, cognitive ethology, and ethics. *The Journal of Ethics* 11: 299–317.

Balcombe, J. 2006. *Pleasurable kingdom: Animals and the nature of feeling good.* New York: Macmillan.

Bekoff, M., Allen, C., & Burghardt, G. M. (eds.) 2002. *The cognitive animal.* Cambridge, MA: The MIT Press.

Hurley, S. and Nudds, M. (eds.) 2004. *Rational animals?* Cambridge: Oxford University Press.

Panksepp, J. 2005. Affective consciousness: Core emotional feelings in animals and humans. *Consciousness and Cognition,* 14: 30–80.

Colin Allen

CONSERVATION ETHICS, ELEPHANTS

He was orphaned at three when he saw his family shot to death. In the ensuing chaos, he was grabbed and taken to a compound several hours away. There, he was able to join up with a few others sharing similar fates and, despite the odds, managed to survive. However, the early brutality and losses experienced left a legacy. When he and two peers reached teenagehood, a violent rampage began that ended in the eventual killing of over one hundred. The fact that all victims were from a completely different ethnic group raised fears about the beginnings of a civil war. Authorities were called in and, within weeks, the three youths were gunned down. The incident appeared quelled, but soon it became evident that the incidents were part of a wider disturbance.

A scene that could fit any number of places around the world—Belfast, Jerusalem, a Native American reservation, or

---

Kosovo, to name only a few. Yet however poignant, this tragedy does not seem relevant to conservation. Not so. Indeed, the three youths are African elephants, and their victims threatened white and black rhinoceros. They have been diagnosed with posttraumatic stress disorder (PTSD), a condition that often develops in humans who have experienced the trauma of war or abuse. The young bulls' story and their diagnosis signal the dramatic change taking place in science, society, and conservation.

We have entered a new paradigm that brings humans and all other animals under the same scientific and ethical umbrella. Theories and data have finally accumulated to the point where even skeptics agree that the differences between humans and other species are small relative to what we share. Down to neuronal detail, human and elephant minds and brains function in much the same way. Mammalian cortico-limbic structures and mechanisms are highly conservative evolutionarily. From rodents to humans, we all share specific areas in the brain responsible for coordinating stress-response behavior, analysis of visual coding and processing, and auditory, somatosensory, and memory-sensory integration. While there are species-specific differences, all mammals share the same generalized emotional brain that includes the prefrontal cortex, cingulate cortex, amygdala, insula, hypothalamus, brainstem and associated physiological (e.g., autonomic, cardiovascular, immunological, analytical), psychophysiological, and behavioral traits (e.g., extinction learning, fear conditioning; attachment and social bonding, pain, aggression; anxiety, and facial recognition). Furthermore, cerebral lateralization of a variety of adaptive capacities has been documented in diverse vertebrates: fish, amphibians, reptiles, birds, and mammals.

Like the human brain, the elephant brain is large. In adulthood, the elephant brain attains a weight of 5000g (.17 ounces), compared to 1400g (.03 ounces) for humans. Elephants exhibit processes reflective of social brain structure and functions found in other highly social animals. Again, similar to those of humans, primates, and dolphins, elephant brains at birth are only a fraction of the size they attain in adulthood. At birth, an elephant's brain is approximately 35 percent adult size and characterized by a high encephalization quotient, a well-developed neocortex, a large complex temporal lobe and, significantly, a cortico-limbic system which includes the fornix, hippocampus, inferior parietal lobe, amygdala, ventromedial thalamic nucleus, and the gyrus cinguli and orbitofrontal cortices that are involved in social attachment, the processing of social, emotional, and visual information, long-term memory, empathy, and stress regulation. A variety of cognitive capacities including tool-use, exceptional long-term and episodic memory, intentionality, complex chemosensory and auditory communication, contextual learning, reasoning, problem-solving capabilities, and the ability to perform premeditated acts have been found in elephants.

Similarities extend to culture and social processes. All young mammals depend on adults, which means that their brains are very sensitive to the surrounding environment as they mature. When the young bulls saw their families killed and had to learn how to survive on their own, their brains were affected. What they experienced is not that much different neuropsychologically from what occurs in human children who experience a similar violent loss of family and community and face an uncertain future alone. Elephant and human brains are rightfully described by a unitary model of brain, behavior, and mind.

None of this is really new. Tragically for animals, this kinship has been exploited. Countless victims have suffered in captivity and at the hands of biomedical experimentation, testing, and research because of the similarities between their physiology, behavior, and brains and those of humans. On one hand, the full spectrum of the animal kingdom—planaria, fruit flies, chimpanzees, sting rays, cats, rats, frogs, and myriad other species—has been considered sufficiently comparable to humans to use as experimental surrogates. On the other hand, ethical parity has been denied. Elephant grief, tool-use, vocal learning, and self–recognition—all capabilities that define what it means to be human—have been ignored when it comes to recognizing elephant rights. However, the new trans-species science compels a parallel revision of ethics, research and conservation based in animal rights.

In the past, animal rights and conservation have been separate. Animal rights insist on a parity among species where animals are deserving of the same ethical and legal considerations given to humans. Protectionism does not generally support the sacrifice of individual wellbeing in favor of the larger group, or of one species for another based on an artificial hierarchy of value. In contrast, conservation focuses on populations and species, and has tended to be more anthropocentric, in the sense that reasons for protecting species are shaped by Euro-American cultural priorities and the structure of the great chain of being. But this disciplinary and political separation is literally killing wildlife.

An elephant strolling through the dusty Tsavo East National Park. (AP Photo/Karel Prinsloo)

The efforts of many notwithstanding, conservation is failing. In 2008, predictions were that 25–36 percent of all mammals were threatened with extinction. As former editor of the premier conservation science journal, *Conservation Biology,* Reed Noss states that it is not for lack of science but of political willpower to put into practice what we know. Elephants are a case in point.

Today, only remnants of elephant society are found in Africa. It is well-known that human-caused starvation, hunting, mass culls, and poaching have reduced elephant numbers in Africa from over 10 million to less than a few hundred thousand. Human encroachment has shrunk elephant habitat to a fraction of its original continental expanse. Such extensive and intensive impacts have even caused genetic change. Elephants lacking tusks have been documented outside Uganda. In 2002, out of the 174 elephants at Addo National Park, South Africa, 98 percent of the females were without tusks.

The threat of elephant extinction is very real, in terms of pure numbers, and in consideration of the degree to which land and animals are pressed to change. And there is something more dire. In Kenya, the heart of elephant lands, the human population has jumped from 8.6 million in 1962 to over 30 million in 2004, and between 1973 and 1989 elephant numbers plummeted from 167,000 to 16,000. As a result, there are no places in Africa or Asia that can claim elephant herds even remotely resembling those of two centuries ago. Even in Amboseli, Kenya where, at a very localized scale, social structure is relatively intact, elephant family life cannot what it used to be at all. Gone is the continent-extending octopus of anastomosing elephant groups. Habitat fragmentation and the pressures of poaching, on the heels of widespread genocide during the first hundred years of colonization, impose an ever-present sense of impending death.

In North Luangwa National Park, Zambia, 93 percent of the population has been killed, and traditional herds composed of mothers and allomothers are virtually absent. According to studies by Delia and Mark Owens, females reproduce at much younger ages (48 percent of births were by females less than 14 years, compared with a normative mean age of first birth at 16 years). Thirty-six percent of groups have no adult females, one quarter of the units consist only of a single mother and calf, and seven percent of groups are sexually immature orphans. In Mikumi, Tanzania, 72 percent of the population was similarly affected, and in Uganda, elephants live in semi-permanent aggregations of over 170 animals, with many females between the ages of 15–25 years having no familial association or hierarchical structure. Infants are largely reared by inexperienced, highly stressed single mothers without a detailed knowledge of the local plant ecology, leadership, and support that a matriarch and allomothers provide. Disoriented teenage mothers raise families on their own without the backbone of elephant society to guide them. They are expatriates in their own land, lacking even the meager protection that refugee camps can sometimes afford to their human neighbors. Parks, which might be considered the equivalent of refugee camps, offer no sanctuary from marauding soldiers and villagers hungry for ivory and machine-gun sport. Like the majority of remaining elephant habitat in Africa, in all of Asia, the total population is estimated to be as low as 35,000 and dwindling fast.

Yet despite the science, the recognition of widespread elephant genocide, and the charismatic spell that elephants cast, conservation continues to use methods that undermine elephant wellbeing and culture. Culling or systematic killing has now been reinstated in South Africa after a 10-year moratorium, even though a scientific team gathered to make a formal assessment could not provide statistically significant evidence for either elephant-caused human deaths, crop damage, or a threat to ecosystem integrity. Human overpopulation and overconsumption have pushed elephants and other wildlife into tiny remaining parcels of land.

Decreased protection and a revival of the ivory market have led to an increase in elephant massacres and displacement. Along with South Africa, Namibia, Zimbabwe, and Botswana are participating in a United Nations-sanctioned auction of more than 100 tons of stockpiled ivory for exclusive purchase by Japan and China. The sale started on October 28, 2008 with one million tons from Namibia and continued into November of that year. The decision to permit the sale was approved by the Convention of Trade of Endangered Species (CITES). Conservationists predicted that this would escalate the dramatic increase in illegal poaching that is decimating herds as a result of more widespread impoverishment and chaotic violence in elephant habitat lands. One prediction estimates that elephants may very well be extinct in about a decade. Without improved conservation and enforcement of anti-poaching laws, the majority of large elephant populations are predicted to be extinct by 2020. When an international ban on ivory was introduced in 1989, African elephant mortality rate from poaching was 7.4 percent. Today, it is eight percent. Even if numbers are ignored, the fact is that elephant culture is already bending to the point of breaking socially and psychologically. The present and future for elephants is dire.

If, as science dictates, elephants and all animals have hearts, minds, and brains like ours, then culls, translocation, and forced confinement in zoos and restricted parks are the same as genocide, relocation, incarceration, and ghettos, with the same effects no matter what the species. Elephants have sustained nothing less than repeated genocidal ethnic cleansings since colonial occupation. We know how violence transmits across generations in humans neurobiologically and culturally. Now we realize that animals are vulnerable, too. The angry young bulls are only the tip of a deep and broad iceberg. What is happening to elephants is happening to all wildlife. Already cougars, moose, deer, lions, and bears have shown symptoms of trauma. A new conservation is needed.

Along with cognitive ethology, trans-species psychology is a new field that has begun to lay down the foundation for an animal protection conservation. Trans-species psychology is the formal study of how animals think, feel, and behave. In contrast to conventional psychology, the neologism signifies that a common model of psyche applies for all animal species, including humans. It uses the same language and concepts used to study and achieve human wellbeing for all species. How do we move this new animal protection conservation forward?

The first essential step is the creation of an elephant conservation based on principles of human rights and health. This entails abolishing culling and eliminating the social engineering of elephant society, as for example by translocation. The new trans-species science brings ethical and legal standards to bear on elephants and other animals to match those granted to humans. Elephant conservation is designed to protect all aspects of elephant life—that is, the psychological, social, physical, and emotional wellbeing of individuals and society.

Second, a multistep strategy needs to be developed to help move toward the ultimate goal: humans and elephants, indeed all wildlife, learning how to coexist as they once did before colonial occupation. A number of intermediate solutions have been suggested. While not perfect, they would take the immense pressure off elephants: well-protected connecting corridors between parks, expansion of parks to mega-parks, education and financial incentives for people who are most affected by or live off poaching. This will also help elephant groups that exist in scattered pockets across the continent to reconnect and restore their former culture.

Philosophically and politically, animal protection conservation is also tied to indigenous human rights. Traditionally, many tribes, such as the Acholi in Uganda whose totem is the elephant, related very differently to elephants and other wildlife. The rich diversity of wildlife that existed before European colonization is testimony to this. Further, many tribes have, like elephants, been forced from their traditional lands and suffered war and severe disruptions to their cultures. Steve Best, a professor at the University of Texas, speaks about the parallels between human and species apartheid in South Africa:

Apartheid was a brutal system of class and racial domination maintained by repression, violence, and terror, whereby a minority of wealthy and powerful white elites exploited and ruled over the black majority . . . . [but] under the pseudo-progressive guise of progress, rights, democracy, and equality, leftists, communists, democratic humanists, black nationalists, and community activists murder animals no different than white, racist, Western, capitalist, imperialists. Consider, for instance, the Zimbabwe "Campfire Conservation Association" that lobbies the United States Congress for funds to kill elephants for community benefit. Through a blatant discourse of objectification, Campfire member Stephen Kasere unashamedly reveals his speciesist outlook: "We just want the elephant to be an economic commodity that can sustain itself because of the return it generates. Ivory is a product that should be treated like any other product." (Best, http://www. drstevebest.org/Essays/TheKilling Fields.htm)

Animal protection conservation includes a restoration of these ancient bonds and ways of interacting. As Nelson Mandela wrote, "A new society cannot be created by reproducing the repugnant past, however refined or enticingly repackaged" (Mandela, 2003, p. 510). Neither can a new conservation be merely a repackaging. Conservation, like human rights, needs to be decolonialized in concept and practice to create an ecocentric elephant conservation—one based on trans-species ethical parity and service to one another.

A trans-species vision and way of life is simpler than it might first appear. Human rights activists Myles Horton and Paolo Freire put it best: "We make the road by walking," in this case, by envisioning what we need ourselves to be able to live in peace with friends and family, for other animals. But there is also another important example. Dr. Daphne Sheldrick, D.B.E., who has rescued over 80 orphaned elephants writes:

During the 50 plus years that I have been intimately involved with Elephants in Africa, and the rearing of over 80 orphans, I am astounded about how forgiving they are, bearing in mind that they are able to recollect clearly that their mother, and sometimes entire family, have perished at the hands of humans. Our Elephants arrive wanting to kill humans but eventually protect their human family out in the bush, confronting a buffalo, or shielding their surrogate human family from wild, less friendly peers. That is why I say that they are amazingly forgiving, because there can be nothing worse in life for an Elephant than witnessing the murder of those they love. And since Elephants never forget (which is a fact), they demonstrate a level of forgiveness that a human would in all likelihood have difficulty in achieving.

To conserve a vital elephant culture, we needn't look farther than the elephants we seek to save. To recreate a comparable vital human culture, we only need to start thinking like an elephant.

Further Reading

Best, S. 2007. The killing fields of South Africa: Eco-Wars, species apartheid, and total liberation. Accessed from: http://www.drstevebest.org/Essays/TheKillingFields.htm.

Bradshaw, G. A., & Finlay, B. L. 2005. Natural symmetry. *Nature* 435: 149.

Bradshaw, G. A., & Sapolsky, R. M. 2006. Mirror, mirror. *American Scientist.* November/December. 487–489.

Bradshaw, G. A., & Schore, A. N. 2007. How elephants are opening doors: developmental neuroethology, attachment, and social context. *Ethology,* 113: 426–436.

Bradshaw, G. A., Schore, A. N., Brown, J., Poole, J., & Moss, C. J. 2005. Elephant breakdown. *Nature.* 433: 807.

Bradshaw, G. A., & Watkins, M. 2006. "Trans-species psychology; theory and praxis." *Spring,* 75, 69–94.

Horton, Myles, Freire, Paolo, Bell, Brenda, and Gaventa, John. 1991. *We make the road by walking.* Philadelphia: Temple University Press.

Mandela, Nelson, Asmal, Kader, Chidester, David, and James, Wilmot Godfrey. 2003. *Nelson Mandela: From freedom to the future: Tributes and speeches.* Johannesburg, South Africa: Jonathan Ball.

Sheldrick, Daphne, personal communication, quoted in G. A. Bradshaw, *Elephants on the edge.* New Haven, CT: Yale University Press (in press).

G. A. Bradshaw

COSMIC JUSTICE

According to the principle of *cosmic justice,* humans are not the only beings in the world whose fortunes matter in considerations of fairness. Proponents of cosmic justice argue that human beings have obligations to treat other living beings such as animals, and perhaps plants and ecosystems, justly. According to critics of the idea of cosmic justice, even if human beings have some moral obligations toward animals and perhaps toward some nonsentient living beings such as plants, the notion of justice applies only to human beings and it makes no sense to say that human beings can do anything unfair to nonhuman beings.

The controversy surrounding the principle of cosmic justice is a controversy over whether nonhuman animals and perhaps some other living beings are the kinds of beings that can be said to merit inclusion in our considerations of fairness. Traditionally, justice has been conceived as a sphere of relations among rational beings who have rights and obligations in relation to other rational beings. To be rational is to be able to reflect on one's own interests, the interests of other beings, the potential conflicts that can arise between these interests, and the appropriate means of resolving these conflicts. Beings that are capable of reflecting on these sorts of matters are considered to be agents, in the sense that they take an active role not only in thinking about the various rights and responsibilities that they and other agents possess, but also in resolving the inevitable conflicts that arise between the interests of different agents in the community.

In the traditional view, the community is conceived as consisting primarily of rational agents, those beings who can take an active part in the process of reflection and in the making of choices that have implications for justice, where justice is understood in terms of fairness. In the traditional view, only human beings are rational; hence only human beings are genuinely agents, and all other living beings are excluded from the sphere of justice. The only exceptions to the traditional view are so-called marginal cases, human beings such as infants, comatose individuals, and the severely mentally impaired, who lack the necessary rational capacities but who by virtue of being human are nonetheless included in the sphere of beings toward whom obligations of justice are assumed to be owed. Agents are able, when principles of morality or justice call for it, to subordinate their own personal interests to the interests of other individuals or to the community. For example, it may be in my interest to take someone else's property without permission when I am in need and unable to acquire the property through legal means, but principles of justice impose on me an obligation to respect the other person's ownership of the property, and to seek to obtain that property by obtaining the consent of the owner. To be an agent is to be able to recognize and respect the rights of other agents, as well as to recognize and seek to protect my own rights.

In this traditional view, nonhuman beings such as animals are excluded from the sphere of justice on the grounds that they are fundamentally incapable of the rational reflection needed to consider and evaluate actions and their consequences and the rights and obligations that characterize justice relations. Some philosophers have argued that because nonhuman beings such as animals cannot be rational agents, they cannot enter into contractual relationships. Nonhuman beings cannot give the consent that is required of any party to a legally or morally binding agreement, hence such beings can properly be said neither to have taken on any obligations toward others nor to have entered into the sort of reciprocal relationship of rights and obligations that would entitle them to assert rights against humans or any other beings. At best, in the traditional view, nonhuman beings are moral patients rather than moral agents, in the sense that they can be affected in ways that are wrong or unfair but cannot be held responsible for their behavior. Such beings can be wronged or treated unfairly, but they cannot act rightly or fairly. In the traditional view, because reciprocity between moral agents and moral patients is lacking, moral patients cannot be said to be beneficiaries of justice. Strictly speaking, nothing we do to moral patients, which is to say nothing we do to nonhuman beings, can be considered to be unjust in the sense of being unfair.

Throughout the history of Western thought, and particularly in recent years, philosophers have challenged this traditional viewpoint and have argued that animals and perhaps some other nonhuman beings are indeed owed obligations of justice even though they cannot take on reciprocal obligations toward moral agents. Proponents of justice toward nonrational (nonhuman) beings argue that agency is neither necessary nor sufficient for membership in the sphere of justice, and that what must be focused on instead is the capacity of living beings to flourish according to their natures or realize their natural potential. Where justice has traditionally been conceived in social terms, that is, as a set of relations that prevail among rational beings in human society, proponents of cosmic justice argue that nature, the world of living beings as a whole, is the proper unit of measure for considerations of morality and fairness. Where the cosmos rather than human society is construed as the sphere within which relations of justice arise, a new dimension is added to our considerations of fairness, namely the rights enjoyed by moral patients and the obligations that moral agents have toward moral patients.

The principles of cosmic justice require human beings, when reflecting on possible choices and their consequences, to take into consideration the interests not only of other human (rational) beings, but also the interests of beings such as nonhuman animals and perhaps also nonsentient beings such as plants and ecosystems. For example, when considering the material consequences and the fairness of destroying a forest to build a housing tract, or when reflecting on the consequences and the fairness of converting a wetland area into a spot for human recreation, we must take into consideration not only the welfare and the rights of all the human beings involved, including, for example, those of any humans who might own the land to be used and who might be personally opposed to selling it or handing it over to the government, but also the welfare and the rights of all the nonhuman beings involved as well, in particular the fortunes of the animal species that will be displaced and possibly rendered extinct, and perhaps also the fortunes of nonsentient living beings such as the indigenous plant species whose lives would be disrupted and possibly destroyed by such human activities.

Debates surrounding the ideal of cosmic justice focus in particular on three key questions. The first is whether it makes sense to attribute any kind of moral status to nonsentient beings such as plants or ecosystems. In maintaining that moral agents are not the only beings that merit consideration in matters of justice, proponents of cosmic justice implicitly raise the question whether a given being need possess *any* kind of consciousness in order to qualify as a beneficiary of justice. For if a being need not be able to reflect on itself as a self with specific interests in relation to other specific selves who possess their own interests, why suppose that a given being must be capable of any awareness of its interests in order to deserve protection of those interests? Why not accept the proposition

that a being can have interests without being aware of those interests, and that it deserves to have its interests protected just as any conscious being does? Some opponents of cosmic justice argue that it simply makes no sense to attribute interests to beings that are incapable of being aware of those interests, and hence that such beings are not proper objects of concern, inasmuch as justice is a mechanism for protecting those interests of individuals which may come into conflict with the interests of other individuals. Any animal that cannot grasp its interests as interests cannot really be said to have interests, any more than, say, a car engine or a sewing machine can be said to have interests and, given that, to the best of our knowledge, all beings in the plant kingdom categorically lack consciousness, plants absolutely cannot be said to have interests. If a being has no interests, then there is no way in which that being can be harmed; hence that being is properly a beneficiary neither of moral consideration nor of justice.

Proponents of cosmic justice argue that the ability to grasp one's interests explicitly as objects of contemplation is not necessary for inclusion in considerations of justice. All that is required is that the being possess the capacity to flourish in accordance with its nature. All such beings are susceptible to harm or interference, and the requirement that a being be conscious of its interests in order to be a beneficiary of justice is simply an anthropocentric, speciesist prejudice that privileges the capacities and the interests of human beings in the sphere of justice.

Proponents of cosmic justice differ on the question of whether a being must be sentient in order to be a beneficiary of justice. Some argue that sentient beings, those beings capable of sense experience and in particular of experiencing pain, can be harmed in ways that are qualitatively different than the ways in which nonsentient beings can be harmed. Those who think along these lines see a special significance in the capacity to suffer; thus they are willing to include sentient animals in considerations of justice while excluding nonsentient animals such as oysters, which have no central nervous system and hence seem to be incapable of experiencing states such as pain, and all plant life. Other proponents of cosmic justice see values such as environmental integrity and biodiversity as values that are worth protecting, not simply because protecting such values benefits human beings, but because doing so protects nature as a whole. In this view, nature itself and its various parts, such as particular ecosystems, are beneficiaries of justice. Even among these sorts of proponents of cosmic justice, there are disagreements regarding whether rationality entitles human beings to any special status in matters of justice, or whether rationality simply confers on human beings a stewardship role and hence obligations to protect and conserve nature.

A related question is whether individual organisms or entire species are the proper objects of concern in considerations of cosmic justice. According to the traditional view of justice as a social relation, individuals are the proper beneficiaries in considerations of fairness, inasmuch as only an individual can have an interest, in the sense of being aware of it. Some proponents of cosmic justice retain a hint of this reasoning in arguing that only sentient beings matter in considerations of justice. It is possible to injure a sentient individual, but nonsentient beings can merely be damaged. Hence we need have no compunction about using

nonsentient nature to satisfy human desires. Other proponents of cosmic justice resist what they consider to be the speciesist, anthropocentric character of this reasoning and maintain that all living beings are susceptible to injury and hence deserve full consideration in matters of cosmic justice, whether or not they possess any capacity for conscious awareness.

A third question in debates surrounding the idea of cosmic justice is how we are to understand the relationship between moral obligations and the notion of justice. Is it possible to have moral obligations toward beings but not have obligations of justice toward them? Morality has traditionally been construed to concern itself with matters of right and wrong, while justice has traditionally been construed to pertain to matters of fairness in situations in which the respective interests of different beings come into conflict with one another. In the human sphere, there is a great deal of overlap between the two spheres. In dealings between human and nonhuman beings, some thinkers have argued that we may have moral obligations toward animals but that we have no obligations of justice toward them. One ancient thinker to argue along these lines was Plutarch, who early in his life argued that we have obligations of justice toward animals, but who softened his position later in life and argued that we have no obligations of justice toward animals, although we do have moral obligations of compassion or pity toward them. The contemporary philosopher John Rawls argued along similar lines in maintaining that animals have no part in considerations of justice, inasmuch as animals are incapable of entering into the sorts of contractual obligations that would bind them together with humans in the sphere of right. Nonetheless, Rawls argued, we may well conclude that we have moral obligations to feel compassion toward animals. Thus we might consider ourselves morally obligated to treat animals humanely, but considerations of fairness would not form part of the basis for such humane treatment. Strong proponents of cosmic justice reject Plutarch's and Rawls's reasoning, and argue that considerations of fairness demand that we extend equal consideration to the interests of humans and animals alike, and perhaps to nonsentient living beings as well.

Further Reading

Bekoff, Marc. 2002. *Minding animals: Awareness, emotions, and heart.* Oxford: Oxford University Press.

Carruthers, Peter. 1992. *The animals issue: Moral theory in practice.* Cambridge: Cambridge University Press.

Kohak, Erazim. 1984. *The embers and the stars: A philosophical inquiry into the moral sense of nature.* Chicago/London: University of Chicago Press.

Hargrove, Eugene C. (ed.) 1992. *The animal rights/environmental ethics debate.* Albany: State University of New York Press.

Rolston III, Holmes. 1994. *Conserving natural value.* New York: Columbia University Press.

Rowlands, Mark. 2002. *Animals like us.* London/New York: Verso.

Steiner, Gary. 2005. *Anthropocentrism and its discontents: The moral status of animals in the history of Western philosophy.* Pittsburgh: University of Pittsburgh Press.

Steiner, Gary. 2008. *Animals and the moral community: Mental life, moral status, and kinship.* New York: Columbia University Press.

Taylor, Paul W. 1986. *Respect for nature: A theory of environmental ethics.* Princeton: Princeton University Press.

Wenz, Peter S. 1988. *Environmental justice:* Albany: State University of New York Press.

Gary Steiner

CRUELTY TO ANIMALS AND HUMAN VIOLENCE

The belief that one's treatment of animals is closely associated with the treatment of fellow humans has a long history, but despite the popular acceptance of this concept, until recently there have been few attempts to systematically study the relationship between the treatment of animals and humans. In the early 1900s, case studies by Krafft-Ebbing and Ferenczi began to explore sadistic behavior toward animals associated with other forms of cruelty. However, single case histories do not provide much insight into the origins of animal abuse and its connections to other violent behavior. In 1966, Hellman and Blackman published one of the first formal studies of animal cruelty and violence. Their analysis of the life histories of 84 prison inmates showed that 75 percent of those charged with violent crimes had an early history of cruelty to animals, fire setting, and persistent bed wetting. Several subsequent studies looked for this triad of symptoms in other violent criminals, with mixed results. Later research found that these three behaviors by themselves do not necessarily predict future violence, unless the animal abuse is particularly aggressive and includes some or all of the following features:

- The child is directly involved in the perpetration of the animal abuse, not just a witness
- The child is impulsive and shows no remorse following the abuse
- The child engages in a variety of acts and victimizes different species
- The child is cruel to valued animals, such as dogs

The concept became more widely appreciated within law enforcement circles following a number of studies of criminal populations. FBI interviews of serial killers and other sexual homicide criminals initiated in the 1970s by Ressler and his colleagues found that 36 percent of these violent criminals described instances of participating in animal mutilation and torture as children, and 46 percent described such activities in adolescence. Prevalence rates of early animal cruelty of 25–50 percent have been described in several detailed retrospective studies of aggressive prison inmates, female offenders convicted of assault, convicted rapists, and convicted child molesters. Questions regarding animal maltreatment have now become standardized in many investigations of violent crime and juvenile fire setting. A major study conducted by the Massachusetts SPCA examined the criminal records of a large sample of 153 animal abusers and a matched control sample of nonabusers over a 20-year period, finding that the animal abusers were significantly more likely to be involved in a variety of crimes, including violent crime, theft and drug offenses. The study supported a notion of deviance generalization in the animal abusing population, rather than an escalation from crimes against animals to crimes against people.

In the 1980s, additional attention began to be given to instances of animal cruelty as part of the dynamics of child abuse and domestic violence. A review in one community in England of 23 families with a history of animal abuse indicated that 83 percent had also been identified by human social service agencies as having children at risk of abuse or neglect. A report on 53 pet owning families in New Jersey being treated for child abuse or neglect indicated that at least one person had abused animals in 88 percent of the families with physical abuse. In two-thirds of these cases the pet abuser was the abusive parent. Recently, several studies have examined the incidence of animal cruelty in families of women seeking protection in shelters for battered partners. In one such survey in Utah, Ascione found that 71 percent of the women with pets who sought shelter reported that their male partner had threatened to kill or had actually killed one or more of their pets. Similar results have been obtained from other surveys throughout the United States and Canada.

Recognition of the significance of the interconnections between violence against animals and violence against people has led to a number of significant changes. A growing number of states have escalated extreme forms of intentional animal cruelty from misdemeanor to felony offenses. Larger fines, longer jail terms, and/or required counseling have become more commonplace in animal cruelty cases. Many areas have begun to train animal care and control officers in the recognition and reporting of child abuse, and some animal shelters have begun to work closely with women's shelters to provide emergency housing for the pets of women and children at risk.

The concept of a link between animal cruelty and other forms of violence has not been without critics. For example, Piper and Myers urge a cautious and critical approach to reviewing the literature before it is applied to public policy, particularly in child protection.

Many advocates for animals and others hope that a better understanding of how cruelty to animals is related to other forms of violence may help in developing tools for prevention and intervention.

See also Cruelty to Animals: Enforcement of Anti-Cruelty Laws; Cruelty to Animals: Prosecuting Anti-Cruelty Laws

Further Reading

Ascione, F. R. 1993. Children who are cruel to animals: A review of research and implications for developmental psychopathology. *Anthrozoos* 6(4): 226–246.

Ascione, F. R., and Arkow, P. 1999. *Child abuse, domestic violence and animal abuse: Linking the circles of compassion for prevention and intervention.* West Lafayette, IN: Purdue University Press.

Currie, C. 2006. Animal cruelty by children exposed to domestic violence. *Child Abuse and Neglect,* 30(4): 425–435.

DeViney, E., Dickert, J, and Lockwood, R. 1983. The care of pets within child abusing families. *International Journal for the Study of Animal Problems* 4(4): 321–329.

Felthous, A. R., and Kellert, S. R. 1986. Violence against animals and people: Is aggression against living creatures generalized? *Bulletin of the American Academy of Psychiatry and the Law* 14: 55–69.

Flynn, C. P. 1999. Animal abuse in childhood and later support for interpersonal violence in families. *Society and Animals,* 7: 161–172.

Hellman, D. S., and Blackman, Nathan. 1966. Enuresis, firesetting and cruelty to animals: A triad predictive of adult crime. *American Journal of Psychiatry* 122: 1431–1435.

Luke, C., Arluke, A., and Levin, J. 1997. *Cruelty to animals and other crimes: A study by the MSPCA and Northeastern University,* Massachusetts Society for the Prevention of Cruelty to Animals.

Merz-Perez, L., and Heide, K. M. 2003. *Animal cruelty: Pathway to violence against people.* Walnut Creek, CA: Altamira Press.

Piper, H., and Myers, S. 2006. Forging the links: (De)Constructing chains of behaviours. *Child Abuse Review,* 15(3): 178–187.

Quinn, K. M. 2000. Animal abuse at early age linked to interpersonal violence. *Brown University Child & Adolescent Behavior Letter,* 16(3): 1–3.

Ressler, R. K., Burgess, A.W., Hartman, C. R., Douglas, J. E., and McCormack, A. 1986. Murderers who rape and mutilate." *Journal of Interpersonal Violence* 1: 273–287.

Randall Lockwood

CRUELTY TO ANIMALS: ENFORCEMENT OF ANTI-CRUELTY LAWS

Special police departments devoted to enforcing animal cruelty laws strike many as a very modern concept, but they have 19th-century origins. Creating animal police forces followed the development of humane societies in Boston and New York. After George Angell founded the Massachusetts Society for the Prevention of Cruelty to Animals (MSPCA), and Henry Bergh the American Society for the Prevention of Cruelty to Animals (ASPCA) in 1866, they both successfully lobbied for anti-cruelty laws.

Enacted in 1868 and revised in 1909, the Massachusetts animal protection law primarily focused on the abuse of horses. Although somewhat antiquated today, the code still stands. To enforce this law, and its parallel in New York, the MSPCA and the ASPCA created small police departments within their organizations. Little is known about the nature of early animal police work other than what has been recorded in the annual reports of humane societies having such departments. For the most part, these brief records only note the numbers and kinds of cases prosecuted by officers. Humane agents, empowered as police officers, primarily investigated cruelty to horses, since the urban infrastructure required these animals to be well tended and healthy. One typical entry catalogued the ASPCA's work in New York, saying that agents carried out 768 prosecutions, of which 446 involved the mistreatment of horses, with offenses such as beating, abandoning, starving, overloading, driving until they fell dead, and working sick, lame, or worn-out horses. Other prosecutions

involved dog and cockfighting, rat baiting, feeding cows swill and garbage,
keeping cows in filthy conditions, refusing to relieve cows with distended udders,
cruelty to cattle, dogs, cats, and poultry,
and maliciously killing, mutilating, and
wounding animals with knives and other
instruments. The only other information
is the rare commentary about the work
of humane law enforcement agents. In
one cases, the ASPCA report noted how
discouraging it was for agents to be criticized for overzealousness.

By the middle of the 20th century, the
makeup and organization of humane law
enforcement departments in cities like
Boston and New York resembled their
present day form. The MSPCA's department is made up of 16 staff members,
including 11 investigative officers, a
consulting veterinarian, two dispatchers,
a director, and assistant director. Except
for the dispatchers, all have been ap-
pointed as Special State Police Officers
by the State of Massachusetts, although
they are restricted to the enforcement
of animal protection laws and regulations. They do, however, conduct investigations, obtain and execute search
warrants, make arrests, and sign and
prosecute complaints. Officers are assigned throughout the state to investigate
whether individuals and, less often, organizations, have been cruel or neglectful.
The bulk of their cases involve everyday
animals—the strays, pets, vermin, and
small-farm livestock—that are neglected
or sometimes deliberately mistreated by
individuals. These officers also visit and
inspect stockyards, slaughterhouses, race
tracks, pet shops, guard dog businesses,
hearing ear dog businesses, horse stables
that rent or board horses, kennels, and
animal dealers licensed by the U.S. Department of Agriculture. During a typical

year, MSPCA officers conduct approximately 5,000 investigations and 1,000
inspections involving more than 150,000
animals. Since such complaints are also
lodged with other organizations in the
state, estimates of abuse complaints easily surpass 10,000 annually in Massachusetts, and show evidence of steadily
rising over time. Of course, this increase
may be due to growing public sensitivity to animal welfare, greater visibility of
humane law enforcement departments, or
simply improved record keeping.

According to the MSPCA's official job
description, the primary purpose of officers' work is:

> to prevent cruelty to animals, to
> relieve animal suffering, and to ad-
> vance the welfare of animals when-
> ever and wherever possible. Such
> purposes are to be achieved through
> the pursuit and implementation of
> a combination of activities, includ-
> ing, but not necessarily limited to,
> the enforcement of Massachusetts
> anti-cruelty and related laws, and
> the dissemination of animal protec-
> tion/welfare related information.

To do this work, prospective employees are expected to have a number
of skills, the first of which is "humane
sensitivity, with affinity for, and ability to
empathize with animals and respond with
compassion and objectivity."

When investigating cruelty complaints, rookie officers think of themselves as a brute force, because they
believe that they have legitimate authority to represent the interests of abused
animals. They see themselves as a power
for the helpless, a voice for the mute, representing and speaking for animals when
their welfare or lives are in jeopardy.

With more time on the job, this view changes. Although they are expected to represent the animal's side when investigating cruelty complaints, officers encounter a number of problems that make it difficult to do this. For the rookie officer fresh from training, these problems can be confusing and discouraging. They are hired in part because of their humane sensitivity; this strong concern for animals plus their recent police training creates a number of expectations in them. Rookies expect to handle complaints against animals that violate the legal definition of cruelty as well as their own standards, to observe animals to ascertain the nature and extent of cruelty, to counsel respondents or perpetrators when necessary to improve the treatment of their animals, to prosecute those who commit egregious acts of cruelty or who do not comply with advice, and to be understood and respected as both police and humane officers. These expectations are quickly shattered as rookies begin investigating complaints.

First, professional identity is a problem. Rookie officers experience a disparity between how they see themselves and how others see them. On the one hand, officers see themselves as professional law enforcers and animal protectors. As one officer said of the department's general job expectation: "They want you to be a humane officer, but have the authority or the presence of a police officer. It's hard to do both." On the other hand, one reason why it is "hard to do both" is that friends, family, strangers, and other professionals are often confused by this combination, and either have no idea what humane officers do, or relegate them to the level of dogcatcher.

Second, officers must enforce a problematic law. Massachusetts, like other states, has an anticruelty code specifying

that animals should not be deliberately mistreated. The law prohibits many types of abuse and neglect that threaten the safety and well-being of animals, including but not limited to beating, mutilating, or killing them, as well as failing to provide them with proper food, drink, and protection from the weather. Those convicted of violating this law can be fined up to $1,000 and imprisoned for as long as one year or both. Newer animal protection laws have classified cruelty as a felony, thereby increasing the maximum prison sentence to as much as five years.

Officers find it difficult to enforce the law, because of vague use of terms such as neglect, abuse, proper care, necessary veterinary care, and suffering. Nor can officers fall back on more general cultural conceptions of suffering, since these, too, are vague and contested by different groups. This problem forces officers to interpret the meaning and application of the law on a case by case basis, a point made by Walter Kilroy, the former director of the MSPCA's humane law enforcement department, who noted the "continuing absence of a widely accepted definition of cruelty to animals. Every activity that threatens the well-being of animals . . . must be challenged and overcome on a largely individual basis."

Third, there is a problem with evidence. The best witness to the abuse of humans is the victim; their testimony certainly facilitates, although it does not guarantee, successful prosecution. Yet animals obviously cannot report or articulate their harm. Rookies must learn how to figure our whether an animal has been mistreated, relying on indirect evidence in order to tell the story of an act of abuse. Rookies discover that a large part of this indirect evidence comes from investigating humans. In fact, this human side of animal cruelty often becomes the deciding factor in handling and resolving complaints.

Finally, there is a problem with enforcement and prosecution. Rookies encounter very few clearcut cases of animal cruelty that lead to prosecution and punishment. Instead, they encounter respondents whose behavior toward their animals does not violate the law, but falls short of what officers would prefer to see. Without a technical violation of the cruelty law, officers feel that they have little, if any, authority to force respondents to improve their treatment of animals. When they meet respondents whose acts violate the law, officers see their advice ignored. Rather than giving up entirely at these times, rookies must learn how to get their message across to respondents and, if necessary, take them to court. This final option can also be particularly frustrating, especially for rookies, as they encounter a judicial system that seems indifferent or hostile to the concerns of animals.

Most officers learn to cope with these problems by developing an attitude of humane realism. With little legitimate authority to enforce the law, officers become humane educators who try to make abusers, or others they meet on the job, into responsible animal owners. With few victories in court, they discover alternative ways to be effective in their fight against cruelty, and, in the face of public confusion about, or derision for, the role of humane law enforcement, they emphasize the police side of their work without forgetting their commitment to animal protection.

See also Cruelty to Animals: Enforcement of Anti-Cruelty Laws

Further Reading

Alexander, L. 1963. *Fifty years in the doghouse: The adventures of William Ryan, Special Agent No. 1 of the ASPCA.* New York: G.P. Putnam's Sons.

Arluke, A. 2004. *Brute force: Animal police and the challenge of cruelty.* West Lafayette, IN: Purdue University Press.

Arnold Arluke

CRUELTY TO ANIMALS: PROSECUTING ANTI-CRUELTY LAWS

Animal cruelty prosecutions have become daily events that attract widespread public and professional interest. Several trends demonstrate the increasing focus on enforcement of anti-cruelty laws:

- Television shows such as "Animal Precinct," which highlights the efforts of the Humane Law Enforcement division of The American Society for the Prevention of Cruelty to Animals (ASPCA) in New York City, are extremely popular, with numerous spin-offs showcasing similar efforts in Houston, Detroit, Miami, San Francisco, Philadelphia and elsewhere

- The number of law schools offering courses in animal law rose from 9 in 2000 to 92 in 2008

- The American Bar Association (ABA) and many state bar associations now have active animal law committees

- Prosecutors in many jurisdictions have established task forces to work with a variety of local agencies to specifically address crimes against animals

- The number of states with felony-level penalties for some forms of animal cruelty has grown dramatically in the last two decades from 5 in 1988 to 43 in 2008

Systematic prosecution of animal cruelty cases did not begin until there were well-defined laws protecting animals, as well as agencies with the authority to enforce these laws. In England, the first comprehensive animal protection law was the Act to Prevent the Cruel and Improper Treatment of Cattle in 1822, which also protected horses, sheep, cows and mules, providing for fines of up to five pounds and up to three months in prison for mistreatment of such livestock. The Society for the Prevention of Cruelty to Animals (SPCA) was founded in England in 1824 to ensure that this legislation would be enforced. It funded its own constables and eventually earned the support of the Queen, becoming the Royal SPCA in 1840.

Inspired by the success of the RSPCA in England, Henry Bergh and his associates founded the American SPCA in 1866 to promote the enforcement of new laws in New York similar to those in England. The animal cruelty law was revised in 1867 to apply to any living creature, a major move away from concern only for animals with commercial value and the first step in protecting pets and wildlife from cruelty. The law was applied regardless of ownership of the animal, recognizing that people are capable of cruelty to their own animals. The list of illegal acts was expanded, until it looked very much like most state anticruelty laws today. It also made all forms of animal fighting illegal for the first time, including bull, bear, dog, and cockfighting. The law comprehensively addressed neglect, and imposed a duty to provide "sufficient quality of good and wholesome food and water," and empowered any persons to enter premises to provide for these needs. Most significantly, the law gave the ASPCA arrest powers to enforce these provisions. Bergh himself acted as a special prosecutor, successfully bringing many cases to court.

In the United States, it has been difficult to assess the impact of the rapid increase in the number of stronger laws on the actual number of prosecutions, since there is no centralized tracking of animal cruelty arrests. In some states where data have been available, rising arrest rates have been related primarily to stronger animal fighting statutes. As of 2008, dogfighting is now a felony in every state in the United States

Successful prosecution of crimes against animals often requires specialized knowledge not only of the relevant laws, but also of veterinary medicine, veterinary forensics, animal care, and the practices used against animals in organized crime, such as dog-fighting and cockfighting. Animal care and control agencies, humane societies and SPCAs, and veterinary associations are important allies to prosecutors in successfully investigating and pursuing animal cruelty cases. These cases are given an unusually high degree of scrutiny by the general public. Prosecutors often receive tens of thousands of letters in support of the prosecution of high-profile animal cruelty crimes.

The effective prosecution of animal abuse has many benefits. It can provide an early and timely response to those who are, or who are at risk of becoming, a threat to the safety of others. It can provide an added tool for the protection of those who are victims of family violence. Finally, it can provide an opportunity for prosecutors to develop new, strong, and helpful allies in the protection of their communities and in helping build a truly compassionate society.

Further Reading

Favre, D., and Tsang. V. 1993. The development of anti-cruelty laws during the 1800s. *Detroit College of Law Review* 1: 35.

Frasch, P. D. 2008. The impact of improved American anti-cruelty laws in the investigation, prosecution and sentencing of abusers. In F.R. Ascione (ed.), *The International Handbook of Animal Abuse and Cruelty*. West Lafayette, IN: Purdue University Press.

Lockwood, R. 2006. *Animal cruelty prosecution: Opportunities for early response to crime and interpersonal violence*. Alexandria, VA: American Prosecutors Research Institute.

Sinclair, L., Merck, M., and Lockwood, R. 2006. *Forensic investigation of animal cruelty: A guide For veterinary and law enforcement professionals*. Washington, DC: Humane Society Press.

Randall Lockwood

D

DEEP ETHOLOGY

The term deep ethology carries some of the same general meaning that underlies the term deep ecology, in which it is asked that people recognize that they are not only an important part of nature, but also that they have unique responsibilities to nature as moral agents. Deep ethological research pursues a detailed and compassionate understanding of the unique worlds of nonhuman animals in order to learn more about their points of views—how they live, what they want, and how they experience various emotions, pain, and suffering. The development of what are called species-fair tests take into account the different sensory worlds, emotional lives, and cognitive abilities of animals, and allow humans to learn more about how all animals deal with their social and nonsocial environments, including pleasurable and painful or stressful stimuli. Recognizing animals as sentient beings or beings with intrinsic or inherent value will allow for an expansion of our compassion footprint.

Further Reading
Bekoff, M. 1997. Deep ethology. In M. Tobias and K. Solisti (eds.). *Intimate relationships, embracing the natural world.* Stuttgart, Germany: Kosmos.
Bekoff, M. 2002. *Minding animals: Awareness, emotions, and heart.* New York: Oxford University Press.
Bekoff, M. 2007. *Animals matter—A biologist explains why we should treat animals with compassion and respect.* Boston and London: Shambhala.
Bekoff, M. 2007. *The emotional lives of animals.* Novato, CA: New World Library.
Bekoff, M. 2008. Increasing our compassion footprint: It's simple to make changes to accrue compassion credits. *Human Ecology Review* 16, 49–50.
Bekoff, M. 2008. Increasing our compassion footprint: Some reflections on the treatment of animals. *Zygon (Journal of Religion and Science)* 43, 771–781.
Solisti, K. and Tobias, M. 2006. (eds.) *Kinship with animals.* Tulsa, OK: Council Oaks Books.

Marc Bekoff

DEVIANCE AND ANIMALS

Social scientists typically understand deviant behavior in two ways. Deviance, on the one hand, is a characteristic of how people act. If the behavior violates social norms—the basic guidelines for behavior that are known and obeyed by well-socialized members of a society—then it is, by definition, deviance. In contrast, some sociologists speak of deviance as a subjective or personal phenomenon. From this view, a behavior is deviant or not depending on who does it, for what reason, and who finds out about it.

Deviant animals are usually displayed in the media in much the same way as deviant humans. At times they are shown to be threatening and dangerous because they are innately evil, like, for example, the shark in *Jaws*. At other times, animals are presented in the media as behaving in deviant ways because they are mad (e.g., the dogs in *Cujo* and *Man's Best Friend*) or because they have been trained by humans to do evil things (e.g., the rats in *Ben* or the guard dog in *White Dog*). Like the human deviants portrayed in the media, deviant animals are easy to recognize because they are slimy, foam at the mouth, bare their teeth, or in other ways physically display their malevolence. It is likely that the fear that many people have for pit bull terriers, bats, snakes, and other definably ugly animals has its roots in our cultural connection of appearance and deviance.

Another common connection between animals and deviance is seen in the tendency for animal terms to be used in most, if not all, cultures as labels that diminish the importance of the person so labeled. In our society, for example, a person can be degraded by calling him or her such things as "animal," "pig," "chicken," "snake," or "dirty dog." ' These animal labels are intended to demonstrate that those to whom they are applied are less than real human beings.

Related to this use of animal terms to label certain individuals as inferior, the symbolic connection of animals to entire groups of people in order to cast them as being outside the bounds of social normality—and, therefore appropriate objects of discrimination—has been common. For example, Fine and Christoforides (1991) describe how in the mid-19th century the English sparrow was used by American politicians and in the media as a metaphorical stand-in for immigrants.

According to this construction, the birds were dirty, foreign, in competition with native birds, and should be excluded from association with American birds. In short, nativists linked the English sparrow to the presumed deviant characteristics of foreign immigrants and the social problem some saw immigration presenting at the time.

From the Middle Ages until the 18th century, it was common in Europe for nonhuman animals to be seen as being able to choose how they behaved. This meant that animals were often put on trial for such things as murder, assault, and destruction of property. If they were judged guilty, the animal defendants were usually executed. One writer recorded 191 judicial proceedings involving such animal defendants as bulls, horses, pigs, dogs, turtledoves, field mice, flies, caterpillars, and bees.

Bestiality is one type of behavior involving people and animals that is seen as a serious violation of the norm. A far more common and less controversial example of the relationship of animals and deviance is seen in the everyday lives we share with companion animals. In some ways, training a dog or breaking a horse may be seen to be forms of socialization. We typically teach animals to abide by certain rules—not to relieve themselves in our homes, not to jump up on visitors, not to make unnecessary noise, and so forth. As is the case with humans, animal companions often break the rules we would like them to obey. When this happens, their misbehavior is usually either ignored or steps are taken to control the deviant behavior.

One study by Sanders (1994) focused on how doctors in a veterinary clinic defined and responded to violations by their animal patients. Typically, the misbehavior of animals was not seen as being their fault, but as being caused by the stress of being in the clinic or the pain the animals were experiencing. While patients' unruliness usually was not interpreted as being due to moral failings, veterinarians were rarely as charitable in their evaluations of owners. The bad behavior of patients was commonly seen as the fault of bad (ignorant, weak, overly permissive) clients.

Social control—the mechanisms employed in order to maintain individual behavior within the bounds of social norms—is directly related to the issue of deviance and is associated with the relationships between people and animals. Dogs, horses, and other animals have been, and continue to be, used in law enforcement as tools or weapons to assist in the maintenance of social order. In a study of K-9 officers and their patrol dogs, Sanders (2006) stresses the ambivalence of this relationship, as officers are torn between regarding their dogs as tool or weapons (and thereby expendable) and as friends and partners in crime control.

Further Reading

Dekkers, Midas.1994. *Dearest pet: On bestiality*. London: Verso.

Evans, E. P. 1987. *The criminal prosecution and capital punishment of animals/1906*. Boston: Faber and Faber.

Fine, Gary Alan and Lazaros Christoforides. 1991. Dirty birds, filthy immigrants, and the English sparrow war: Metaphorical linkage in constructing social problems. *Symbolic Interaction* 14(4): 375–393.

Hearne, Vickie. 1991. *Bandit: Dossier of a dangerous dog*. New York: HarperCollins.

Laurent, Erick. 1995. Definition and Cultural Representation of the Ethnocategory *Mushi* in Japanese Culture. *Society and Animals* 3(1): 61–77.

Sanders, Clinton. 1994. Biting the hand that heals you: Encounters with problematic patients in a general veterinary practice. *Society and Animals* 1(3): 47–66.

Sanders, Clinton. 2006. The dog you deserve: Ambivalence in the K-9 officer/patrol dog relationship. *Journal of Contemporary Ethnography* 35(2): 148–172.

Clinton R. Sanders

DISASTERS AND ANIMALS

Any catastrophic event that affects people on a large scale will also affect animals. Pets, wildlife, livestock, and captive animals face risks from floods, hurricanes, and earthquakes. Fire, drought, and disease can affect wild animals. Animals also face risks in technological disasters such as nuclear accidents, oil spills, terrorist attacks, and chemical leaks. In addition, large-scale disease outbreaks, such as avian flu, SARS, and foot-and-mouth disease, can devastate livestock populations and local economies. Moreover, many diseases are zoonotic, meaning they can spread between humans and animals. The intensive agriculture practices widely used today present ideal environments for the rapid spread of livestock disease. The close confinement and transportation of birds and animals destined for slaughter means that a disease outbreak in one facility can quickly escalate into a regional or national disaster that devastates the economy. Animal stakeholders of all kinds, including pet owners, breeders, zoo keepers, farmers, veterinarians, and others face unique challenges in planning and response.

The difference between a disaster and an emergency is a matter of scale. In both cases, the response begins locally. In an emergency, the existing local authorities, such as police and fire departments, can take action and meet the immediate needs created by the event. In contrast, a disaster overwhelms local resources and often makes it difficult for outside help to arrive. A request for assistance activates a network of government and nonprofit agencies at the federal, state, and regional levels. The response to a disaster that affects animals will usually begin within the local framework and involve animal control departments, animal shelters, veterinary associations, and livestock organizations. Local animal control and law enforcement agencies often seek the help of national nonprofit animal welfare groups that have disaster response programs, such as the American Humane Association and the Humane Society of the United States. In most events, large numbers of volunteers donate time and money.

Depending on the type of incident and the numbers and species of animals affected, various government agencies may assist with the response. In a large-scale incident within the United States, the Federal Emergency Management Agency might activate Veterinary Medical Assistance Teams to assist when a disaster compromises an area's veterinary infrastructure. The Department of Agriculture and the Fish and Wildlife Service each have many branches that play roles when animals are involved. The Department of Health and Human Services, which oversees the Centers for Disease Control and Urban Search and Rescue, could also participate. At the state levels, offices of emergency management and departments of agriculture and wildlife can enter the picture. However, state and federal agencies get involved only after requests from the local level.

In disasters, animal issues are associated with matters of public safety, the human-animal bond, public health, the economy, and ethical and moral issues.

Public Safety

People will risk their lives to protect their pets, horses, and livestock. They will consequently jeopardize the lives of others by refusing to evacuate or by reentering evacuated areas. A common reason for evacuation failure (along with fear of looting) is the inability or unwillingness to evacuate animals. When people remain in unsafe buildings or reenter them to rescue pets, emergency responders often have to rescue them, using time and resources that are always in short supply during a disaster. This public safety risk is not limited to pet and horse owners, but occurs with those who own and work with livestock as well.

Numerous issues surround the evacuation of animals, including property rights, contamination, evidence preservation, and infrastructural hazards. In 2005, following Hurricane Katrina, rescuers entered many properties without permission to rescue stranded pets. Some homeowners objected to what they saw as breaking-and-entering. Moreover, rescuers encountered sewage, oil, gas leaks, and other chemical hazards because of their efforts to save stranded pets, who were also contaminated. After a disaster, the scene must be maintained for insurance documentation. When people enter damaged areas, they can compromise the integrity of the evidence needed for insurance claims through their movement and by moving debris.

A dramatic example of the public safety risk when people reenter evacuated areas

U.S. Army flight surgeon Capt. Devry C. Anderson, of HHC 2-4 Aviation, 4th Infantry Division out of Fort Hood, Texas, holds a small dog named Chip after he was rescued with his owner, Friday, September 2, 2005 in New Orleans. (AP Photo/Haraz N. Ghanbar)

comes from a chemical spill in Weyauwega, Wisconsin. Early in the morning on March 4, 1996, 35 cars of a train derailed while passing through the town. Fifteen of the cars carried propane, and five of these caught fire. At 7:30 AM, residents of 1,022 households were ordered to evacuate because of the risk of explosion. Emergency managers anticipated that the response would take several hours. The effort instead took over two weeks, reflecting the unpredictability of disaster response. Half of the 241 pet-owning households left their pets behind. Others who were not at home at the time had little choice. Shortly after the evacuation, pet owners began to reenter the evacuation zone illegally to rescue their pets, at considerable risk to their own safety. Following protocol, emergency managers prevented residents from entering

their own homes. In response, a group of citizens made a bomb threat on behalf of the animals, which directed considerable negative media attention at the response. Four days after the evacuation, the Emergency Operations Center organized an official pet rescue, supervised by the National Guard and using armored vehicles.

The Human-Animal Bond

Approximately 70 percent of American households now include pets, which exceeds the numbers that include children. The majority of pet owners consider their pets members of the family. Thus, the human-animal bond is a powerful presence in our society. Interaction with animals has positive effects on people's mental health and physical well-being.

During disasters, the human-animal bond can be either a source of support for the victims of disaster or a source of significant stress, anxiety, and even depression. Failure to consider this bond in disaster response creates substantial concerns among the public. Consequently, disaster planning at all levels must take animals into account. In 2005, Hurricane Katrina brought this need to public attention, which called attention to the importance of including pets in evacuation plans.

One year after Hurricane Katrina, President George W. Bush signed the Pets Evacuation and Transportation Standards Act into law, which requires that state and local emergency planners address the needs of individuals with household pets and service animals in their disaster preparedness efforts. When Hurricane Gustav struck the Gulf region in late August 2008, plans provided for the housing of animals and transporting evacuees with their pets. The aftermath of Gustav offers a dramatic and positive contrast to that of Katrina.

Public Health

The roles that animals play in public health seldom come to mind when people think of disasters. However, animal and human health issues are closely connected. Many diseases, known as zoonoses, can affect both humans and animals. Some, such as rabies, are transmitted directly through human contact with an animal. Others, such as Hendra virus, require reservoir hosts, such as bats, who suffer few if any symptoms. Other examples of zoonoses include Lyme disease, Nipah virus, sleeping sickness, West Nile virus, Severe Acute Respiratory Syndrome (SARS), avian flu, HIV, and monkey pox. In addition, some animal diseases,

such as anthrax and plague, could serve as weapons of mass destruction. Animal diseases could also become weapons in agroterrorism, in which an agricultural disease outbreak causes economic damage and loss of citizen confidence in authorities. Many experts say the new strain of avian flu, H5N1, has the potential to be much worse than SARS.

In addition to natural disasters, animals face hazards of other sorts.

The Economy

Animal issues in disasters have economic impact not only because of the costs of the recovery efforts themselves, but also because of the role of animals in the economy. For cxample, when wildfires and drought affect wildlife, local and state economies feel the impact in their tourism, hunting, and fishing industries. The economic impact is particularly notable with livestock disasters, such as widespread disease outbreaks. In the United States, livestock production directly contributes over $100 billion to the economy annually, and multiple times that value indirectly. Disease threats to livestock, either accidental or intentional (as in agroterrorism), could devastate the economy and the nation's food supply. Great Britain serves as an example of the impact of livestock disease. Britain's first cases of foot-and-mouth Disease (FMD) appeared in 2001, only five years after the outbreak of bovine spongiform encephalopathy (BSE) or mad cow disease.

The 2001 outbreak of FMD paralyzed Britain's agricultural infrastructure and cost the equivalent of 12 billion U.S. dollars. The outbreak resulted in the killing of over four million cows, pigs, and sheep, the majority of whom lived in the affected areas but did not have the disease. The economic impact included direct costs such as lost animals, carcass disposal, and response and eradication efforts. Slaughterhouse workers lost jobs. The outbreak also affected peripheral industries. Hauling companies reporting a large downturn, and the rendering industry, which had previously produced economically valuable raw materials, essentially became a waste disposal industry in response to the massive slaughter. The outbreak also caused significant indirect costs to tourism and trade in Britain and Western Europe. Travel was significantly restricted to control the spread of the disease. Many small businesses, such as pubs and inns, in the affected areas closed down. In addition, the outbreak brought significant nonmonetary and moral consequences. Some herds in Great Britain were legacy herds, raised by particular families for generations. The outbreak meant the loss of lifestyle. Many farm families were ostracized within their communities, and over 80 suicides were reported among farmers and other animal stakeholders affected by the outbreak of FMD.

Today, severe economic problems for people in the United States and elsewhere, especially when they lose homes due to foreclosure, are forcing some either to give pets up to shelters or to abandon them. Compounding problems for shelters and other rescue organizations are the decreases in donations to these groups, as people lose jobs or take lower-paying jobs.

Ethical and Moral Issues

Ethical and moral issues enter into disaster response, because humans are responsible for animals in so many ways. We bring them into our homes and include them in our families. Therefore, they depend on us when they are in danger. We house food animals in extremely crowded conditions with little chance of escape when barns catch fire, collapse, or become flooded. On a more basic level, human beings are responsible for bringing many species of animals into existence in the first place. When we domesticated animals, we took on the responsibility for their care.

The impact of oil spills illustrates the ethical issues involved in disasters and the ensuing response efforts. Estimates indicate that 380 million gallons of petroleum make their way from various sources into the world's oceans each year. Oil is so toxic that many animals die from ingesting it. Oil is also carcinogenic to fish, birds, and mammals. Seals and sea lions often drown because of the weight of oil on their coats. Some of the high-profile spills illustrate the scope of the issue. In 1978, the tanker *Amoco Cadiz* ran aground and split in two off the coast of Brittany, spilling 223,000 tons of heavy crude oil into the Atlantic Ocean. Rescuers recovered 20,000 dead birds. Marine life in the area suffered tremendous mortality. In 1989, the *Exxon Valdez* spill killed an estimated quarter of a million birds, as well as countless sea otters, harbor seals, salmon, and creatures in the supporting food chain. In 1999, the tanker *Erika* broke in two and sank off the French coast, affecting an estimated 77,000 birds. In 2000, the freighter *MV Treasure* sank off the coast of South Africa, contaminating over 20,000 African penguins, whose worldwide numbers are estimated at only 180,000. In 2002, the sinking of the crude oil tanker *Prestige* off the coast of Spain and Portugal topped the *Exxon Valdez* as the worst spill and possibly the worst ecological disaster in history. As many as 300,000 sea birds died as a result.

Although most spills result in massive efforts to rescue, clean, and rehabilitate birds and animals, the effort might not always pay off. The birds and animals experience stress during the rescue and cleaning, in addition to the trauma and injury due to the oil itself. Studies of sea birds found that most did not survive the rehabilitation efforts. Others found that cleaned birds died soon after release back into the wild. One study of sea otter rehabilitation efforts following the *Exxon Valdez* spill determined that the cost of capture and rehabilitation was $18.3 million, or $80,000 per otter. The high costs and low survival rates raise questions about what we should do for wild birds and animals affected by oil.

Planning for Disasters

Disaster planning on a large scale takes place at the governmental level, but individual households must also make plans. Animal stakeholders such as veterinary clinics, breeding facilities, boarding kennels, shelters, and farms must also have plans in place. Whereas some disasters require evacuation, others necessitate "sheltering in place," or staying put until the risk has passed. Depending on the disaster, animal stakeholders might have to evacuate their facilities or take in evacuated animals. Consequently, preparations must consider various scenarios.

Disaster planning begins with assessing the risks in a given area. A region that is vulnerable to hurricanes and flooding probably faces little risk of blizzards and ice storms. Wildfires do not threaten urban areas. The type of response necessary will depend on the potential threat. However,

there are many equal opportunity risks. For example, railroad tracks intersect most regions, and there are numerous homes within a mile of tracks. Trains regularly transport hazardous chemicals, posing risk in the event of derailment, such as the incident in Weyauwega.

Planning at the household level usually means anticipating the needs of pets. Experts suggest designating a cupboard, shelf, or container for emergency supplies for pets. At minimum, households should have sufficient food, water, litter, bedding, and other necessities to last at least 72 hours. Pets should have up-to-date identification and vaccinations. A waterproof plastic bag can hold copies of vaccination records and any licenses. It can be helpful to include one or two photos of the pets, ideally with family members, in the emergency supply kit. If an animal is lost, the photo can supplement a description and also verify that a found animal belongs with a particular family. If the incident requires evacuation, rather than sheltering in place, dogs' leashes must be easily located. Cats and smaller dogs must have travel carriers. An adequate supply of any medications must accompany the animals. Because most emergency shelters do not allow pets to be housed with people, animals and their guardians will most likely be separated during the evacuation period. This highlights the importance of up-to-date identification.

Horses and livestock bring additional issues to consider in planning. Experts note that owners must have sufficient, operable trailers and transporters. Moreover, horse owners should practice loading their horses into trailers so that they can do it quickly and safely when necessary. Horses and livestock are often evacuated to local farms and ranches, but they are also housed at fairgrounds and similar facilities that have barns. The need for identification also arises with livestock. Brands on livestock and tattoos on horses link owners with animals, and all owners should ensure that their animals have current identification.

In addition to preparing to shelter in place, individuals and families should locate animal-friendly accommodations outside the immediate area in case emergency managers call for evacuation. Knowing where to find pet-friendly motels before the incident occurs, or having friends and family who can house pets, can save lives and prevent separation from pets.

Further Reading

Convery, I., Bailey, K., Mort, M., & Baxter, J. 2005. Death in the Wrong Place: Emotional Geographies of the UK 2001 Foot and Mouth Disease Epidemic. *Journal of Rural Studies* 21, 99–109.

Heath, S. E., Beck, A. M., Kass, P. H., & Glickman, L. T. 2001. Risk factors for pet evacuation failure after a slow-onset disaster. *Journal of the American Veterinary Medical Association* 218, 1905–1910.

Heath, S. E., Beck, A. M., Kass, P. H. & Glickman, L. T. 2001. Human and pet related risk factors for household evacuation failure during a natural disaster. *American Journal of Epidemiology* 153:659–665.

Heath, S. E., Voeks, S. K., & Glickman, L. T. 2001. Epidemiological features of pet evacuation failure in a rapid-onset disaster. *Journal of the American Veterinary Medical Association* 218, 1898–1904.

Irvine, L. 2007. *Animals in disasters: Responsibility and action.* Ann Arbor MI: Animals and Society Institute.

Mead, C. 1997. Poor Prospects for Oiled Birds. *Nature* 390, 449–450.

Sharp, B. 1996. Post Release Survival of Oiled, Cleaned Seabirds in North America. *Ibis,* 138:222–28.

Leslie Irvine

DISASTERS AND ANIMALS: LEGAL TREATMENT IN THE UNITED STATES

In the United States, the law treats non-human animals as personal property, making them vulnerable during disasters. Governmental policies for evacuation, shelter and rescue during disasters place priority on saving human lives, with a secondary focus on protection of property. Since nonhuman animals are property, the welfare of nonhuman animals during hurricanes, floods, and other disasters is less important than the welfare of humans. As a result of this status as property rather than as living beings with inherent value, large numbers of nonhuman animals, including companion animals, stray and feral domesticated animals, livestock, and wild animals, are left behind and suffer or die during disasters and their aftermaths.

The treatment of nonhuman animals hinges upon the value that humans place upon the these animals. Household pets, or companion animals and service animals, which have direct bonds with individual humans, are treated differently than livestock and wild animals. Livestock, as commodities that provide food and fiber, are generally considered only for their economic and subsistence value to humans, with disaster policies focusing on issues such as the maintenance of the food supply rather than on the preservation of individual animals. Disaster policies for wild animals, both those captive in zoos and other manmade facilities, and those in natural habitat, primarily discuss these animals in terms of danger to humans during disasters; these animals will either be kept captive or left in the wild to use their own instincts for survival in a disaster. This is not to say that either livestock or captive wild animals are ignored by disaster policies and plans, but these categories of animals are not treated as a priority for evacuation or rescue.

In contrast to livestock or wild animals, companion animals (also referred to as household pets) and service animals have a special relationship with humans and are afforded a measure of protection in disasters that livestock and wild animals are not. Humans will make greater efforts to protect and rescue their companion and service animals than they will livestock or wild animals. However, all animals are faced with the possibility of abandonment, destruction, or removal if considered to be endangering human health and safety during a disaster, whether by possible attacks on humans, by exposure of humans to biohazards, or by use of limited resources including food, water, and space in transportation and shelters.

While wild animals and livestock are handled by entities such as animal control or other governmental entities in disasters, the general approach to the evacuation, rescue, and care of companion animals and service animals during disasters has been one of personal responsibility by animal owners. Rescue organizations and government agencies continue to emphasize the need for humans to be prepared to handle disasters that might require evacuation. Proposed advance plans from the Red Cross and the U.S. Department of Homeland Security include identification of pet-friendly lodging, health certificates for animals so they can be sheltered or transported out of state, a proper carrier, and sufficient food and water for several days.

During Hurricane Katrina in 2005, many people did not have private transportation available to them or funds to pay for shelter; without private vehicles, these people had to rely on public evacuation systems, which did not provide the means to evacuate nonhuman animals. Disaster plans did not offer options for evacuation of companion animals with humans; shelters for people would not accept animals for health and hygiene reasons. As a result of the lack of options available to people in New Orleans in 2005, an estimated 250,000 pets, including dogs, cats, birds, and fish, were stranded in the hurricanes and the flooding.

Prior to Hurricane Katrina, federal and state law had focused on the proper handling of animals in disasters to preserve the health and hygiene of humans. No federal statutes dealt with animal evacuation. Federal laws and regulations that dealt with animals in the event of a disaster focused on proper handling of animal carcasses and biohazards and looked at ways to prevent disease and other harm to humans. For similar reasons, state health and safety regulations prohibited sheltering of animals with humans or the transport of animals with humans. Even transport of animals from the disaster area was limited by laws and regulations prohibiting movement across state lines. While humans were transported over state lines to neighboring states for refuge from the storm and its aftermath, nonhuman animals were not allowed unfettered transport out of the states affected by Hurricanes Katrina, Rita and Wilma. For example, states outside the disaster area, such as Massachusetts (which issued an emergency order concerning importation of animals from Louisiana, Mississippi, and Alabama in September 2005), declined to accept animals from the disaster area for fear of diseases such as heartworm.

Following Hurricane Katrina, the dangers to people of governmental failure to provide for evacuation of companion animals became clear. People who refused to evacuate without their animals endangered themselves by remaining in an unsafe situation. Rescue workers were endangered when trying to save people who had stayed or who tried to return to unsafe areas to protect their companion animals. Rescue workers also faced the dangers of trying to capture abandoned animals, who were often terrified or aggressive and difficult to remove. In addition, animals who were left alone sometimes turned to foraging for food or died in houses or in the streets, causing safety, health, and hygiene problems for people returning to the area.

The dangers to people prompted legislators to make companion animals and service animals the subject of governmental evacuation and rescue laws and policies. While livestock and wild animals are generally still left to their own devices during disasters, with at best minimal efforts at rescue and, at worst, execution to protect humans from potential harm, companion animals and service animals are now the focus of federal and state statutes and regulations affecting evacuation during disasters.

In 2006, Congress passed the first federal legislation to address evacuation issues for nonhuman animals during disasters. The legislation amended the Stafford Disaster Relief and Emergency Assistance Act, which provides for federal government assistance to the states in times of disaster. The Pets Evacuation and Transportation Standards Act of 2006 (PETS Act) requires that state and local emergency preparedness plans take into account the needs of people who have household pets and service animals if the state is to be eligible to receive Federal Emergency Management Agency (FEMA) funds. The Act also allows federal agencies to provide assistance to protect property by providing rescue, care, shelter, and essential needs to individuals with household pets or service animals and to those pets and service animals.

In October 2007, FEMA released Disaster Assistance Policy 9523.19, which sets out the costs related to emergency pet evacuations and sheltering activities by state and local governments that may be reimbursed by FEMA following a declaration of a major disaster or emergency. This policy defines a household pet as a "domesticated animal, such as a dog, cat, bird, rabbit, rodent, or turtle that is traditionally kept in the home for pleasure rather than for commercial purposes, can travel on commercial carriers, and be housed in temporary facilities." Under the policy definition, "reptiles (except turtles), amphibians, fish, insects/arachnids, farm animals (including horses), and animals kept for racing purposes" are not household pets and are not be covered by the policy.

State legislatures also reacted in the aftermath of Katrina, passing their own laws to require inclusion of animals in state disaster plans. In 2006, Louisiana amended its disaster act to require that the Governor's Office of Homeland Security and Emergency Preparedness assist in the formulation of parish emergency operation plans for "humane evacuation, transport, and temporary sheltering of service animals and household pets in times of emergency or disaster." The Louisiana act made a distinction between service animals and household pets. Under the act, provisions must be made for a service animal to be evacuated, transported, and sheltered with the person served, while household pets (defined as "any domesticated cat, dog, and other domesticated animal normally maintained on the property of the owner or person who cares for such domesticated animal") are to be protected by the agency providing assistance in identifying suitable temporary shelters and providing guidelines for admission to those shelters and by enabling, whenever possible, evacuation of pets and pet owners for disabled, elderly, special needs residents, and all other residents when the evacuations can be done without endangering humans. The act also provided that pets in cages or carriers are to be allowed on public transportation during an impending disaster, when doing so does not endanger human life, and that the agency may provide separate transportation for pets that are not allowed on public transportation.

While the PETS Act was the first federal legislation to address evacuation of companion animals, some states had already included animals in their evacuation and rescue planning as a consequence of previous disasters. Following Hurricane Floyd, which hit the east coast of the United States in September 1999, North Carolina developed a public/private interagency animal response team model that provides coordination of efforts to address animal-related issues during disasters. The State Animal Response Team (SART) model, which is dependent upon cooperation among local, state, and federal agencies and private organizations, has been or is being adopted by about one-half of the states, many of them having adopted or begun to develop a version of the model since Hurricane Katrina. Unlike the PETS Act, the SART model does not restrict coverage

and consideration to companion animals. Under the section on animal protection in North Carolina's March 2008 Emergency Operations Plan (NCEOP), the stated purpose is to "protect domesticated and wild animal resources, the public health, the food supply, the environment, and to ensure the humane care and treatment of animals during disasters." The protection is "aimed at all animals (whether owned, stray, domestic) that may need help during disaster situations." Although the purpose and scope of the plan note the goals of protection, humane care, and treatment of animals during disasters, the stated policies in the EOP place priority on "saving human lives and protecting property, in that order" and place responsibility for sheltering and protection of companion animals and livestock on their owners, while wild animals are to be "left to their own survival instincts." Wild animals who pose a threat to themselves or to humans will be handled by local animal control or wildlife management personnel and returned to their natural habitats.

While the legislation and policies that are currently in effect may be aimed at the protection of humans rather than at the protection of nonhuman animals, the legislation does give those interested in the welfare of nonhuman animals a seat at the table and an opportunity to offer recommendations in the planning process that will affect animals in future disasters. State and local disaster planning processes now include representatives from animal welfare organizations, and offer the possibility of more careful thought regarding the needs of animals, and not just the needs of humans affected by the inability to take their pets and service animals with them in the event of a disaster.

Further Reading

Blum, S., and Silver, R. C. Why is it important to allow people to evacuate disaster areas with their pets? available at APA Online Public Policy Office, *Psychological Research on Disaster Response*, at http://www.apa.org/ppo/issues/katrinaresearch.html.

"Katrina's Animal Rescue." 2005. *Nature.* Thirteen/WNET New York and National Geographic Television, Inc.

Louisiana Homeland Security and Emergency Assistance and Disaster Act, La. Stat. Anno. R.S. 29: 726(E)(20)(a) and (c); North Carolina Emergency Operations Plan (March 2008).

North Carolina State Animal Response Team, at http://nc.sartusa.org.

Pets Evacuation and Transportation Standards Act of 2006, Pub. L. 109–308, 120 Stat. 1725 (Oct. 6, 2006).

Marsha L. Baum

DISNEYFICATION

The Disneyfication of animals refers to the assignment of some human characteristics and cultural stereotypes to the animals. Although this practice is best shown by the way cartoon characters and animals are pictured in Walt Disney movies, it is not restricted to the Disney Corporation, but is widespread as a marketing strategy. The most noticeable human characteristic projected onto animals is that they can talk in human language. Physically, animal cartoon characters (and toys styled after them) are also most often deformed in such a way as to resemble humans. This is achieved by showing them with human-like facial features (eyebrows, expressive lips) and altered forelimbs to resemble human hands (although with a smaller number of fingers). In more recent animated movies, the trend has been to depict the animals in a more natural way.

However, they still use their limbs like human hands (for example, lions can pick up and lift small objects with one paw), and they still talk with an appropriate facial expression. A general strategy that is used to make the animal characters more emotionally appealing, both to children and adults, is to give them enlarged and distorted childlike features.

Probably the most significant aspect of Disneyfication of animals is the projection of cultural stereotypes onto animal behavior. The members of the animal kingdom are often used as a means of presenting male-dominated societies with stereotypical gender roles. Racist attitudes are subtly conveyed not only through the choice of the physical characteristics of bad animal characters, but also through the use of language with accents and characteristic expressions indicative of racial or ethnic background. In Disney's 1994 best selling *The Lion King*, the members of the royal family speak with British accents, whereas the voices of hyenas resemble those of urban black and Latino populations.

Disneyfication is widely used in popular visual culture, including everything from video games, television, and film to amusement parks and shopping malls. Its effects on the formation of the individual and collective identities of children and youth are not yet fully understood. One of the direct effects of misrepresentation of animals is that animals and their behavior tend to be misinterpreted by children, sometimes with tragic consequences. Objectification of animals promotes the pet industry and the view of animals as goods to be bought. This strategy may lead to the formation of adult personalities incapable of functioning outside of stereotypical frameworks modeled after their childhood experiences.

Further Reading

Complete Details on Disney's Animal Kingdom. 1995. *Orlando Sentinel,* June 21, 1995, A1, A6

Giroux, H. A.. 1994. Animating Youth: The Disneyfication of Children's Culture. *Socialist Review* 24(3): 23–55.

Noske, B. 1989. *Humans and other animals.* London: Pluto Press.

Oswald, Michael. 1991. Report on the Potentially Dangerous Dog Program: Multnomah County, Oregon. *Anthrozoos* 4(4): 247–254.

Thompson, W. I. 1991. Disney's world: The American replacement of culture. *The American Replacement of Nature.* New York: Doubleday.

Slavoljub Milekic

DISSECTION IN SCIENCE AND HEALTH EDUCATION

Medical and veterinary schools have largely phased out the practice of having students dissect animals, and yet animal dissection continues across the United States as a widespread practice for children in intermediate school science classes. This practice was introduced in the 19th century at the same time that it became a national goal to provide science education with laboratory experience for all children in the United States. When science education became universal, it was modeled on the teaching style that had been used for hundreds of years for medical students. Medical students had typically been provided experience with human cadavers, but providing hands-on experience to all children required a shift to animal bodies for laboratory instruction. The emphasis shifted somewhat to animal biology, reflecting that the dissection focused on the frog, cat, or guinea pig. Perhaps it seems paradoxical that laboratories for medical and veterinary education have shifted to newer methods, whereas pre-college instruction has not changed and still emphasizes animal dissection.

History of Human and Animal Dissection and Science Education

Dissection was used in the Middle Ages as a method for illustrating Galen's ancient texts, and later became a method for discovering the anatomical and physiological aspects of humans and other mammals. Human dissection was most informative and productive for learning and teaching, with other mammals used for supplementary work reflecting a shortage of human cadavers. Demonstrations of human dissection were conducted in a theater setting as a special occasion, typically in winter when the cold slowed the rate of decomposition of the cadaver. Vesalius in the 16th century, and those following him, began using dissection to investigate the human body and also to make anatomical and physiological discoveries. The use of human bodies for dissection was controversial, and violated religious concerns regarding the need to be resurrected with an intact body. Even when dissection became an accepted part of medical education, gaining access to a sufficient number of bodies was challenging, sometimes resulting in grave robbing. Furthermore, being dissected was considered to be an even worse fate than hanging.

With the widespread establishment of science education for children in the 1850s, for which laboratory exercises were a valued part of instruction, it became common practice to use dissection of small animals to support the teaching.

Dissection of animals was adopted as a convention to illustrate and provide children with hands-on experience of the body systems as a surrogate for the human body, a practice that remains common today. This practice, like the dissection of human bodies, has been controversial, at times having both strong advocates and dissenters.

National and State Standards, Plus Legislation and Regulation, for Science and Health Education

Learning goals and objectives for the course material that is to be taught at various grade levels is officially defined by national and state education standards, and further spelled out in individual frameworks. The content related to body systems is addressed in the 7th grade science standards and appears again with a more physiological emphasis in the standards for high school biology. The health standards include some discussion of certain diseases and practices affecting health and offer prescriptive recommendations for maintaining good health.

Although standards are defined at the national and state level, schools are considered to be locally governed. State and local legislation may constrain the content to be taught, or add specific requirements for what must be taught. The teaching of health is particularly subject to regulatory and policy requirements, such as mandatory instruction on the use of alcohol and tobacco. In some cases, local or state laws specify the minimum or maximum classroom time to be spent on certain content such as reproduction or sexual activity, or state a requirement for parental permission for children to participate in instruction on certain topics. Teaching of

health can be less than optimal when not presented within a biological framework by teachers who have majored in science. Teaching certificates for health are a part of the physical education curricula rather than biology coursework, and health teachers may have limited backgrounds in basic science content. Integration of biological science and health, with an emphasis on the human rather than non-human, can better prepare children for managing their lifelong health. Most children in the United States have their last biology instruction in the 7th grade, making this an important opportunity to prepare them in biology and health.

Since dissection is a teaching method, not a subject area with informational content, nor pertaining to teaching objectives and goals, it is not discussed within either the national or state standards for science or health, or in the frameworks. There are no official recommendations for teachers concerning the presence or absence of dissection as a laboratory experience, nor is there much discussion of dissection in the professional education literature. Hence, teachers receive little guidance with regard to using dissection as a laboratory exercise, or implementing other resources that could provide similar learning experiences.

Testing and Funding

National funding for schools, which is based on the results of required testing, currently sets policy for local school districts and demands that teachers give their primary attention to preparing children for standardized tests. The results of this mandatory testing of children, such as the requirements legislated by the No Child Left Behind Act of 2001, are linked with high-stakes consequences, affecting the funding provided to school districts. Students' capability to perform well on tests affects and can reduce the funds available for teaching resources in a particular school district. Teachers are obliged to devote a significant portion of classroom time to helping children succeed in tests that have far-reaching implications for the district as well as for the students personally.

Challenges for Teachers

Teachers seek to inspire their students, a goal that provides rewards and usually accounts for their choice of teaching as a profession. To be effective, teachers seek out learning opportunities to continually increase their mastery of ever expanding subject matter and incorporate new teaching methods. Most science teachers face various challenges, including small budgets to purchase laboratory equipment and supplies that could enhance their teaching. Many use personal funds to purchase laboratory supplies. They must spend valuable classroom time teaching and testing to national and state standards. Teachers strive to find ways to offer laboratory experiences that will motivate their students. Dissection offers a riveting experience that fully engages students, and is something that students tend to remember, often with some combination of excitement, fear, and revulsion. An additional feature is that dissection is familiar to teachers, is not intimidating, and does not require extensive new learning for them.

Resources for Human Health and Science Education

Medical and veterinary schools have invested during the past couple decades in creating new laboratory teaching resources that draw on new technologies for learning, including computer software, plastination of tissues, and reusable prosections. In contrast, major initiatives have not yet been made to modernize pre-college laboratories. Hence, current biology laboratory curricula in pre-college classes seem more likely to rely on dissection than those in college or professional school laboratories.

Recently, some outstanding software on human biology has become freely available on the web. For example, National Geographic, the British Broadcasting Corporation (BBC), and the Public Broadcasting System (PBS), among others, have produced some fine instructional materials concerning the systems and major organs of the human body. While some of these are fine resources that are visually appealing and informative, they do not fulfill teachers' needs for materials that would stimulate children to solve problems and interact with the subject matter rather than rote learning. The most gifted and motivated teachers are looking for webquests, materials that engage students in interacting with the information.

The technological capabilities evidenced in computer games and Hollywood films have yet to be brought to educating our children concerning their own bodies and health. While some recent web-based resources on the human body are promising, much more can be done to support health promotion and knowledge of the human body by using the full range of web technology to engage children and adults in learning.

See also Alternatives to Animal Experiments; Student Objections to Dissection

Further Reading
American Alliance for Health, Physical Education, Recreation and Dance, Health Education

Standards, accessed on December 15, 2008, http://www.aahperd.org/aahe/pdf.files/standards.pdf.

Balcombe, Jonathan. 2001. Dissection: The scientific case for alternatives. *Journal of Applied Animal Welfare Science* 4:117–126.

DeBoer, George E. 1991. *A history of ideas in science education: Implications for practice.* New York: Columbia University.

French, Roger. 2000. *Ancients and moderns in the medical sciences: From Hippocrates to Harvey.* Aldershot, England: Ashgate.

Hart, Lynette A., and Wood, Mary W. 2005. Mainstreaming alternatives in veterinary medical education: Resource development and curricular reform. *Journal of Veterinary Medical Education* 32:473–480.

Hart, Lynette, Wood, Mary W. and Hart, Benjamin L. 2008. *Why dissection? Animal use in education.* Westport, Connecticut: Greenwood.

Huxley, Thomas H. 1876/1902. On the study of biology. In *Science and Education.* New York: P.F. Collier & Son.

Jukes, Nick, and Chiuia, Mihnea. 2003. *From guinea pig to computer mouse: Alternative methods for a progressive, humane education,* 2nd ed. Leicester, England: InterNICHE.

National Academies Press. 1996/2007. *National science education standards.* Washington, DC: National Academy Press

NORINA (Norwegian Reference Centre for Laboratory Animal Science and Alternatives, NORINA: A Norwegian Inventory of Audiovisuals, accessed on December 15, 2008, http://oslovet.veths.no/NORINA.

Patronek, Gary J. and Rauch, Annette. 2007. Systematic review of comparative studies examining alternatives to the harmful use of animals in biomedical education. *Journal of the American Veterinary Medical Association* 230:37–43.

Singer, Susan R., Hilton, Margaret L., and Schweingruber Heidi A. (eds.). *America's lab report: Investigations in high school science* (Washington, DC: Committee on High School Science Laboratories: Role and Vision, National Research Council), accessed December 15, 2008, http://www.nap.edu/catalog/11311.html.

Vesalius, Andreas, 1543/1964. *De humani corporis fabrica libri septem.* Bruxelles: Culture et Civilization.

Wood, Mary, and Hart, Lynette. *Why dissection? Animal use in education: Resources,* accessed December 15, 2008, http://www.vetmed.ucdavis.edu/Animal_Alternatives/appendices.html.

Lynette A. Hart

DISTRESS IN ANIMALS

Distress denotes mental suffering and may be reflected in a change in molecular receptor binding in the central nervous system (e.g., benzodiazepine, opioid, serotonin, noradrenalin). It may be an integral part of other aspects of suffering. An animal in pain from a broken leg may be fearful of being moved or touched, as well as being distressed by its inability to move normally. Such changes in receptor binding in the central nervous system may lead to stereotypic behaviors.

In a physiological sense, it means that an animal is no longer able to cope with its environment, usually over a long period of time, and is becoming hormonally deranged, that is, homeostasis is lost. Most animals can adapt to short-term minor stressors, and this is an important part of survival and retaining fitness to live and reproduce, but when the stressors are severe or prolonged so that animals are unable to adapt, they can be described as physiological distress.

David B. Morton

DOCKING

Docking refers to the removal of varying amounts of the tail. Docking is done for reasons of fashion (dogs, horses), protection of some animals from diseases where other preventative measures are impracticable (lambs, hill farming of sheep against fly-strike), convenience of the stockperson (dairy cattle swishing their tails in the face of the person milking it), to prevent tail biting in pigs, which is most often caused by poor farming conditions (e.g., overstocking in barren environments). Occasionally, it is done therapeutically for the benefit of the individual animal.

David B. Morton

DOGFIGHTING

The arrest and imprisonment of Michael Vick, star quarterback for the Atlanta Falcons, for dogfighting in 2007, focused international attention on a brutal blood sport which thrives in a netherworld devoted to pain and suffering. As currently practiced, fights feature two dogs attempting to inflict maximum damage on each other for the entertainment and profit of spectators and owners, who frequently bet heavily on the outcome. Ranging in length from minutes to two hours, matches end when one dog can no longer continue due to loss of will, exhaustion, injury or death. Owners are known to kill or simply abandon losers, generally for lacking gameness, the drive or quality that dogfight trainers believe compels a dog to attack its opponent head-on and continue fighting until it is killed or kills. Winners often suffer serious injury and are seldom unscathed.

The Humane Society of the United States estimates that 40,000 people nationally participate in organized dogfighting rings that sponsor high-stakes matches where tens of thousands of dollars are wagered on a single fight. An estimated 100,000 participants, the majority of them disaffected urban youth, fight their dogs opportunistically in less structured matches, often for little more than bragging rights. The total number of participants worldwide is unknown.

But evidence indicates that dogfighting is a global problem. Even where it is legal, or at least officially ignored, dogfighting is tied to issues of caste and class; to urban decay and rural decline; to gangs and other criminal groups, especially those trading in guns and drugs; to gambling; to alcohol abuse; to animal cruelty; to alienation and socially deviant behavior; and to violence against women and children.

Dogfighting is closely related to other blood sports involving animals that are rooted in antiquity and flourished in medieval and Renaissance Europe. Bull and bear-baiting, in which dogs attempt to maul and kill a tethered bull or bear, were popular among commoners and aristocrats. Queen Elizabeth I herself sponsored bull and bear-baiting spectacles. Other animals were baited as well, particularly badgers and wild boars. Hog dogfighting, where a dog is sent to fight a caged boar, is a contemporary variation.

As religious and social reform groups voiced increasingly strident opposition to blood sports involving animals and other atrocities against humans and animals throughout the 17th century, pit dogfighting gained popularity in England and America.

In 1835, with a major push from the newly organized Royal Society for the Prevention of Cruelty to Animals, England became the first country to outlaw dogfighting, bull and bear baiting, and other blood sports, as well as the use of dogs as beasts of burden. English blood sports shifted to dogfighting, which could be staged in a tavern's back room, in barns, or other private spaces, unlike a bear- or bull-baiting, which required

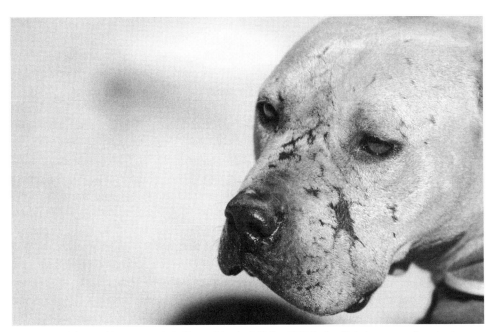

The scarred face of Lucas, a pit bull used in the Michael Vick dogfighting operation, is shown at Best Friends Animal Sanctuary, north of Kanab, Utah, in January 2009. A year after some experts left them for dead—in fact said they should die—many of these dogs are alive and thriving at the Best Friends Animal Sanctuary, rewriting myths about who pit bulls really are and who they can be. (AP Photo/Jae C. Hong)

larger venues. (Rat killing competitions also flourished, until they fell victim to better urban rodent control.) To legitimize their sport, the dogmen established rules dictating the dimensions of the pit in which the fight takes place, how the dogs should engage, how a break is enforced, and a winner determined.

In America, dogfighting flourished among gamblers, grifters, bar hoppers, sportsmen, and gentlemen with and without portfolio through the 19th and into the 20th century. The American Society for the Prevention of Cruelty to Animals was founded in 1866 in New York City, and the next year the City outlawed dogfighting and bull-baiting as part of a campaign to control stray dogs and clean up slaughterhouses. Bull-baiting—using dogs to harass, catch and hold bulls by the nose while the butcher bled them to

death before throwing unwanted scraps into the gutter full of contending dogs—was still common, justified by the belief that an animal terrorized in that fashion produced more tender meat. With each state responsible for its own animal laws, legislation was a patchwork of poorly enforced laws, until dogfighting itself went out of vogue. The United Kennel Club published its last U.K.C. Pit Rules in 1940, and ended its sanctioning of the blood sport.

Still, dogfighting was not outlawed in all 50 states until 1977. In the United States, animal law resides with the states, unless interstate commerce is involved, and the federal government in 2007 strengthened its statutes forbidding the transport of animals for fighting. Most other countries in Western Europe and among former British colonies have followed England's lead over the past two centuries, and even in nations where dogfighting is popularly considered legal, like parts of Central and South America, Russia, Afghanistan, and Japan, it often exists in a netherworld just outside rarely enforced laws against animal cruelty, if its existence is officially recognized at all.

Each culture follows its own dogfighting tradition, including the choice of dogs. Russians and Afghans, for example, use big sheep dogs like the Caucasian *Ovcharka*. Argentineans are said to prefer the *dogo de Argentina*. In Japan, where dogfighting dates at least to the 14th century, large mastiff-like *Tosa inu* are fought, as are pit bulls, reportedly more for honor and prestige than money. Anglo-Americans favor purpose-bred pit bulls.

Following the principles of scientific breeding then coming into vogue, 19th century dogmen created the new pit fighting dogs from terriers and the big, mastiff-like *bandogges* that had been used to guard the Tower of London and in bull- and bear-baiting, according to Johannes Caius, a Cambridge physician, in his 1576 classification of English dogs, *A Treatise of Englishe Dogges*. The butcher's dog, with its shortened, brachycephalic muzzle for catching and holding bulls for slaughter, might also have figured in the mix, as dogmen sought animals that were quick and relentless on the attack but sturdy, possessed of a low center of gravity, great strength, a high tolerance for pain, and an inability to understand canine body language.

The bull terrier, Staffordshire bull terrier, American Staffordshire terrier, American pit bull terrier are all pure breeds with their roots in the 19th century Anglo-American dogfighting, that

through World War II, were also acceptable companion dogs, for men. General George S. Patton, for example, had a bull terrier throughout the war. In the 1980s, when pit bulls purpose-bred to fight became an urban scourge, the bull terrier was deemed safe because it was the same breed as Spuds McKenzie, star of a national Budweiser beer advertising campaign. Thus, when Miami-Dade County in South Florida became the first major metropolitan area in the United States to ban a specific type of dog—the pit bull or pit fighting dog in all its guises—bull terriers were specifically exempted.

Breed-specific bans have proliferated around the world, largely because of the epidemic in dog bites and the association of fighting pit bulls with disaffected urban minority youth and violent criminals. Yet members of the Fancy, who breed the bull terrier, Staffordshire bull terrier, and American Staffordshire terrier, claim they have bred their dogs away from aggression and maintained them as companion animals. Dogfighters do, in fact, maintain their own bloodlines, independent of registration with any kennel club. Defenders of pit bulls, even non-fighters, argue that the dogs can be gentle with people, but their message frequently gets lost in the violence and negative publicity.

In some jurisdictions, legal bans on pit bulls have led dogfighters to adopt other breeds and cross breeds, including the *dogo de Argentina*, Rottweiler, and *Presa de Canario*.

The pit gives the dogs their name. It is a square with sides at least 14 feet long, except when space limitations require it to be smaller, and walls 2.5 to 3 feet high. Any dog jumping out of the pit is disqualified. Scratch lines, behind which the dogs are held, are drawn seven feet from the opposing corners.

Rules governing the preparation of dogs, scratch, and turn, and other aspects of the fight, are intended to make fair an event tainted with the aura of cheating in the form of illegal use of steroids, and of poisons on the coat of the dog. Essentially dogs are expected to cross the scratch line within 10 to 30 seconds at the beginning of the match and after each break, called when one contestant turns its head and forequarters away from the other. The dog who turns first must scratch first, that is, prove itself still game. The dogs alternate all subsequent scratches, regardless of which one turns. Break sticks are used to pry a dog's jaws apart and off its opponent when necessary.

Investigators have identified several types of dogfighters, more than 90 percent of whom are male, according to a Humane Society of the United States survey: professionals who make their livings breeding, training, buying and selling, and fighting dogs at matches that can carry $100,000 prizes. Results are published in magazines devoted to the sport. A dog with five straight wins is a grand champion.

It is not unusual for these individuals to have well over a dozen dogs bound by heavy three-foot chains to a stake or car axle in a dog yard when not traveling, training, or fighting. Professionals often maintain their own bloodlines; forced breedings at rape stands are used to make sure the desired dogs mate. Semiprofessional dogmen participate in organized fighting on a smaller scale and not as a full-time preoccupation.

In the 1980s, pit bulls began to appear in urban neighborhoods, among gang members and street drug dealers, who used them as protection, and eventually among young men embracing through hip-hop and rap music the violent culture that was their home. Where pit bulls went, dogfights followed, abiding by street rules, which were dismissed by professionals.

But street fighting appeared to be a growing international phenomenon among the urban poor, and occasionally one of its former practitioners, like Michael Vick, moved to enter the ranks of major players. He started his own Bad Newz Kennels on property he bought for that purpose in Smithfield, Virginia, bought, bred, trained, and fought his dogs while gambling on them. He and his dog handlers used treadmills, suspended tires, and stray and stolen cats and dogs as training aids. Vick admitted that he participated in killing eight dogs by drowning and hanging because they had failed to show proper gameness when rolled or tested in a fight with an older kennel dog. Others were shot and electrocuted. Vick and three others were convicted of federal charges and imprisoned, and Vick was suspended by the NFL, losing lucrative endorsements as a result. As many of his dogs as possible were taken in by rescue organizations.

Vick's dogfighting kennel may have been unusual because of his financial resources. In other regards, it was like other kennels, a place defined by the cruelty and violence of the dogfighting culture that breeds more cruelty and violence in people and animals.

See also Blood Sports

Further Reading

American Society for the Prevention of Cruelty to Animals. www.aspca.org/site/PageServer?pagename=cruelty_dogfighting.

Derr, Mark. 1997. *Dog's best friend: Annals of the dog-human relationship.* New York: Henry Holt and Company.

Derr, Mark. 2004. *A dog's history of America: How our best friend explored, conquered, and settled a continent.* New York: North Point Press.

Duggan, Paul. 2007. A blood sport exposed: Vick case puts dogfighting culture in the spotlight. *Washington Post,* August 22, 2007.

Gibson, Hanna. 2005. "Dogfighting detailed discussion," Animal Legal and Historical Rights Center, Michigan State University College of Law, 2005. http://www.animal law.info/articles/ddusdogfighting.htm#s2.

Humane Society of the United States Animal Cruelty and Fighting Campaign. www.hsus. org/act/.

Mark Derr

DOGS

Over the years, dogs have been widely used in biomedical research to investigate heart disease, bone injury, hearing loss, blindness, lung disorders, infectious diseases, the effects of lethal poisons, and other conditions that have relevance to human health. They are also used to study the nutritional value of dog food. In the United States, the number of laboratory dogs used peaked in 1979 at 211,000 animals per year. Recently, the numbers have declined so that in 1995 the number had dropped to 89,420 per year. In 2006 more than 87,000 dogs were used for research, a sharp increase from previous years in which 65,000–70,000 dogs were used annually (http://www.aavs. org/researchDogs.html; see also http:// www.hsus.org/animals_in_research/spe cies_used_in_research/dog.html). To put these figures in perspective, dogs comprise a relatively small fraction (less than about one percent) of all animals used for research. Nevertheless, the use of any dogs for research has always been controversial. Dogs are a well-loved species and public sympathy for dogs runs high.

Controversy over the use of domesticated dogs for research has a long history.

In the early days of animal experimentation, the 19th-century French physiologist Claude Bernard encountered fierce public criticism because he performed painful experiments on dogs. On one occasion, he was reported to have experimented on the family pet, which caused his wife and daughter to become antivivisectionists. In those days, there were no commercial breeders of laboratory animals, and it was hard for researchers to obtain suitable animals for their work.

In the 20th century, as the volume of animal experiments increased, researchers found a ready supply of dogs and cats for their work from shelters and pounds. Shelters and pounds are places where lost, stray, and abandoned animals are temporarily housed. By law, shelters have to retain animals in their care for a certain number of days so that owners have an opportunity to reclaim their pets or, alternatively, adoptive homes are sought. If a suitable home is not found, the dogs are often euthanized.

In 1945, a lobbying group for animal researchers was formed whose primary purpose was to work for passage of state laws to permit researchers to have access to unwanted and unclaimed animals in shelters. These efforts persist to this day. However, these efforts are strongly resisted by members of the animal welfare and animal rights movement, who hold that shelter animals should not be used for research. Leading humane societies including the Animal Welfare Institute, the Humane Society of the United States, the American Humane Association, and others have been involved. Currently, state laws are mixed. Some states, notably Minnesota, Utah, and Oklahoma, specifically require shelters to hand over their animals to research, whereas 17 other states prohibit this practice (http://www.

Dogs confined in a very small, but legal cage. Dogs are used in a variety of laboratory experiments. (Shutterstock)

hsus.org/animals_in_research/species_used_in_research/dog.html). In states where there is no law, shelters operated by humane societies usually will not permit their dogs or cats to go to research. But city pounds, whose responsibility it is to keep stray animals off the streets, do not share the same compunction about the eventual fate of former pets, and so are often glad to sell dogs to labs.

The rationales for these opposing viewpoints of researchers and members of the humane movement are as follows: Researchers argue that shelter animals are unwanted and are doomed to die anyhow, so why not use them for a socially useful purpose? Also, the animals are less expensive than animals bred specially for the purpose of research,

thus saving research dollars. The animal welfare/rights view is that human beings have a profound moral responsibility to domesticated animals and this cannot be forsaken at any point in those animals' lives. Shelters should be sanctuaries for animals, and not a supply line for biomedical researchers. From a dog's viewpoint, a humane death may be a better choice than a longer life as the subject of a painful experiment. Animal welfarists hold that overpopulation of pet animals should not be exploited for the benefit of researchers. Animals for research should be a different population of animals than those that were once pets.

This clash of viewpoints has been somewhat lessened by the fact that, since the 1980s, commercial breeders for laboratory dogs have become well established. It is a profitable business. Commercial (Class A) breeders can supply animals who are healthy, and of known age and genetic make-up, and who are more reliable experimental subjects than so-called random source dogs obtained from shelters or from Class B breeders, of which there are about 15 remaining in the United States (http://www.hsus.org/animals_in_research/animals_in_research_news/Class_B_Dealers.html). As of the late 1990s, researchers obtained about half their dogs from commercial suppliers and the other half from shelters. Increasingly, researchers are finding that so-called purpose-bred animals obtained from Class A breeders are scientifically preferable to using random source animals. However, Class B breeders also are used as a source of dogs for research by some research facilities, despite the fact that they often sell dogs that are lost, strays, or have been stolen or obtained from auctions, flea markets, or pound seizures (http://www.hsus.org/animals_in_research/species_used_in_research/dog.html).

Public Health Service Policy protects dogs that are used in federally funded research. Pending legislation entitled The Pet Safety and Protection Act (S. 451) (http://www.hsus.org/animals_in_research/animals_in_research_news/pet_safety_and_protection_act.html) would ensure that any dog or cat used by research facilities was obtained legally. It specifically targets Class B breeders.

Dogs are the only animals required to have exercise under USDA standards. USDA standards also require that dogs housed without sensory contact with other dogs must be provided with "positive physical contact with humans at least daily" (http://www.hsus.org/animals_in_research/species_used_in_research/dog.html). Updated information about the use of dogs in research can be found at these websites: http://www.hsus.org/animals_in_research/species_used_in_research/dog.html; http://www.hsus.org/animals_in_research/animals_in_research_news/Class_B_Dealers.html; and http://www.aavs.org/researchDogs.html.

Further Reading

Festing, M. 1977. Bad animals mean bad science. *New Scientist* 73(1035):130–31.

Giannelli, M. A., 1986. The Decline and Fall of Pound Seizure. In *The animals' agenda.* Monroe, CT. July/August. Pages 10–13, 36.

National Association for Biomedical Research. The use of dogs and cats in research and education. NABR Issue Update (1994).Washington, D.C.

Number of Animals Used by Research From the First Reporting Year (FY1973) to the Present. http://www.aavs.org/images/pdf/animalChart2.pdf

Orlans, F. B. 1993. *In the name of science: Issues in responsible animal experimentation.* New York: Oxford University Press.

F. Barbara Orlans and Marc Bekoff

DOLPHINS

See Whales and Dolphins

DOMESTICATION

In the Western world today, animals are divided into three basic groups, the wild, the tame, and the domestic, but these divisions are fluid and more interchangeable than they seem at first. It is difficult to define what is a wild and what is a domestic animal. A wild animal is usually thought of as one that is fearful of humans and runs away if it can. But this fear of humans is in itself a behavioral pattern that has been learned from experience of human predation over countless generations. A wild animal that has no contact with humans has no fear of them and can be quickly exterminated, as was the dodo on Mauritius. This large flightless bird evolved without any predators, so when Portuguese sailors landed on the island for the first time in about 1507 they only had to knock the dodos on the head to get much-needed fresh meat. However, for perhaps the past 150,000 years, humans have become so supremely successful at killing other species that there are rather few wild animals left on Earth that do not attempt to escape from us as the master predator. On the other hand, it is remarkable how many species of wild animals can be tamed, and taming is not a modern phenomenon. It has probably always been a very important and essential part of human behavior and an adjunct to hunting. Young animals whose mothers were killed in the hunt would have been nurtured and reared by people, and it is not only in modern times that wild animals were captured and tamed as symbols of status, as shown by this anecdote recorded by the Greek writer Diodorus Siculus and written in the first century BC (Oldfather, 1979, pp. 2,187). It is about the capture of a python for King Ptolemy's zoo in ancient Egypt in the middle of the third century BC:

> Observing the princely generosity of the King in the matter of the rewards he gave, some hunters decided to hazard their lives and to capture one of the huge snakes and bring it alive to Ptolemy at Alexandria. . . . They spied one of the snakes, 30 cubits long, as it loitered near the pools in which the water collects; here it maintained for most of the time its coiled body motionless. . . . and so, since the beast was long and slender and sluggish in nature, hoping that they could master it with nooses and ropes, they approached it the first time, having ready to hand everything which they might need. . . . but the beast, the moment the rope touched its body whirled about and killed two of the men.

> Nevertheless the hunters did not give up. . . . They fashioned a circular thing woven of reeds closely set together, in general shape resembling a fisherman's creed and in size and capacity capable of holding the bulk of the beast. . . . and so soon as it had started out to prey upon the other animals as was its custom, they stopped the opening of its old hole with large stones and earth and digging an underground cavity near its lair they set the woven net in it and placed the mouth of the net opposite the opening. . . . And when it came near the opening which had been stopped up, the whole throng, acting together, raised a mighty din and so it was caught.

> When they had brought the snake to Alexandria they presented it to the king. . . and by depriving the beast of its food they wore down its spirit and little by little tamed it, so that the domestication of it became a thing of wonder. (Bk III, p. 36)

The Process of Domestication

In one sense it can be said that a domestic animal is just one which has lost its fear of humans, like that snake, but true domestication involves much more than this.

The process of domestication is subject to two profound overriding and interlocking influences, the biological and the cultural (Clutton-Brock, 1999a). The biological process of domestication begins when a small number of animals are separated from the wild species and become so tame that they have lost all fear of the humans around them and are said to be habituated. For domestication to follow from taming, the animals have to go through a series of morphological and behavioral changes, which in mammals broadly follow the same pattern in succeeding generations, irrespective of the species. In general what happens is that the characteristics of the juvenile animal are retained into the adult state, a process that is known as neotony. Thus domestication of the wolf, the wild cat, the wild sheep, or the wild boar all led in the initial stage to reduction in size of the skull, skeleton, and brain. This was followed in succeeding generations by an increase in the proportion of fat to muscle in the body, to changes in the coat, in the carriage of the ears and tail, and to loss of the wild temperament.

When a small population of animals that has undergone the first stages of domestication is bred over many years in isolation from the wild population, it may form a founder group which is changed both in response to natural selection under the new regime of the human community and its environment, and by artificial selection for economic, cultural, or aesthetic reasons.

Once a species of animal has become fully domesticated, say the domestic dog, *Canis familiaris,* new breeds are produced by further reproductive isolation. The founders of the new breed contain only a small fraction of the total variation of the parent species, and they become a genetically unique population, which continues to evolve under natural and artificial selection. At any point the process can begin again, and further new breeds can be developed by crossbreeding. A breed can be defined as a group of animals that has been bred by humans to possess uniform characters which are heritable, and distinguish the group from other animals within the same domestic species.

There are many anomalies in the interface between the wild and the domestic. For example, domestic rats, mice, and rabbits can be adored animal companions or laboratory animals that are highly valued for medical research, but their wild counterparts are universally treated as vermin and killed on sight.

The Cultural Process of Domestication

The second fundamental side to the process of domestication is the equally important cultural process, which affects both the human domesticator and the animal domesticate. Domestication begins with ownership. In order to be domesticated, animals have to be incorporated into the social structure of a human community and become objects of ownership, inheritance, purchase, and exchange. The relationship between human and animal is transformed from one of mutual trust, in which the environment and its resources are shared, to total human control and domination.

The process of taming a wild animal, whether it is a wolf or a wild goat, can be seen as changing its culture. The term culture has many meanings, but here it can be defined as a way of life imposed over successive generations on a society of humans or animals by its elders. Where the society includes both humans and animals, then the humans act as the elders.

The animal is removed from where, in the wild, it learns from birth either to hunt or to flee on sight from any potential predator. The tamed animal is brought into a protected place where it has to learn a whole new set of social relationships, as well as new feeding and reproductive strategies and, under domestication, this culture is passed down from generation to generation.

A domestic animal is a cultural artifact of human society, but it also has its own culture, which can develop, say, in a cow, either as part of the society of nomadic pastoralists or as a unit in a factory farm. Domestic animals live in many of the same diverse cultures as humans, and their learned behavior has to be responsive to a great range of different ways of life. In fact, so closely do many domestic animals fit with human cultures that they seem to have lost all links with their wild progenitors. The more social or gregarious in their natural behavioral patterns are these progenitors, the more versatile will be the domesticates, with the dog being the earliest animal to be domesticated (around 14,000 years ago), and an extreme example of an animal whose culture has become humanized.

It is not fully understood why the broad domestication of livestock animals, these being sheep, goats, cattle, pigs and equids in the Old World and camelids in South America, occurred progressively from 8,000 years ago, but this was the basis of the so-called Neolithic revolution when the fundamental change in human societies occurred, and groups of hunter-gatherers became farmers and stock-breeders. Archaeologists in the past have hypothesized that there was a natural progression first from generalized or broad-

spectrum hunting in the Paleolithic era, at the end of the last ice age, to specialized hunting and herd following of, for example, reindeer or llama. It was believed that this stage was then followed by control and management of the herds, then to controlled breeding, and finally to artificial selection for favored characteristics. However, the sequence would very rarely have been so smooth, for the social implications of ownership by a social group of hunter-gatherers are a bigger hurdle to domestication than they may seem. Many hunter-gatherer societies that could have domesticated animals never did so, and this was probably for cultural as much as for many other complicated reasons. Why, for example was the bighorn sheep never domesticated in North America?

Tim Ingold has argued that for hunter-gatherer societies there is no conceptual distance between humanity and nature, and the boundary is easily crossed. The animals in the environment of the hunter act with the hunter in mind and present themselves to him. The hunter believes that if he is good to the animals they will be good to him, and if he maltreats them, the animals will desert him. Animals to be hunted are not seen as wild, but as individuals that allow themselves to be taken. The best known survival of this belief is seen among the Ainu of Hokkaido, Japan, who still practice a bear sacrifice in which a bear cub is nurtured for months and then killed in an elaborate and ancient ritual.

In the pre-domestication world, humans and animals lived in mutual trust, but all is changed by the herding of animals and even more so by full domestication. Herdsmen do care for their animals, but it is quite different from the care of the hunter, because equality is lost and domination takes over from trust. By

8,000 years ago, domination of the natural world was already well under way, and by the period of the ancient Egyptians and the capture of the python described above, agriculture and the breeding of livestock were the established foundations of all the ancient civilizations of the Old World. The transformation in attitudes toward the animal world from those of the hunter-gatherer to those of the farmer and stock-breeder was epitomized by Aristotle (384–322 BC) who wrote about more than 500 kinds of animals all of which, he believed, existed for the sake of men (Clutton-Brock, 1999b). This belief that the world exists for the benefit of humans has persisted until the present day, and is imbued in the worldwide sport of hunting. But the wild places and their fauna are shrinking fast and, increasingly in the future, biologists will have to tackle the great problems of their conservation and management. Whether these fauna include African elephants, Asian lions, or giant tortoises, they are all becoming increasingly hedged in. In order to survive, the wild will have to merge with the tame, and as a result of morphological and behavioral changes brought about by human ownership and control, wildlife may even become domesticated.

Further Reading

Clutton-Brock, J. 1999a. *A natural history of domesticated mammals.* 2nd ed. Cambridge: Cambridge University Press/The Natural History Museum.

Clutton-Brock, J. 1999b. Aristotle, the scale of nature, and modern attitudes to animals. In A. Mack (ed.). *Humans and other animals*, 5–24. Columbus: Ohio State University Press.

Ingold, T. 1994. From trust to domination: an alternative history of human-animal relations. In A. Manning & J. Serpell (eds.). *Animals and human society changing Perspectives*, 1–22. London & New York: Routledge.

Oldfather, C. H. (trans.). 1979. *Diodorus siculus.* Cambridge: Harvard University Press, and London: William Heinemann.

Juliet Clutton-Brock

DOMINIONISM

According to one dictionary, the word dominion means "a supremacy in determining and directing the actions of others. . . . the exercise of such supremacy." Dominionism is the West's basic ideology, one that views the world and all of its life forms as God-given property to serve human needs and whims. Dominionism drives science and technology to take ever-increasing power and control over the living world so that some human beings, at least, may have safety, comfort, convenience, longer lives, and other benefits. Dominionism is older than the Judeo-Christian ideology. As farmers, humans stepped up ways to use some plants and animals while they subdued the competition, the plants and animals of the natural world. As farmers, humans learned to take the laws of nature into their own hands. In time, agrarian peoples regarded the living world less as a divinity and more as an enemy. Nature was not to be held in awe; it was to be subdued, outwitted, and controlled. Animals, who had long been regarded as the souls and powers of the mysterious living world, became tools, goods, and pests. With their relegation to inferior status, the much older sense of kinship and continuity with the living world broke up, and the agrarian sense of human supremacy and alienation set in.

See also Anthropocentrism; Evolutionary Continuity; Religion and Animals—Christianity; Religion and Animals—Disensoulment

Further Reading

Collard, Andree, and Contrucci, Joyce. 1989. *Rape of the wild*. Bloomington: Indiana University Press.

Eisler, Riane. 1987. *The chalice and the blade*. San Francisco: Harper and Row.

Shepard, Paul 1991. *Man in the landscape,* 2nd ed. College Station: Texas A&M University Press.

Thomas, Keith. 1983. *Man and the natural world: A history of the modern sensibility*. New York: Pantheon Books.

Jim Mason

DONKEYS

The story of the donkey makes an important contribution to the complex and contradictory history of human and nonhuman animal relationships. Donkeys were one of the earliest animals to be domesticated, and their history with humans is long and close, but it is almost invariably a story of cruel exploitation. It is ironic that the domestic donkey, designed by humans to carry their burdens as cheaply as possible, was relegated to low status and associated with the world's poorest societies. The social construction of donkeys has generally been as unfeeling beasts of burden, ignored, abused, and derided. They have embodied a variety of social, cultural, symbolic, and religious meanings. Donkeys have, in many ways, acted as mirrors to the human condition, standing between us and our sense of ourselves. In *Don Quxiote,* for example, Sancho Panza's beloved donkey is both true companion and humble and steadfast mirror to his master.

The story of the donkey began in Africa and Asia, where they ran free as wild asses before their domestication by humans over 10,000 years ago. As donkeys made the long journey from Africa and Asia to Europe and on to the United States and Australia in the service of humans, their physical journeys were accompanied by their changing fortunes in terms of their treatment by humans. Thus the history of the donkey is irrevocably tied to human history. Archaeologists and anthropologists, for instance, have discovered where and when donkeys were first used by people. They have found that this event marks an important cultural shift from a sedentary lifestyle to a more mobile society that enabled humans to extend their worlds, to travel, and to trade with different cultures. Despite their valuable contribution to human society, however, very little is known about the process of donkeys' domestication or their welfare over time. It is as though they are beneath consideration or interest.

Throughout the world, donkeys have been used for innumerable tasks, mainly as pack animals, during times of peace and of war. By 1000 BCE, donkeys were the main means of transport throughout Egypt and western Asia, as the horse was in the rest of Asia and Europe. Extensive wear on the joints of 5,000-year-old excavated donkey skeletons show that they were used for heavy transport. This was at the dawn of the Egyptian empire, which was built on the backs of donkeys. There were times during their association with humans when donkeys were considered valuable and had a high status. The Egyptians, who exploited donkeys as beasts of burden, for example, were at the same time proud of their large, valuable and graceful white donkeys.

However, donkeys were used for more than practical purposes. They have had religious significance for humans since the start of their domestication. In

Egyptian history, for instance, the donkey is identified with the god Set, a god of the desert depicted with a donkey's head. The worship of the early Christian God was associated with the donkey that Jesus rode into Jerusalem, and there are strong associations with donkeys in the Christian, Jewish and Muslim religions. Despite the connections between donkeys and religion, however, other traditions and customs have had a greater effect on the way donkeys are perceived. Donkeys played an important but often shameful part in the customs of the Middle Ages. Enemies were often placed on donkeys, facing backwards, as a typical form of humiliation. The backwards ride of the criminal to the gallows on a donkey was also used as a form of pre-execution disgrace. Shaming people in this way is evident in many cultures; a recent example was reported from Afghanistan in 1990.

Donkeys came to Europe before the second millennium BCE, most probably to accompany the introduction of viticulture. Their further distribution through Europe took place with the Roman army in the first century BC. The supply trains of the expanding empire consisted in the main of droves of pack donkeys. Later, they were used in agriculture in Roman colonies, and in the new vineyards that the Romans planted as far north as France and Germany. Cruelty to animals in 19th-century Europe was common. Donkeys were abused, starved, and thrashed. They have generally been dismissed as stupid and unfeeling beasts of burden and their very nature—patient, humble, loyal, and accepting—reinforces this perception. Harsh treatment continues in Third World countries today, where donkeys are abused and accorded little care or status.

Donkeys carrying plastic water jugs, on the outskirts of Kandahar, Afghanistan. (AP Photo/ Eugene Hoshiko)

Donkeys arrived in the United States with the Spanish and in Australia with the British. Without the service of these animals, it would have been difficult to colonize these continents. Their hardiness in harsh and inhospitable conditions was invaluable to the pioneers. However, the success of European humans and their animals in colonizing new lands led to many unforeseen consequences. One of those consequences is that some of those animals, the descendants of which are now running wild in vast numbers, having successfully adapted to their environment, are causing problems for the descendants of the humans who brought them there.

When donkeys were no longer considered of any economic value, they are socially constructed as pests, feral, exotic invaders, and even vermin. They are targeted for eradication when they compete with livestock for resources, destroy cultivated or wild environments, and threaten humans economically. Although they were previously shot in the United States, there was an outcry from some sections of the public. In 1952, legislation was passed making it illegal to shoot wild donkeys in Death Valley in California, where the greatest numbers roamed. A sanctuary was also set up for their safety.

In 1971, the US Senate and House of Representatives passed the Wild Free-Roaming Horses and Burro Act, which protected wild horses and donkeys from harassment and death. In fact, wild donkeys have been removed from National Parks, and agencies for and against the donkey still battle in various states; however, it would seem that those who wish to preserve the donkey as an important player in America's history are winning with their "Adopt a wild horse or burro"

scheme, run by the Bureau of Land Management. Despite attempts to revoke the protections afforded by the Act of 1971, it was reaffirmed unanimously in the House of Representatives in May 2006, with the passage of an amendment prohibiting taxpayers' money from being used to sell or slaughter America's wild horses and burros. In Australia, on the other hand, government agencies are intent on the eradication of feral donkeys. There have been public outcries whenever it is reported that brumbies (wild horses) are to be slaughtered, so it would be more hopeful for the remaining wild donkeys if, as in the United States, they were considered equal to wild horses, and recognized as important players in Australia's European history, rather than slaughtered as vermin.

Within government and scientific communities, ethical issues regarding the suffering of donkeys is of secondary consideration to the management of feral animals. Many believe that, as a society, humans must decide the moral standing and significance of nonhuman animals and the duty of care afforded to them. An anthropocentric ethic prevails, where environmental, agricultural, and economic considerations override the value of the individual animal. Questions like these have been asked: Is it morally defensible to assign value to a native animal and death to a non-native animal? Are donkeys, who have served humans for thousands of years, somehow less worthy now that they have become overly abundant when we no longer need them? If we decide that it is ethically defensible to slaughter them, then we must be very sure of our reasons, and ensure that the killing is humane.

Those who are concerned about the mistreatment of donkeys believe that these animals have the right to respectful treatment. The labels assigned to them, such as pest, exotic invader, and feral have no relevance outside human constructions. Many believe that all species and all individual animals, regardless of the value humans may place upon them, positive or negative, have an equal right not to be harmed. Critics of the destruction of donkeys point out that such policies support the premise that humans have the right to destroy elements of nature whenever they choose. Donkeys are caught between a rock and a hard place, literally and philosophically, as they are gunned down from helicopters in isolated rocky outcrops in northern Australia, their bodies left to rot where they fall.

Further Reading

Armstrong, S. and Boltzler, R. eds. 1993. *Environmental ethics: Divergence and convergence.* New York: McGraw-Hill.

Beja-Pereira, A., England, P. R., Ferrand, N., Jordan, S., Bakhiet, A., Abdalla, M., et al. 2004. African origins of the domestic donkey. *Science* 304:1781.

Brookshier, Frank. 1974. *The burro.* Norman: University of Oklahoma Press.

Crosby, A. 1986. *Ecological imperialism: The biological expansion of Europe 900–1900.* Cambridge: Cambridge University Press.

Dent, A. 1972. *Donkey: The story of the ass from east to west.* London: Harrap.

Low, T. 1999. *Feral future: The untold story of Australia's exotic invaders.* Melbourne: Penguin.

Marshall, F. 2000. The origins and spread of domestic animals in East Africa. In *The origins and development of African livestock: Archaeology, genetics, linguistics and ethnography,* edited by Blench, R. M. and MacDonald, K. C., 191–221. London: UCL Press.

Shelton, J. 2004. Killing animals that don't fit in: Moral dimensions of habitat restoration. *Between the Species.* http://cla.calpoly.edu/bts/issue_04/04shelton.htm.

Tobias, M., and Morrison, J. 2006. *Donkey: The mystique of equus asinus.* San Francisco, Tulsa: Council Oak Books.

Jill Bough

DRAIZE TEST

See Toxicity Testing and Animals

E

ECOFEMINISM AND ANIMAL RIGHTS

Ecofeminism, or ecological feminism, is the view that there are important connections between what is characterized as oppression of women and the domination of nature. These connections may be historical (causal), experiential (empirical), symbolic (literary and religious), theoretical (conceptual, epistemological, and ethical), political, and/or practical. One connection, for example, is that Western culture inherits a belief system based on mastery; this in turn supports racism, sexism, and exploitation of animals. Ecofeminists are concerned about broad questions of the ethical and epistemological quality of relationships between humans and other animals, as well as the connections between animal oppression and the oppression of women, people of color, and the natural world. While not all ecofeminists agree about how connections may be drawn, all agree that any feminist theory or environmental ethic that fails to recognize some connection is inadequate.

Some critics of ecofeminism object to drawing connections between women, animals, and the rest of nature. This, they say, appears to essentialize women, ignoring important differences among and between women. Critics also are wary of views that make women

appear inherently closer than men to nature and animals, believing that this connection makes women inferior and less valuable. Ecofeminists reject these dualisms and argue that being close to animals is a problem only if animals are seen as less than human.

A particular strength of ecofeminist writing is its critique of the way that animals are often excluded from environmental ethics and politics, which often focuses on generalized protection of nature. Many environmentalists justify killing individual animals or eradicating entire populations of invasive species in order to protect ecosystems. Some ecofeminists argue that individual animals matter, whether they are domestic animals or nonnative species, and that animal interests should not be ignored in order to promote broader environmental values.

Many ecofeminists have begun to develop theories and practices linking ecofeminism to animal defense. Part of this work involves highlighting parallels between the specific ways that women and animals are oppressed. Two examples are the ways that female farm animals have their reproductive labor exploited, and the way U.S. doctors encourage menopausal women to use the drug Premarin, produced through large-scale exploitation of pregnant horses. Ecofeminist animal defense theories draw on traditional animal defense theories, such as the rights approach of Tom Regan and the utilitarian approach of Peter Singer, and emphasize the importance of animal suffering. However, ecofeminists are cautious about using traditional concepts of rights, which are based on a dualistic notion of rationality, understood as distinct from emotions. Accordingly, some ecofeminists stress the need to develop an ethic of care towards other animals that emphasizes attentiveness to animals' interests, feelings, pain, and ability to flourish.

Further Reading

Adams, Carol J. 1990. *The sexual politics of meat: A feminist-vegetarian critical theory.* New York: Continuum.

Adams, Carol J. 1994. *Neither man nor beast: Feminism and the defense of animals.* New York: Continuum.

Adams, Carol J., and Donovan, J. (eds.) 2007. *The feminist care tradition in animal ethics.* New York: Columbia.

Birke, Lynda. 1994. *Feminism, animals, and science: The naming of the shrew.* Philadelphia: Open University Press.

Gaard, Greta (ed.). 1993. *Ecofeminism: Women, animals, nature.* Philadelphia: Temple University Press.

Kheel, Marti. 2008. *Nature ethics: An ecofeminist perspective.* Lanham, MD: Rowman and Littlefield.

Plumwood, Val. 1993. *Feminism and the mastery of nature.* London: Routledge.

Sturgeon, Noel. 1997. *Ecofeminist natures: Race, gender, feminist theory and political action.* London and New York: Routledge.

Lori Gruen and Lynda Birke

ECOLOGICAL INCLUSION: UNITY AMONG ANIMALS

The Concept of Ecological Inclusion

Ecological inclusion as a concept is an evaluative process to better review exploitative practices, to gauge the degree of disconnectedness that results from those practices, spatially, temporally and contextually, and to formulate moral, ethical, and practical responses to how exclusion may be overcome or, if not, largely minimized. The overriding objective of ecological inclusion is to establish the foundations for a new interrelationship between animals and humans that is more respectful and caring on the part of humans, and one that may help alleviate some of the ecological problems that result from practices and policies of exclusion.

Ecological inclusion is both a simple prescription and an alternative worldview that all humans, individually and/or collectively, can potentially apply to all interrelationships with nonhuman animals. It allows humans to apply appropriate inclusive solutions to various exclusionary interrelationships in an effort to eliminate, or at least alleviate, exploitation. Solutions may need to be justified on certain ecological criteria and by looking at the total environment, as well as on moral and ethical grounds, such as offered by animal welfare or rights theory.

Furthermore, the practical application of the concept of ecological inclusion will always be dependent upon place, time, and context and, therefore, solutions will vary. For example, killing domesticated animals that have escaped and established themselves in ecologically destructive nonendemic wild populations should only occur if it can be justified scientifically, culturally, ethically, and morally. That justification is dependent on the protection of, for example, an endangered species in an area where that species has little chance of survival, and only upon ensuring that the nonhuman animals killed would not suffer in any way. Taking the life of any individual is in reality a denial of their intrinsic value, and denying such value in any individual should not be taken lightly. Making decisions regarding the life of individuals is also potentially reductionist and mechanistic. An inclusive framework seeks to address such conundrums.

Ecological inclusion involves reviewing those exclusionary practices that humans have imposed on nonhuman animals over time with a view to ultimately providing a holistic and less discriminatory worldview. The concept involves gauging the degree of exclusion that results from exploitative practices and formulating ethical and practical responses as to how humans might establish better interrelationships and enhance the greater whole within which all reside.

Given the heavy reliance upon nonhuman life, it is perhaps not surprising that human utilization and dependence upon other forms of life has become so over-exploitive and exclusionary of nonhuman animals. If humans are to become less exploitive and more inclusive of nonhuman animals, then new interrelationships need to be established. A number of fundamental ethical and moral principles and precepts have developed over the centuries which, to varying degrees, have attempted to do just that. That has not been enough; there too many barriers have been erected that prevent humans from better engaging with other life forms. The overall objective of ecological inclusion as a concept, therefore, is towards overcoming those barriers by acting and thinking inclusively, with both individuals and collectively, in all our relationships with nonhuman animals.

The sciences of biology and ecology have allowed humans to develop an understanding that what links human and nonhuman animals is their shared evolution and the overall interconnectivity of planetary life. This understanding is implicit within the concept of ecological inclusion, that is, that the entire planetary ecology is invariably holistic. Some theocentric and atomistic concepts were devised to provide better outcomes for nonhuman animals and for the interrelationships between human and nonhuman animals. However, such concepts more often than not fail to give enough credence to the complexities associated with evolution and ecology, and are either bound to fail, or are not about seeking better outcomes for nonhuman animals at all, but in fact about providing ongoing justification for the *status quo*. Concepts such as stewardship, utilitarianism, rights and duties, based as they are on individualistic foundations, can only ever offer piecemeal solutions to much more complex problems. Nonetheless, they do offer some solutions.

Fundamental Philosophical Principles of Commitment, Respect, and Compassion

Aldo Leopold's land ethic, Albert Schweitzer's reverence for life, and Charles Birch's postmodern ecological worldview offer important insights for a worldview that is more ecologically inclusive. Leopold stressed a deeper, more holistic approach to nature. He extended collective moral considerability to the entire land community. Whether it was Leopold's intent to extend moral considerability beyond the realm of the individual and thereby capture all individual nonhuman animals within that consideration, or to extend it to include all individual members of wild nature, is a point that has led to much debate.

Undoubtedly, Leopold recognized that individual nonhuman animals are active members of the land, a "community of interdependent parts" (1970, p. 239), and stated that humans can be "ethical only in relation to something we can see, feel, understand, love, or otherwise have faith in" (1970, p. 251). Leopold undoubtedly believed that humans should seek to understand and love individual nonhuman animals, as well as have a high regard for their value, in order to be truly ecological.

Whereas Leopold believed that humans must think and act ecologically, Schweitzer extended the notion that it is human nature to revere life. Schweitzer believed that all human actions that affect life must be judged on necessity, that all life forms are morally or ethically considerable and deserve a sense of reciprocity. Schweitzer believed humans need to act in ways conducive to the overall maintenance of life itself; as Schweitzer said, "If I save an insect from a puddle, life has devoted itself to life, and the division of life against itself is ended" (Schweitzer, 1987, p. 313).

In the pursuit of the good, Schweitzer believed that everyone must adhere to principles similar to that of the Jaina principle of ahimsa a total worldview of right thought, word, and deed (1959). Schweitzer's simple prescriptions and basic principles, contained within the Jaina ecologically-imbued ethic and lifestyle, should lie at the center of any human thought or intended action involving nonhuman animals.

Birch's philosophy is built on strong Christian and panentheist precepts (1991, 1993). He maintains that God is both within and independent of nature, and that all human and nonhuman animals are subjects that belong to a community of individual beings. Yet Birch's postmodern ecological worldview should not be seen as limited to a Christian or stewardship foundation; this would be far from Birch's intention. He strongly contends that humans need to bridge the gap between their inner intentions and outer acts, that a "sense of at-one-ment" (1991, p. xvi), or wholeness, between humans and the rest of the universe should be a constant objective. Such a holistic worldview is also implicit in the ideas of Leopold, and in other philosophies such as deep ecology.

What makes Birch's ethic inclusive is his insistence that life, and other components of nature that are not living but critical to life, be afforded respect. It is only when humans feel love and compassion for both humanity and nature that they can act in ways and be committed to lifestyles that are truly inclusive. The concept of ecological inclusion supports Birch's view, but widens his idea of inclusiveness to encapsulate Leopold's holism, which incorporates the entire ecological community of "soils, water, plants, and animals, or collectively: the land" (Leopold, 1970, p. 239).

Further, in the spirit of Schweitzer, if humans are committed to protecting or caring for the land beyond the human self, then humans should revere all life. Not to revere all life, or to love and respect it, as Birch suggests, would mean that any action to correct the damaging results of exploitative actions, or to justify the continuance of actions aimed at achieving a greater good, would more than likely fail. In revering all life, humans could then negotiate ways of engaging with nonhuman life and live ecologically-imbued lives, and acknowledge their place within nature, that they are subjects within a community of countless individuals.

Ecological inclusion is thus both a metaphysical and a practical response to the human interrelationship with nonhuman animals and is applicable to all life. If a human has reverence for life, then that human has the potential to commit to an ecologically-imbued lifestyle, a respectful and compassionate interrelationship with all individual life forms, and the whole of the environment. That interrelationship is thereby inclusive of the life of all individuals and the ecological integrity of the whole.

Thinking and Acting Inclusively

Implicit within an ecologically inclusive worldview is the recognition that, no matter what perceptions of nature may be held by any human individual, there is an overarching oneness or unity within nature, and all life forms have an inherent worth or intrinsic value. Such worth or value is an importance that cannot be quantified in any human sense as a degree of usefulness to humanity. Further, all individuals can potentially contribute to the fecundity and well-being of the whole; appositely, their inherent worth or intrinsic value also relates to their potentiality as fecund and/or contributive individuals within the greater whole.

Every individual human and nonhuman animal, whether they be cognizant of it or not, is also in and of themselves ecologically significant to the integrity and maintenance of the whole. Thus, all individuals should have the right to have their inherent worth or intrinsic value upheld, whilst having the ability to pursue their individual ecological and evolutionary paths, as long as that does not impinge

on the biological and ecological integrity of the greater whole.

Yet, human lives and their interrelationships with nonhuman animals are more temporally and spatially complex than any prescription for exclusionary practices can possibly remedy. Humans perceive nature and engage with nonhuman animals in countless ways, from the most caring and respectful of relationships to the most destructive and exclusionary forms of exploitation. Yet, if humans acted in ways that were less exploitative and exclusionary, then survival rates for threatened individuals and habitats would improve.

Further, humans act in ways which equate individual nonhuman animals with their species, as environmentalists and conservation biologists do when they stress the need to eradicate the inappropriately-labeled invasive species; such ideas justify continued exploitation and exclusion. As ecological inclusion as an alternative worldview does not support any action that excludes or exploits nonhuman animals, then it should not automatically advocate such reductionist and mechanistic prescriptions as eradicating invasive species. However, problems do occur with some forms that become overly abundant and/or ecologically destructively. In such cases, and after careful consideration of all possible alternatives, eliminating a life form from a certain locale, by death or relocation, might be justified if the survival of a particular life form is under direct and serious threat from the overabundance of another. Other examples of ecosystemic reductionism include whether nonhuman animals should be utilized in invasive medical research or as food.

As human/nonhuman animal interrelationships are so extraordinarily complex, so must the solutions to the most problematic of situations be. The responses that humans undertake to impact upon nature are by implication complex, and are also time and scale dependent. They also need to be based on the context in which problems need resolution. Hence, to be ecologically inclusive, humans must take great care and consideration as to the possible impacts that actions may have over smaller and larger scales, and over time.

Any small action that humans take to enhance human interrelationships with nonhuman animals might result in massive changes to the way humans behave towards nonhuman animals. Such actions could result in a welcome positive reengagement with our nonhuman animal kin, or result in tremendous harm. To achieve or enhance the possibility of positive outcomes, humans need to think and act in certain ways to which they are not accustomed, and that includes thinking and acting in ways that are ecologically inclusive. Humans have a personal and societal obligation to commit to a worldview that is both respectful of and compassionate towards nonhuman animals and the total environment. Also, there must be an underlying commitment by all humans to view all components of nature as morally considerable.

See also Anthropocentrism

Further Reading
Birch, L.C. 1991. *On purpose.* Repr. ed. Sydney: New South Wales University Press Ltd.

Birch, L.C. 1993. *Regaining compassion for humanity and nature.* Sydney: New South Wales University Press.

Leopold, A. 1970. *A sand county almanac, with essays on conservation from Round river.* First Ballantine Books edition. New York: Random House.

Schweitzer, A. 1959. *The animal world of Albert Schweitzer: Jungle insights into reverence for life.* Translated and edited, with an introduction by C.R. Joy. Beacon Paperback Number Seventy. Second Printing. Boston: Beacon Press.

Schweitzer, A. 1987. *The philosophy of civilization.* Originally published in English by the Macmillan Company in 1949. Translated by C.T. Campion. New York: Prometheus Books.

Rod Bennison

ELEPHANTS

See Conservation Ethics, Elephants

EMBRYO RESEARCH

The study of nonhuman animal embryos has provided a wealth of information about normal embryonic development. A variety of questions has been asked concerning how sperm fertilize eggs, how early embryonic nervous systems develop, and how arms and legs develop. This basic research has important clinical relevance. For example, research on fertilization in sea urchins and mice has provided the data needed to develop methods for *in vitro* fertilization in humans. This technique is used by many infertile couples to allow them to have children. Studies of the development of the nervous system in frogs have permitted researchers to identify the processes involved in a serious human birth defect, spina bifida, in which the spinal cord does not form normally. Limb development is another developmental process that has been extensively studied in nonhuman animal models. Basic research on chicken embryos first identified the importance of retinoic acid in limb formation. These studies made it clear that drugs containing forms of retinoic acid, often used in formulations designed to treat acne and wrinkling of the skin, are potentially dangerous to the unborn fetus.

As mentioned above, a wide range of organisms is used in embryological studies, ranging from invertebrates such as sea urchins and fruit flies to vertebrates including frogs, chickens, mice, and primates. The choice of animal model for a particular embryological question depends on several factors. For example, fruit flies are an excellent model for examining how genes control the formation of the basic body plan, and for asking questions such as where the head will be and where the dorsal and ventral will be located. On the other hand, sea urchins have been widely used for studies of fertilization, because in them the processes are easily visualized. Later studies were then able to confirm that many of the same mechanisms are used in mammals, and identify the specific processes that are different. The advantage of using invertebrates such as fruit flies and sea urchins is that they are available in large numbers at low cost, they are small in size, and relatively easy to house in a laboratory. On the other hand, the disadvantage is that the relevance of the mechanisms used in invertebrate embryonic development to those used in humans is not always immediately clear. The use of vertebrates, and particularly mammals such as mice and primates, has the advantage that the results are likely to be more directly relevant to human development. However, smaller numbers of embryos are typically available, they are larger in size, and cost more to maintain. As a result, research is often first carried out in animals that are less closely related to humans. Once mechanisms are understood there, then more targeted research can be carried out on vertebrates and, finally, mammals.

The ethics of using nonhuman animal embryos in research has not been widely discussed. This is most likely because the vast majority of embryonic research takes place in the newly fertilized egg and early embryo, because most of the major organ systems are developed very early in embryonic development. Therefore, the stages studied most often occur before the nervous system is functional, so that neither pain nor consciousness are an issue. In contrast, the question of whether human embryos should ever be used in research has generated a great deal of controversy. However, even here, most people agree that prior to neural tube closure, even human embryos are "too rudimentary to have interests or rights and thus cannot be harmed when used in research" (Robertson, 1995).

Further Reading

Robertson, J.A. 1995. Symbolic issues in embryo research. *Hastings Center Report* 25: 37–38.

What research? Which embryos? 1995. *Hastings Center Report.* 25: 36–46.

Anne C. Bekoff

EMPATHY WITH ANIMALS

Empathy is a term used to describe the tendency that most people have to be emotionally affected by witnessing the emotions, for example, suffering or distress, of another person. On the whole, the more empathetic we are, the more likely we are to show compassion and concern, and to offer help to someone in distress.

Psychologists studying empathy have long assumed that people who are strongly emotionally affected by the distress of another human being will also be strongly emotionally affected by the distress of a nonhuman animal, and this has recently received some confirmation from researchers. A study which showed participants video clips of humans, primates, other mammals, and birds in victimized circumstances found that those who were highly empathetic with humans also showed more empathetic responses to the suffering of animals, and a questionnaire survey found positive correlations between people's self-reported empathy with humans and with animals. However, this association was not as strong as might have been expected. There were still plenty of people who showed high empathy with humans but low empathy with animals, and others who were very concerned about animals but showed no greater concern than average about people. So although there does appear to be some association, feeling empathy with or compassion for animals seems to be a process that is not entirely the same as feeling empathy with or compassion for people.

From a developmental perspective, there is a traditional belief that children who are brought up to love and care for animals will develop into adults who love and care for people as well. This is exemplified by stories told about public figures such as Florence Nightingale, the famous British nurse, who cared for injured cats and dogs as a child, before graduating to caring for sick and injured humans in later life. The notion seems to be that looking after someone smaller, weaker, and more dependent than oneself during childhood will instill an enhanced sense of empathy or compassion that can later be applied to the weaker and more dependent individuals in human society. However, the mere existence of a few well-known tyrants and mass murderers (for example, Hitler) who were also pet lovers seems to weaken the idea that keeping pet animals alone can lead inevitably to enhanced empathy with humans.

Another popular hypothesis is the idea that people who keep pets will tend to show greater empathetic concern for the welfare of all animals, not just the species kept as companions. Historically, there has certainly been a correlation between the rise of pet keeping and the rise of concern for animal welfare and animal rights. And a number of questionnaire-based studies have supported the idea that an association between pet owning and concern for animal welfare continues to exist today. One survey of university students found that a lower proportion of pet owners than non-owners found the use of animals in biomedical research acceptable. Another found that students who had pets in their childhoods that they considered to have been important to them (mostly the more interactive pets such as cats and dogs), showed significantly more concern about a variety of animal welfare issues. The mechanisms by which these associations arise has not yet been elucidated, but it seems probable that by living with animals in the home setting, we are more likely to classify them as like us, or as members of our in-group, thereby granting them a more humanlike moral status as creatures deserving of our empathy and compassion.

Further Reading
Davis, M. H. 1996. *Empathy: A social psychological approach.* Westview, CO: Westview Press.
O'Malley, I. B. 1933. Florence Nightingale and animals. *British Journal of Nursing.* June, 170.
Paul, E. S. 2000. Empathy with animals and with humans: Are they linked? *Anthrozoös* 13.
Paul, E. S., and Serpell, J. A. 1993. Childhood pet keeping and humane attitudes in young adulthood. *Animal Welfare* 2: 321–337.

Elizabeth S. Paul

ENDANGERED SPECIES ACT

Out of concern for native plants and animals imperiled "as a consequence of economic growth and development untempered by adequate concern and conservation," the 93rd Congress of the United States created the strongest species protection statute in the world, the Endangered Species Act (ESA). Congress was inspired to action by the nation's brushes with species loss: the disappearance of the passenger pigeon, and the near-extinction of the whooping crane, black-footed ferret, gray wolf, and American alligator. President Richard Nixon signed the ESA into law on December 28, 1973, describing the rich array of animal life as a vital part of the country's natural heritage (Rosmarino, 2002).

The ESA's explicit purpose is to conserve imperiled species and the ecosystems upon which they depend. To achieve this, the law directs the federal government to classify imperiled species as endangered or threatened, to designate critical habitat for listed species, and to develop recovery plans that actively conserve and restore listed plants and animals. The law requires federal agencies to proactively conserve endangered and threatened species, to avoid jeopardizing them or adversely modifying their critical habitat, and to protect listed species from take (for example, killing and harassment) by private individuals and public agencies (Rosmarino).

The ESA has many precautionary facets, erring on the side of protecting wild flora and fauna in the face of scientific uncertainty. A wide variety of life forms are eligible for protection, including species, subspecies and, for vertebrate species, distinct population segments. The ESA not only protects wildlife on the brink of extinction—endangered species—but also those on the road to becoming endangered—threatened species. Moreover, the law provides for plants and animals to be listed based on the best available science, rather than mandating a higher threshold of scientific certainty. Acting on the best available data makes addressing suspected risks to species the priority, rather than allowing species to languish during largely unachievable quests for perfect knowledge.

Congress authorized citizen enforcement of the ESA in the event that the federal government violated the law or failed to enforce it against nongovernmental violators. This is particularly important when administrations are hostile or indifferent to species protection; citizens can step in, holding the government accountable to the ESA's charge.

The need for a strong ESA is clear, considering the diversity of life in the United States and the threats to this diversity. Scientists have documented 200,000 species existing in the nation, and the actual number may be twice this amount. This diverse tapestry of life derives from the nation's large size and its varied terrain and ecosystems (Stein et al., 2000). The United States includes more biome and ecoregional types than any other nation. Richly varied plant and animal life forms find niches in these diverse habitats. However, rapid development, including massive urban sprawl and fossil fuel extraction, is taking its toll on these life forms, as is climate change, continued widespread livestock grazing, crop agriculture, logging, mining, recreation, and over-allocation of rivers.

The U.S. biodiversity crisis is a microcosm of the global human-caused Sixth

Extinction, with current extinction rates at least 100–1000 times higher than natural rates of extinction. Possible causes of the first five mass extinctions include volcanic eruptions, climate change, and asteroids colliding with the Earth. This time, the extinction crisis is human-caused (Leakey and Lewin, 1995). Extinction rates during some of the previous mass extinctions topped 75 percent. The current extinction rate is the highest it has been in 65 million years (Leakey and Lewin, 1995), and there is a growing international scientific consensus on biological catastrophe resulting from lost biodiversity (Ehrlich and Ehrlich, 1996). In the United States, the ESA can provide a defense against the Sixth Extinction.

Endangered Species Act Implementation

Passed almost unanimously by Congress in 1973, the ESA continues to be very popular. University researchers have documented that 84 percent of the American public supports the current or an even a stronger ESA (Czech and Krausman, 1999). Scientists have also reported that the law is effective. Research shows that 227 species might have gone extinct had it not been for ESA protection (Scott et al., 2006). Only nine species (of more than 1,350 listed) have gone extinct after being listed under the ESA (USWFS, 2009), which means this law is over 99 percent effective in preventing extinction.

Despite the ESA's popularity and efficacy, it has been enshrouded in controversy since the late 1970s. Controversies involve private and public lands. Private property rights groups have continually claimed that the law erodes property rights by restricting actions that harm species on private land. Despite the rhetoric, the ESA has been only lightly applied to private lands (Rosmarino, 2002). However, the law's reach to private land is fundamental to the goal of preventing species extinction. The former General Accounting Office (now the Government Accountability Office) issued a report estimated that 75 percent of listed species find the majority of their habitat on private land, and some 90 percent find a significant portion of their habitat on private land. In addition, the ESA can curtail ecologically destructive activities being permitted on federal land or by federal agencies by requiring that federal agencies not permit or engage in actions that result in jeopardy to listed species or adverse modification of critical habitat (Rosmarino, 2002).

Other critics of the ESA include industry interests, state wildlife agencies, federal agencies, and pro-industry administrators and politicians at all levels. Given the law's capacity to curb ecologically destructive economic activities, there have been sustained efforts to weaken the ESA for more than a decade, and these efforts have reached a crescendo in recent years (Goble et al., 2006).

Industrial interests have long been hostile to the ESA's purpose of tempering economic growth with adequate concern and conservation, claiming that it harms economic growth.

While it has endured many sets of amendments over the past three decades, the ESA has emerged fairly intact (Rosmarino, 2002).

The most focused attack in Congress has been on the critical habitat provisions of the ESA. Given that 85 percent of species listed under the law are at least partially imperiled due to habitat degradation (Wilcove et al., 1998), the ESA's strong safeguards for critical habitat have resulted in significantly increased rates of species recovery. Plants and animals with such designations are twice as likely to be recovering as those without (Taylor et al., 2005), yet most listed species in the United States lack critical habitat designations (*Ibid.*).

Because citizens can and have used the ESA to protect endangered species, federal agencies and federal administrators themselves often seek to weaken the law. The George W. Bush administration added only eight listings per year, all as the result of citizen lawsuits. This is compared to 58 listings per year under President George H. W. Bush and 65 per year under President Bill Clinton (Eilperin, 2008). Meanwhile, as of February 2009, over 300 species awaited listing as candidates and proposed species (USFWS, 2009).

Some species have waited on the candidate list for decades, and delays in protection have led to the extinction of dozens of species (Greenwald et al., 2006). Moreover, in the 2000 book *Precious Heritage,* 6,460 species in the United States were identified as imperiled or vulnerable (Stein et al., 2000), and the majority of these were not even candidates for listing. Ironically, the majority of endangered species in the United States are not protected under the nation's Endangered Species Act.

The bottleneck on listings under George W. Bush was second only to that under the Reagan administration, which was also generally opposed to regulatory protections for endangered species. Administrative hostility to the ESA was evident in the George H.W. Bush Administration as well (Rosmarino, 2002). George H. W. Bush characterized the law as a "sword aimed at the jobs, families,

and communities of entire regions" (Houck, 1993). And while the number of species listed per year was higher in the Clinton Administration than in all other administrations after Jimmy Carter's, most listings still resulted from litigation, and listing delays were common (Greenwald et al., 2006).

Other ways the George W. Bush Administration weakened the ESA administratively include cutting funding for endangered species programs, designating far less critical habitat than biologists have recommended, political appointees overruling biologists' recommendations to list imperiled species, and approving regulations that dramatically reduce the ability of federal wildlife agencies to protect endangered species from federal projects (See, e.g., Winter, 2009).

Citizen Enforcement of the ESA

Citizens have long played an important role in ESA enforcement, including petitioning for species to be listed, for critical habitat designations to be revised, and to sue any party that violates any section of the ESA. An early high-water mark for controversy over the ESA was the Tellico Dam issue in the late 1970s. A citizen lawsuit stopped this dam from being completed on the Little Tennessee River due to its threat to critical habitat of the snail darter. In the 1978 landmark opinion by the U.S. Supreme Court in *Tennessee Valley Authority v. Hiram Hill* (437 U.S. 153 (1978)), the majority of the justices found that, given the incalculable worth of endangered species, it would be inappropriate for the court to weigh the economic costs of protection against the value of protecting a species. Consequently, the Supreme Court affirmed the injunction against the Tellico

Dam, a $100 million dollar project that was 90 percent complete, because its completion would jeopardize the snail darter, a three-inch long fish, in its critical habitat on the Little Tennessee River (Rosmarino, 2002).

As a result, many in Congress were up in arms about the ESA. While the statute itself remained essentially intact, the Tellico Dam was ultimately completed, via an appropriations rider heard for a mere 42 seconds on the House floor. A similar end-run around the ESA and citizen participation unfolded in the 1990s in the context of the threat to the Northern spotted owl from logging in the Pacific Northwest. A nondescript rider was attached to a Senate general appropriations bill, and it briskly passed through Congress and overrode a court injunction that had stopped the logging of old-growth forests on U.S. Forest Service land. This was one of several riders used by the Pacific Northwest delegation to avoid logging prohibitions intended to minimize threats to the owl (Rosmarino, 2002),

More recent examples of citizen enforcement of the ESA include efforts to address the impact of livestock grazing on listed species inhabiting federal lands in the Southwest, such as the Mojave Desert tortoise and the Mexican spotted owl. Citizen groups are also pushing for better river management to address water needs of endangered fish in U.S. rivers, for example, salmonids in the West and the Rio Grande silvery minnow in the Southwest. In the Rocky Mountains, citizens continue to challenge logging and ski resort expansions because of their impact on the Canada lynx and other forest wildlife. In the Intermountain West, citizens have sought the listing of the greater sage grouse, to protect the bird and its

habitat in the Sagebrush Sea. Hotspots of species imperilment include California, Hawaii, and Florida, and active citizens' campaigns are working to protect species, their habitat, and native ecosystems in these states.

The importance of citizen ESA enforcement is underscored by the fact that federal agencies are not just allowing land uses and actions by private parties that jeopardize imperiled species and harm their habitat, they are also themselves committing actions that are harmful to species on the brink. For instance, a division within the U.S. Department of Agriculture, misleadingly named Wildlife Services, kills millions of animals every year, both wild and feral. In 2007, this agency killed 2.2 million animals, including gray and Mexican wolves, which are listed under the ESA (WildEarth Guardians, 2009).

The ESA in Perspective

The ESA was visionary when Congress passed it almost unanimously 41 years ago, and it remains at the vanguard today. The law's architects and supporters argued for a strong biodiversity statute based on moral, ecological, and utilitarian reasons, and from the perspective that imperiled species represent unwilling canaries in a coal mine. Most of Congress in 1973 agreed that we ignore the onward march of species extinction at our own peril (Rosmarino, 2002).

That warning still rings true. Two-time Pulitzer Prize winner E. O. Wilson argued in *The Future of Life* that we are literally mortgaging the Earth by continuing down the path of unsustainable economics. Rather than merely living off the interest that the Earth's natural capital provides, we are drawing down the capital, and our bank account will soon be empty (Wilson, 2002).

Continuing on this path ensures both economic and ecological collapse. Economists estimate that intact natural systems provide us with $33 trillion annually in ecosystem services (Costanza et al., 1997). Whether it is the maintenance of the atmosphere, the creation of clean air, recycling of rainfall by forests, pollination by insects and animals, or myriad other functions, these are the processes of nature that make the Earth habitable to humans.

Yet estimates of the monetary value of a living planet are likely to be gross underestimates. We generally cannot replace ecosystems once they are in tatters (Ehrlich and Wilson, 1991). Monetary measurements also do not address the intangible aesthetic, spiritual, and moral rationales that are important components of support for endangered species protection. The ESA honors these widespread attitudes.

As John Muir put it, "When we try to pick out anything by itself, we find it hitched to everything else in the universe" (Muir, 1911). By requiring caution when economic growth and human activities overstep nature's bounds, the ESA protects the diverse plants and animals in the United States and the ecosystems of which they are a part, and can guide us to a more sustainable future.

Further Reading

Costanza, R., d'Arge, R., de Groot, R., Farber, S., Grasso, M., Hannon, B. et al. 1997. The value of the world's ecosystem services and natural capital. *Nature* 387:253–260.

Czech, Brian, and Krausman, Paul R. 1999. Public opinion on endangered species conservation and policy. *Society and Natural Resources* 12(5): 469–479.

Ehrlich, Paul R., and Wilson, E. O. 1991. Biodiversity studies: Science and policy. *Science* 253:758–62.

Ehrlich, Paul R., and Ehrlich, A. H. 1996. *The betrayal of science and reason: How anti-environmental rhetoric threatens our future.* Washington, DC: Island Press.

Eilperin, Juliet. 2008. Since '01 guarding species is harder: Endangered listings drop under Bush. *Washington Post.* March 23, 2008. P. A1.

Goble, Dale D., Scott, J. Michael, and Davis, Frank W., eds. 2006. *The Endangered Species Act at thirty.* Volume I. Washington, DC: Island Press.

Greenwald, D. Noah, Suckling, Kieran F., and Taylor, Martin. 2006. The listing record. In Goble, Dale D., J. Michael Scott, and Frank W. Davis, eds. *The Endangered Species Act at thirty.* Volume I, Chapter 5, 51–67. Washington, DC: Island Press.

Houck, Oliver A. 1993. The Endangered Species Act and its implementation by the US departments of Interior and Commerce. *University of Colorado Law Review* 64(2) 277.

Leakey, Richard, and Lewin, Roger. 1995. *The Sixth extinction: Patterns of life and the future of humankind.* New York: Doubleday.

Muir, John. 1911. My first summer in the Sierra. in *The Wilderness Journeys* (1996), 91. Edinburgh: Canongate Classics.

Rosmarino, Nicole J. 2002. Endangered Species Act: Controversies, science, values, and the law. Ph.D. Dissertation, University of Colorado at Boulder.

Scott, J. Michael, Goble, Dale D., Svancara, Leona K., and Pidgorna, Anna. 2006. By the numbers. In Goble, Dale D., Scott, J. Michael, and Davis, Frank W.. eds. *The Endangered Species Act at thirty.* Volume I, Chapter 2, 16–35. Washington, DC: Island Press.

Stanford Environmental Law Society. 2000. *The Endangered Species Act handbook.* Stanford: Stanford University Press.

Stein, B. A., Kutner, L. S., and Adams J. S., eds. 2000. *Precious heritage: The status of biodiversity in the United States.* New York: Oxford University Press.

Taylor, M., Suckling, K., and Rachlinski, J. J. 2005. The effectiveness of the Endangered Species Act: A quantitative analysis. *BioScience* 55(4): 360–367.

U.S. Fish and Wildlife Service. 2009. Threatened & Endangered Species System. Online database, http://ecos.fws.gov/tess_public. Visited 2/1/2009.

Wilcove, David S., Rothstein, David, Dubow, Jason, Phillips, Ali and Losos, Elizabeth. 1998. Quantifying threats to imperiled species in the United States. *BioScience* 48(8): 607–615.

WildEarth Guardians. 2009. War on wildlife: The U.S. Department of Agriculture's "Wildlife Services." Report issued in 2009.

Wilson, E. O. 1992. *The diversity of life.* Cambridge, MA: Harvard University Press.

Wilson, E. O. 2002. *The future of life.* New York: Alfred E. Knopf.

Winter, Allison. 2009. Endangered species: Bush admin's ESA changes face all-fronts assault. E&E Publishing, LLC. Article dated January 15, 2009.

Nicole Rosmarino

ENDANGERED SPECIES AND ETHICAL PERSPECTIVES

Few persons doubt that humans have obligations to endangered species. People are helped or hurt by the condition of their environment, which includes a wealth of wild species, many of which are currently under threat of extinction. Whether humans have duties directly to endangered species is a deeper question, part of the larger issue of biodiversity conservation, but many believe so. The United Nations World Charter for Nature states that, "Every form of life is unique, warranting respect regardless of its worth to man." The Biodiversity Convention affirms "the intrinsic value of biological diversity." Both are signed by over a hundred nations.

Many endangered species have no resource value, nor are they particularly important for the usual humanistic reasons: medical, industrial, agricultural resources, scientific study, recreation, ecosystem stability, and so on. Many environmental ethicists believe that species are good in their own right, whether or not they are good for anything. The duties-to-persons-only line of argument leaves deeper reasons untouched.

Questions are at two levels: (1) facts (a scientific issue, about species), and (2) values (an ethical issue, involving duties). Sometimes species can seem questionable, since some biologists regularly change their classifications as they attempt to understand and classify nature's complexity. From a more realist perspective, a biological species is a living historical form, an ongoing lineage expressed in organisms and encoded in the flow of genes. In this sense, species are objectively there—found, not made up.

Responsibility to species differs from that to individuals, although species are always exemplified in individuals. When an individual dies, another replaces it. As it tracks its environment, the species is conserved and modified. Extinction shuts down the generative processes, as a kind of superkilling. This kills forms (*species*) beyond individuals, and kills collectively, not just distributively. To kill a particular animal is to stop a life of a few years or decades, while other lives of such kind continue unabated; to superkill a particular species is to shut down a story of many millennia, and leave no future possibilities.

A species lacks moral agency, reflective self-awareness, sentience, or organic individuality. An ethic that features humans or sentient animals may hold that specific-level processes cannot count morally. But each ongoing species defends a form of life, and these forms are, on the whole, good.

The wrong that humans are doing, or allowing to happen through carelessness, is shutting down the life stream, in the most destructive event possible. One argument is that humans ought not play

Sumatran tigers, unique to the Indonesian island Sumatra, are smaller than Indian tigers. Because some forms of Asian medicine prize tiger body parts, the species, despite being endangered, continues to be hunted. (Photos.com)

the role of murderers or superkillers. The duty to species can be overridden, for example, by pests or disease organisms. Increasingly, humans have a vital role in whether these species continue. The duties that such power generates no longer attach simply to individuals, but are duties to the species lines, kept in ecosystems, because these are the more fundamental living systems, the wholes of which individual organisms are the essential parts. In this view, the appropriate survival unit is the appropriate level of moral concern.

It might seem that for humans to terminate species now and again is quite natural. Species go extinct all the time. But there are important theoretical and practical differences between natural and anthropogenic (human-generated) extinctions. In natural extinction, a species dies when it has become unfit for its habitat, and other species appear in its place; this is a normal turnover. By contrast, artificial extinction shuts down speciation. One opens doors, the other closes them. Humans generate and regenerate nothing in this extinction; they dead-end these lines. Relevant differences make the two as morally distinct as death by natural causes and murder.

Humans appear late in the scale of evolutionary time. Even more suddenly, they have increased the extinction rate dramatically. What is wrong with such conduct is the maelstrom of killing and the insensitivity to forms of life that it creates. What may be required is not just prudent preservation of resources, but principled responsibility to the Earth.

Further Reading

Cafaro, Philip J., and Primack, Richard B. Ethical issues in biodiversity protection. In *Encyclopedia of biodiversity,* Vol. 2, 593–607. San Diego: Academic Press, 2001.

Norton Bryan G. ed., 1986. *The preservation of species*. Princeton, NJ: Princeton University Press.

Rolston, Holmes, 1988. *Environmental ethics*. Philadelphia: Temple University Press.

Rolston, Holmes. 1994. *Conserving natural value*. New York: Columbia University Press.

Wilson, Edward O. 2002. *The future of life*. New York: Alfred A. Knopf.

Holmes Rolston, III

ENRICHMENT AND WELL-BEING FOR ZOO ANIMALS

Environmental enrichment may be defined as the actions taken to enhance the wellbeing of captive animals by identifying and providing key environmental stimuli. Well-being is a notoriously slippery concept that is difficult to define and measure, but it generally includes good health and biological functioning, the ability to maintain physiological homeostasis, and—the most difficult to measure—good psychological health.

Early in the 20th century, zoo professionals were the first to express concern about what today would be called *psychological well-being,* noting that animal behavior in zoos often seemed abnormal compared with that observed in the wild. The practice of enrichment to address these problems started in zoos and later spread to more intensively managed captive settings, such as farming and animal laboratories. Largely due to public concern for animal welfare in all these settings, governments began to legislate minimum standards for animals held in captivity, many of which involve enrichment.

In comparison with the wild, captive environments are often unchanging, that is, lacking novelty, spatially limited, stimulus-poor or lacking in complexity, and generally provide the inhabitant with little control over its environment. The result is animals with a great deal of time with nothing to do. Without opportunities to engage in species-typical natural behaviors, many animals show signs of poor wellbeing, such as stereotypy—highly repetitive behaviors, invariant in form, with no obvious function. Pacing is the stereotypy most frequently seen in many mammal species.

Types of Enrichment

Visitors to zoos are likely to see enrichment in action when, for example, they witness cheetahs sprinting after mechanical rabbits, monkeys sifting through piles of straw for food tidbits, or a bear endeavoring to extract peanut butter from crevices in a log. Other forms of enrichment do not necessarily involve food rewards, relying instead on the animal's natural curiosity to explore novel and interesting changes in their environment. Something as simple as a burlap bag stuffed with straw can keep a giant panda entertained for hours. But successful enrichment strategies involve much more than tossing a random mix of interesting items into an animal's enclosure.

Types of enrichment have taken many different forms in the literature. In one prevalent schematic, enrichment may be divided into five categories:

1. Occupational enrichments are those efforts that try to keep the animal busy, for example, encouraging the animal to work for food or providing some sort of exercise equipment.

2. Physical enrichment attempts to improve the quality of the enclosure through permanent changes or temporary introduction of novel objects. The enclosure may be enlarged or made more complex, and climbing structures, water pools, soft substrate, or vegetation may be added. In one of the most highly visible types of enrichment, animals are given novel toys that encourage exploration and play, which in addition to providing psychological benefits to the animal are sure to entertain the zoo-going public.

3. Sensory enrichment can be similar to novel object enrichment, but the

aim is to activate the senses with visual, auditory, olfactory, or other stimuli.

4. In nutritional enrichment, animal caretakers attempt to introduce more natural variation to diet and feeding schedules. Rather than plopping down a bowl of processed, quickly consumed food once a day, they scatter feedings at various times throughout the day and present the food in ways that encourage the animal to use its natural foraging behavior, for example, by hiding it in the crevices of logs and rocks. They may also provide a more varied diet and include food items that are more challenging to consume. Giraffes may be given browse (tree branches) instead of hay, and lions may be given bones or whole carcasses instead of ground meat.

5. Finally, social enrichment can provide endless opportunities for challenge and change, and can meet species needs for social interaction. Overlap among these types of enrichment is inevitable, but these distinctions are useful when devising a well-rounded, holistic enrichment program.

Why Use Enrichment?

Why is enrichment necessary and when is it used? Most often, unfortunately, enrichment is introduced or improved when animals in our care show signs of poor wellbeing. Sometimes poor reproduction or health, or physiological signs of stress, alert zoo animal caretakers to the possibility of poor well-being, but most often it is the readily observable abnormal behaviors such as stereotypies that key us into a developing problem. Stereotypies can take several forms. Reviews of the literature on zoo animals suggest that pacing is the most common, followed by oral stereotypies such as tongue-flicking, and other repetitive movements, for example head bobbing. Although we may never fully understand the subjective experience of another species, the scientific evidence is clear that stereotypies are more often than not associated with poor well-being. Sometimes stereotypies can continue as a scar from past poor environments even after improvements have been made, so stereotypic behavior is not a foolproof measure of an animal's current psychological state or the quality of its environment. Thus, it is recommended that stereotypies alone not be used to infer psychological well-being, though in actuality they often are. One interpretation of stereotypies, with some supporting evidence, is that they are used to cope with suboptimal environments. Thus, the goal of management should not be to prevent the stereotypy itself, but to recreate the environment to meet the animal's needs and obviate its reliance on stereotypy as a coping mechanism.

Documented Benefits of Enrichment

Does enrichment really work or does it just make us humans feel better about keeping animals in captivity? In fact, a great deal of science has shown clearly that animals do benefit from enrichment. Much of this research has taken place outside the arena of zoos, because zoo researchers often cannot achieve the level of experimental control necessary to rigorously test the effects of enrichment, and because they typically avoid research methods that are invasive and potentially harmful to the animals.

Several studies have demonstrated a variety of positive developmental effects on brain function. Animals reared in more enriched environments have heavier brains with more synaptic connections between neurons and enhanced levels of neurotransmitters, all indications of a more effective and efficient brain. These animals are better learners, adapt to change more readily, and show less hormonal evidence of stress. The benefits of stress reduction are significant, because stress can suppress immune system function and reproduction. Thus, enriched animals are less prone to disease and reproduce better. Enriched environments also promote a greater diversity of species-typical behaviors and fewer abnormal behaviors. Literature surveys of published zoo enrichment studies indicate that the typical enrichment program reduces stereotypies by more than half. However, the ultimate goal of completely eliminating stereotypies, once developed, has not yet been fulfilled.

The weight of evidence from these and other studies also suggests a clear role for enrichment in maintaining animals in captivity for conservation purposes, one of the main goals of today's zoos. In addition to creating a better atmosphere for conservation education, enrichment promises to increase successful mating and rearing of offspring, and promote the development of more behaviorally competent candidates for reintroduction to the wild. In fact, enrichment is playing an increasing role in specifically preparing captive-bred endangered species for release back to the wild.

Studying Enrichment and Well-being

How we employ science can facilitate or compromise our goal of discovering the secrets of optimal animal well-being. The answers we get are only as good as the science we use to address the questions. The zoo environment provides exceptional challenges to carrying out good science, but with greater effort zoo research can approximate that found in the more controlled settings of the laboratory. Zoo environments often offer little experimental control. Researchers need to work more closely with animal care personnel to reduce the number of confounding variables that may affect the results. Where possible, husbandry practices should be held constant during the course of an enrichment study. Sample size, essential for legitimate statistical analysis and interpretation of results, is another problem plaguing zoo research. When a zoo has only a few members of a species, researchers may need to collaborate with other zoos to obtain a sufficient sample size.

One frequent failing of zoo enrichment research is the tendency to use the everything-but-the-kitchen-sink approach. Here enrichment practitioners make so many changes to the environment at the same time, with the reasonable hope that at least something will help their animals, that it is impossible to determine which changes had beneficial effects. In these cases, we learn little about the underlying motivation that led to improved well-being, and we are ill equipped to provide guidance to others who need to know which enrichments work best. For scientific purposes, one thing needs to be changed at a time, so that we can draw a conclusion about its effects. However, this does not mean that animals should be kept in barren environments and given one measly enrichment item at a time. Enrichment programs should always be holistic, but scientific studies need to measure the effects of simple changes to the program that may or may not provide incremental improvements to well-being.

Another important point to consider when testing enrichment strategies is that the goal should be to understand the motivational factors underlying poor and good attempts to improve well-being. By understanding motivation, we are better able to predict when well-being will suffer elsewhere, and we will be better able to address it with appropriate enrichment. This has consequences for the experimental design used to test enrichment efficacy. For example, an enrichment study may find a significant reduction in stereotypic behavior, but this may result from the simple fact that the animals used the enrichment and therefore had less time left over to perform stereotypies. If, after the animal tires of the enrichment, it returns to stereotyping at the same rate, have we really affected its motivation or done anything to enhance its well-being? The goal of enrichment is to reduce the *need* to perform stereotypic behavior by making the environment less aversive. A careful experimental design, such as measuring the rate of stereotypy in the aftermath of an interaction with enrichment, can help us determine whether the motivation to perform stereotypies has been reduced.

The importance of detailed descriptions of the forms of stereotypy, the enrichment, and the behavioral response to enrichment, cannot be overemphasized. Too often researchers omit these details from their publications, leaving the reader guessing. These details are important for a variety of reasons. Significantly, the form of the stereotypy can provide insight into the cause that motivates it; for example, oral stereotypies may be related to a thwarted desire to forage, whereas pacing may be related to the need to express natural ranging behavior. It is also important to describe precisely the enrichments used and even attempt to quantify the properties of enrichment, because it is the properties that determine how animals will use enrichment. For example, with novel object enrichment it is important to know if it is moveable, manipulable, destructible, or stimulating to the senses. Is the object complex, does it respond unpredictably, and does it allow the animal to exercise control? These factors and more will determine whether and how the animal uses the enrichment. Moreover, each enrichment item may evoke a unique suite of behaviors that map on to its properties, underscoring the importance of also recording the behavioral details of how the animal interacts with the enrichment. This information is important not just for academic purposes, but essential to understand why some enrichments work better than others, and how enrichment can be designed to more effectively target the animal's needs. If properly understood, the right combination of enrichments may act together to meet all of an animal's psychological needs.

There are ways to tackle animal welfare problems other than enrichment. Low-stereotyping animals may be bred to produce a population free of stereotypy, but does this address the underlying problem of insufficient well-being, or does it eliminate one of the animal's most important mechanisms for coping with an aversive environment? Similarly, drugs can be used to eliminate stereotypic behavior, obstacles can be placed to prevent the behavior, or animals can be trained not to rely on stereotypy. The advantage that enrichment has over all these alternative methods is that it addresses the root cause of the observable behavioral problem, not the symptoms. In fact, many enrichment advocates argue that enrichment is *the* key concept for maintenance of captive animals, on a par with food, water, and shelter.

See also Stereotypies in Animals; Zoos: Roles; Zoos—Welfare Concerns

Further Reading

Kleiman, D. G., Thompson, K. V., & Baer, C. K. (eds.). 2009. *Wild mammals in captivity*, 2nd ed. Chicago: University of Chicago Press.

Mason, G. J., Cooper, J., & Clarebrough, C. 2001. Frustrations of fur-farmed mink. *Nature*, 410, 35–36.

Mason, G. J., & Rushen, J. (eds.). 2006. *Stereotypic animal behaviour: Fundamentals and applications to welfare*, 2nd ed. Wallingford, UK: CAB International.

Moberg, G. P., & Mench, J. A. 2000. The biology of animal stress: Basic principles and implications for animal welfare. Wallingford, UK: CAB International.

Shepherdson, D. J., Mellen, J. D., & Hutchins, M. 1998. *Second nature: Environmental enrichment for captive animals*. Washington, DC: Smithsonian Institution Press.

Swaisgood, R. R. 2007. Current status and future directions of applied behavioral research for animal welfare and conservation. In R. R. Swaisgood (ed.), *Special Issue: Animal Behaviour, Conservation and Enrichment. Applied Animal Behaviour Science*, 102, 139–162.

Swaisgood, R. R., & Shepherdson, D. J. 2005. Scientific approaches to enrichment and stereotypies in zoo animals: What's been done and where should we go next? *Zoo Biology*, 24, 499–518.

Ronald R. Swaisgood

ENTERTAINMENT AND AMUSEMENT: ANIMALS IN THE PERFORMING ARTS

Defenders of the use of animals in entertainment must contend with one argument beyond all those that arise in the context of other uses of animals: that this use is utterly without necessity or utility. Unlike the use of animals in science and agriculture, where clear benefits to humans, and sometimes even to animals, can be claimed, animals who are made to perform in any genre or venue whatsoever are being used purely to fill leisure time, to provide human beings with amusement and distraction and, some would argue, to implicitly assert human dominion over other species. No human or animal lives are saved or even demonstrably improved by capturing wild animals and teaching them to jump through flaming hoops. No diseases are cured or hungers reduced by having stallions prance in unison or elephants rear up on their hind legs.

Several common rejoinders are made to this charge of inutility and human self-indulgence. First, it is pointed out that the use of animals in entertainment is an age-old and universal practice and, as such, is deeply entwined with cultural values and identities. In recent years, this argument has frequently been offered, for example, in defense of bull-fighting, with the implication that opposition to this sport is tantamount to cultural insensitivity or even disrespect.

There is no doubt that animal performance has a long history, expansive geography, and deep cultural roots. While organized circuses, dating at least as far back as imperial Roman times, are the best-known form of animal performance, more specialized animal acts are found all over the world, ranging from such admired institutions as the Lipizzaner Stallions to the often desperate dancing bears, drumming monkeys, and undulating cobras used in the street performances of organ grinders and snake charmers. A similarly vast range of venues and presentation styles makes animal performance difficult to conceptualize and analyze coherently. From the cosseted biblical animals of Radio City Music Hall's annual Christmas Spectacular to the listless alligators forced to behave threateningly for tourists in Florida's

Everglades; from the white lions and tigers in Siegfried and Roy's glitzy Las Vegas magic show to the inscrutable felines in such bizarre and idiosyncratic acts as the Moscow Cats Theatre, animals appear in every genre and level of performance. Whether this ubiquity in itself is testimony to the value, naturalness, or necessity of animal performance is an open question. Certainly the reform or rejection of such practices, while it might alter some traditional formulations of national identity, could hardly hurt humans more than it will help animals.

A second defense of the use of animals in entertainment is the claim that human beings, especially children, who witness the awe-inspiring prowess or amusing

Elephants from the Ringling Brother's circus walk through the streets of Manhattan on their way to Madison Square Garden in New York City. Elephant trainers use chains and metal-tipped prods called bull hooks to train and control the elephants in this and other circuses. (AP Photo/Seth Wenig)

antics of creatures they had previously dismissed as instinct-driven and uncreative will begin to understand, appreciate, and respect animals in new ways. From this point of view, animal performance is far from being a useless exploitation of animals; rather, it is a long-delayed recognition and most salutary celebration of their extraordinary capacities. This argument is similar but not identical to the familiar defense of zoo as a means of fostering public knowledge of and respect for animals.

The third, most ingenious and therefore most challenging defense of the use of animals in performance is the claim that animals actually enjoy participating in many of the activities we make them do for our entertainment. Far from being victims of our selfish need to amuse ourselves, they are, according to this perspective, benefactors of our expertise as their artistic collaborators. The idea that animals are often willing and eager partners to their human handlers and trainers was given philosophical credibility by the late Vicki Hearne, who made a persuasive case for human-animal collaborations based on her observations of horse- and dog-training practices. Hearne's insights have recently been creatively extended by philosopher of science Donna Haraway into a challenging new conception of human-animal relationship she terms companion species. However, both Hearne and Haraway are careful to restrict their theorizations only to those animal species, mainly dogs and horses, whose biological histories are deeply entwined with ours and whose behaviors undeniably evince an interest in interacting with us.

The most common objection to the use of animals in performance is simply that such use frequently involves cruelty and enjoins unnecessary suffering. In recent years, animal advocacy organizations have uncovered a host of specific abuses in the training, housing, maintenance and care of performing animals, including the use of goads and prods, painful restraints, undersized cages, and poor nutrition. In the case of wild animals, there is growing consensus that the very fact of captivity is cruel, and there is a steady growth of legislation all over the world banning the use of wild animals for entertainment. That performing animals suffer stress will come as no surprise to any human who has ever had to perform, but the boredom that defines the life of performing animals between acts is unimaginable to their human counterparts. Finally, there is the issue of degradation and indignity that arises when one group, in this case, humans, gets to dictate and choreograph the performance behavior of another group. The cigar-smoking apes, tutu-wearing elephants, and boxing kangaroos that have amused generations of spectators have also implicitly reinforced the anthropocentric conviction that nonhuman animals are our inferiors, and as such they are ours to do with as we please.

If practices that degrade animals without physically hurting them are within the purview of animals ethics—because it can be argued that such practices contribute to speciesist thinking and behavior by derogating animals—then inquiry into the ethics of using animals in performance must move beyond those instances in which actual animals are or were present when the work was created. The performing arts frequently bypass the living animal and instead make use of its likeness, either as subject-matter, or imagery or, as often in performance, through costume and physical imitation. The ethical stakes of this kind of representation, that is, the ethics of what we could call virtual, as distinct from actual, animal performance, are very hard to determine, because its capacity for harming or improving the lot of actual animals is mediated through such variable and subjective factors as the artist's intention and skill, and the spectator's response and interpretation. Theorizing the existence of an animal apparatus in performance, Michael Peterson writes:

"How are animals made to *perform*"? Collars reins, bits, whips, good, treadmills are part of this apparatus [. . .] "How are animals made to *mean*?" is a more difficult question, and beyond basic questions of semiosis, this involves abstractions like wildness, nature, freedom, servility, and even "the great chain of being," or the concept of the soul.

Thus the ethics of animal performance must be understood in terms of a vast continuum ranging from such unambiguously pro-animal pieces as Rachel Rosenthal's *The Others,* which featured 40 actual animals and their human companions, to such obviously destructive or sacrificial pieces as Kim Jones's *Rat Piece,* in which three live rats were burnt to death in front of an audience. Between the two extremes one finds everything from the magnificent, soul-stirring horses and other animals of the French company Zingaro, or Martha Clarke's *Endangered Species,* to the mischievous dog performances of Oleg Kulik, to such deeply disturbing and morally essential human-animal tragedies as Peter Shaeffer's *Equus* and Edward Albee's *The Goat.*

Performance intersects most directly with animal rights and animal welfare in the area of pro-animal activist performance. Perhaps the best known examples of this genre are the actions of PETA, the international animal advocacy organization known for its inventive and sometimes outrageous use of costumes, props, and human bodies to get its message across. PETA activists deploy both the revelatory and the confrontational powers of performance: the former to unveil and expose the hidden abuses of animals in zoos, circuses, factory farms, and laboratories, and the latter to awaken people to the consequences of their unquestioning acceptance of many cultural animal practices. For example, in a recent action described on the PETA website, "one of our activists lay on a 'grill' on a busy sidewalk, her skin painted to mimic the charred flesh that some people still happily consume at barbecues." Such performances may seem superficial, but they are in fact brilliantly encapsulated uses of such key elements of performance as embodiment, presence, live interaction in actual space and time, and imaginative engagement with persons, situations, and stories. As such, they can inspire conventional and mainstream artists to address animal issues in their performances.

Debates around the representation of animals in performance of both actual and virtual kinds often center around a phenomenon that has also recently sparked new and important debates in the sciences: the issue of anthropomorphism. While developments in ethology have shaken the longstanding scientific prohibition against ascribing mental states and emotional responses to animals, the arena in which such attributions have always been customary, if not obligatory, that of the arts, has of late evinced growing awareness of the fact that anthropomorphism is often an innocent-seeming foundation for anthropocentrism and speciesism. This awareness parallels the critique of philosophical anthropomorphism that underlies the new interdisciplinary academic field of Animal Studies. A seminal text of that field, John Berger's "Why Look at Animals?" initiated the reappraisal of anthropomorphism in art and culture, characterizing Disney's talking animals as the vanguard of modernity's colonization of actual animals by such monuments to their disappearance as cartoons, toys, pets, and zoo animals. More recently, Steven Baker's writings on the animal figure in contemporary art have drawn attention to a style he calls botched taxidermy, deliberate distortions of animal form that have the effect of exposing and disrupting the sentimentalism or smugness that often lurks behind traditional anthropomorphic representations. Theater groups like Forced Entertainment and individual performance artists like Edwina Ashton and Nina Katchadourian combine botched form with talking animals to explore new, more complex, critical, ironic and, most importantly, non-anthropocentric, modes of anthropomorphism.

The call for an enlightened anthropomorphism to be reflected in and encouraged by the arts comes partly from recent developments in continental philosophy as well as phenomenology. In the last decades of life, the late French philosopher Jacques Derrida applied his signature method of analysis, deconstruction, to the human-animal binary, which he regarded as the tenacious foundation of all hierarchical systems of thought and of the oppressive practices they enabled. The link to performance comes with Derrida's notion of the animal as an interruptive encounter in real time and space, an event, as Derrida says, rather than as the conveniently malleable concept it has always been in Western thought. The idea of animality as encounter resembles the idea of animality as process

found in another abstruse but academically popular theoretical formulation, the Becoming-Animal of French philosophers Gilles Deleuze and Felix Guattari, a neo-Nietzschean and vitalist affirmation of the transformative potential of human and nonhuman animality. Offering a nonmimetic and nonliteral mode of animal performance, the idea of becoming-animal has been widely influential among contemporary artists, critics, and curators. Another recent formulation which holds equally great promise, especially for the performing arts, is phenomenologist Ralph Acampora's notion of corporal compassion, an interspecies ethos based not on sympathy but on symphysis, the felt commonality of bodily being shared by humans and animals.

These philosophical developments could point to a heightened role for performance in the next phase of human-animal relationship. If the renewal of the ancient and shamefully frayed bonds between human and nonhuman animals are to be healed, individuals and cultures will need to employ modes of inquiry and attention that go beyond the rationalist reliance on thought and language. They will have to devise more intuitive, emotional, and embodied explorations of animal life, and in this endeavor the performing arts and performance in general could play a significant role.

Further Reading

Acampora, R. 2006. *Corporal compassion: Animal ethics and philosophy of body*. Pittsburgh: University of Pittsburgh Press.

Baker, S. 2000. *The postmodern animal*. London: Reaktion Books, 2000.

Berger, J. 1991. Why Look at Animals? In *About looking*. New York: Vintage International, 1991.

Chaudhuri, U. 2001. Special Issue of *TDR: The Journal of Performance Studies on Animals and Performance*, T93, 51:01.

Derrida, J. 2008. *The animal that therefore I am.* New York: Fordham University Press.

Haraway, D. 2008. *When species meet.* Minneapolis: Minnesota University Press.

Peterson, M. 2001. From to an ethics of animal acts. In *TDR: A Journal of Performance Studies,* T93, Spring 2001(51): 01.

Read, A. 2000. Special issue of *Performance Research on Animals and Performance,* 5 (2) Summer 2000.

Thompson, N., ed. 2005. *Becoming animal: Contemporary art in the animal kingdom.* Boston: MIT Press.

Una Chaudhuri

ENTERTAINMENT AND AMUSEMENT: CIRCUSES, RODEOS, AND ZOOS

Animals have entertained people for centuries. From the gladiatorial contests of ancient Rome, to modern day circuses, rodeos, and zoos, they have been held captive for the amusement of crowds.

Animal Circuses

There can be few people who are not aware of the controversy caused by the use of animals in circuses. From Australia to Norway, Russia to Peru, animal rights campaigners have highlighted the conditions these animals may endure in the name of entertainment. Increasingly, governments are responding to public demand: several countries, including Austria and Israel, have banned the use of all wild animals in circuses.

To some, a circus is not a circus if it has no animals. The majestic elephants walking across podiums and powerful tigers jumping through hoops may provide the only opportunity many people have to see these creatures. But does a few minutes in the ring justify the hours that elephants—who in the wild may walk up to about 43 miles a day—spend chained by their legs, or tigers—solitary predators—are confined to a cage on the back of a truck?

Some say that an elephant born in the circus knows nothing different and does not need a large herd of other elephants; rather she is happy with her human trainers. The same people argue that big cats, primates, and bears caged on the backs of large trucks are fine as it means they have the comfort of being transported in their own homes.

How comfortable is life for these performers? One study revealed that elephants spent between 72 percent and 96 percent of their time chained, big cats were confined in cages for 75–99 percent of their time, and horses spent up to 98 percent of their time closely tethered in a stable tent (Creamer and Phillips 1998).

Research used frequently by defenders of animal circuses actually supports the view that animals suffer in these conditions. It found that all species examined showed abnormal behavior patterns, indicative of prolonged stress or suffering. For elephants, this behavior occupied up to 25 percent of the animals' time; with bears, prolonged or undirected pacing occupied 30 percent of their time (Kiley-Worthington 1990).

Circuses claim to train animals through reward and repetition, and by having "trust and a personal relationship with the animal" (ECA 2004). Yet, undercover investigations of circuses around the world show that animals are whipped, kicked, and hit with sticks on a daily basis. When famous animal circus trainer Mary Chipperfield was prosecuted for cruelty after being exposed by undercover investigators, the industry rallied to her support. Despite viewing film of a crying young chimpanzee being kicked and thrashed with a stick, and a sick elephant being whipped, another circus director, appearing as a defense witness in court, said he saw nothing wrong with this and would do the same thing himself (ADI 2006).

People often show greater concern for the elephants, lions, and bears than domestic animals. However, horses and dogs are subjected to the same constant transportation, restricted movements and training as their co-performers. As Lord Hattersley (2006) said: "I would be opposed to circuses exploiting performing animals [even] if every dog which ever walked round a ring on its hind legs lived in conditions approved by a joint committee of the RSPCA and Dogs Trust with Saint Francis of Assisi in the chair. Animal acts are demeaning—not to the animals which perform them but to the grown men and women who enjoy the spectacle."

While the ethics of zoos have been the subject of a great deal of discussion, less has been written about circuses from an academic perspective. Moral philosopher Dr. Elisa Aaltola (2008) suggests that "this is possibly because animal circuses are seen to be so blatantly at odds with animal welfare and value that it is not even necessary to point out that they would have negative implications on the way we conceptualize and treat nonhuman animals."

Rodeos

Beginning in 1869 as a skill contest between cowboys, rodeos are billed as "showmanship and hard work," showcasing a contestant's "skill with a rope or his ability to ride a bucking animal" (PRCA 2008).

Described by participants as "man versus beast" (Rodeo Productions 2008), electric shock devices, bucking straps, and spurs are all used to assist the human competitors.

Quintessentially American, rodeos actually take place around the world, including Brazil, Australia, New Zealand, and parts of Asia and Europe.

A range of events make up a rodeo, usually consisting of chasing and catching animals or riding bucking bulls. Calf-roping is considered one of the cruelest: three- to four-month-old calves are lassoed around the neck while running at very high speeds. Sometimes they are pulled over backwards in what rodeos call a "jerk-down," a brutal snapping back of their heads. They are then picked up and slammed to the ground, stunning them while their feet are tied together.

Other events in the rodeo game involve crashing a steer to the ground by twisting his horns and roping a speeding steer in such a way that the 500- to 600-pound animal flips over in the air and smashes to the arena floor (IDA undated).

Most animals used in rodeos—bulls, steers, and calves—are completely domesticated and not naturally aggressive, and those used in bareback riding often need some extra encouragement to buck. Electric prods, spurs and bucking straps are used to irritate animals and put on a good show for the crowds; 6,000 volts into the body certainly makes an animal buck.

Bucking (or flank) straps are leather straps tightened around the lower abdomen, a very sensitive area on a horse's body. For bulls it is tightened across the urethra, adding to the pain (TVT 2005). A veterinary study reported that many horses on whom these straps were used "showed stressed facial expressions," and "the flank strap has to be seen as a cause of suffering and as a potential cause of pain" (TVT 2005). Horses stop bucking once the strap is removed.

Injuries are common, including paralysis from spinal cord injuries, severed tracheas, as well as broken backs and legs (MFA 2003), despite organizers' assurances that great care is taken to provide for the animals' welfare.

Rodeos are now being challenged wherever they exist. In the United States, major companies have pulled their sponsorship of rodeos; in New Zealand, campaigners have prevented individual events taking place; and in Portugal, activists have disrupted attempts to promote rodeos by taking legal action under animal welfare laws.

Zoos

"Ambassadors for their species" is how zoos refer to their animals, as if they had volunteered to be put in cages on display to visitors. Similar euphemisms are used in relation to other animals in entertainment, which to some observers hides the realities in which these animals live. Joan Dunayer (2001) writes, "deceptive language perpetuates speciesism, the failure to accord non-human animals equal consideration and respect."

The managers of zoos maintain that they are about more than entertainment, that they are essential for conservation and education. Yet, Dale Jamieson (2003), Professor of Environmental Studies at New York University notes that "zoos are still more or less random collections of animals kept under largely bad conditions." The vast majority of animals in zoos are not threatened species (only 11% are, according to a UK study) (Casamitjana and Turner 2001). Critics have also pointed out, however, that conservation is not about how many individual animals are in captivity, but about the protection of ecosystems so that species can survive in their natural habitats.

Captive breeding is heralded by zoos as essential for the protection of threatened species but is considered by some conservation scientists to be a diversion from the reasons for a species' decline, giving "a false impression that a species is safe so that destruction of habitat and wild populations can proceed" (Snyder et al 1996).

Redirecting money to protecting natural habitats, instead of confining animals, benefits all species of fauna and flora. Measures to protect giant pandas' habitat, for example, also helps support hundreds of species of mammals, at least 200 birds, dozens of reptiles and over half of the plants known to exist in China (Viegas 2007).

Captivity is increasingly offered as the better alternative to the dangers of the wild. "Some species do absolutely great in zoos—they get great food, they get it every day, they have great veterinary care, says one leading zoo scientist, adding "for some species, the zoo trumps the wild" (Stern 2008). Yet many scientists point out that wild animals are uniquely adapted to their own environment and occupy specialized places in their ecosystems, so are they really better off in a cage? "Who needs the wild when we have zoos?" seems to be the message given here.

In their natural habitats, animals face infinite challenges that opponents of zoos claim cannot be provided by a cage. Zoo enclosures are tiny compared to natural home ranges (those for polar bears are one million times smaller) (Sample 2003).

Many animals spend their time pacing up and down or rocking backwards and forwards, abnormal behaviors indicating the boredom and frustration captivity brings. Those lions who pace in zoos spend 48 percent of their time doing it (Mason and Clubb 2004).

Animals in zoos live longer than in the wild, zoo supporters state. Critics wonder if this is an adequate justification for a caged life, or if it is true.

Forty percent of lion cubs die before one month of age—in the wild only 30 percent of cubs are thought to die before they are six months old and at least a third of those deaths are due to factors that are absent in zoos, such as predation. (Mason and Clubb 2004). Elephants in zoos live on average 15 years—half the age of those in timber camps and less than a quarter of the life expectancy in the wild (Clubb and Mason 2002). It is not only the length of life, but the quality of life that is important, and the latter is increasingly questioned.

Zoos describe themselves as "excellent centers in which to inform people about the natural world and the need for its conservation" (WAZA 2005). One critic, novelist Terence Blacker (2008), replies that "the idea that children are educated by gawping at miserable wild animals is an insult to the intelligence. If anything, all they learn is that it is fine to treat wild animals as a show."

The circuses, rodeos, and zoos of today may seem very far removed from the animal baiting of the Romans, or the traveling "freak shows" of the early 1800s, but are they? Or are they just a continuation of what Professor of English Randy Malamud (1998) calls "spectatorial attractions with a related heritage"?

Defenders of animal use in entertainment believe they should have freedom of choice, whether to watch a bear balancing on a ball in the circus, see a three-month-old calf slammed to the ground by a rodeo contestant, or to stare at a gorilla in a zoo cage. But those who question the use of animals in entertainment wonder why humans' freedom of choice, when it results in the suffering or death of another sentient being, should be any more important than the freedom of the subjects of their entertainment.

Further Reading

Aaltola, E. 2008. *The ethics of animal circuses"* Captive Animals' Protection Society. http://www.captiveanimals.org/news/2008/ethics.html.

ADI. 2006. The Mary Chipperfield trial. Animal Defenders International. www.ad-internatio nal.org/animals_in entertainment/go.php?id=236&si=1&ssi=10.

Blacker, T. 2008. Zoos show us little more than our own cruelty. *The Independent* (London). August 22, 2008.

Captive Animals' Protection Society. Animal Circuses and Zoos. www.captiveanimals.org.

Casamitjana, J. & Turner, D. 2001. Official Zoo Health Check 2000. Born Free Foundation.

Clubb, R. & G. Mason. 2002. A review of the welfare of zoo elephants in Europe. RSPCA (Royal Society for the Prevention of Cruelty to Animals). http://www.rspca.org.uk/serv-let/Satellite?blobcol=urlblob&blobheader =application%2Fpdf&blobkey=id&blobtab le=RSPCABlob&blobwhere=10244737184 78&ssbinary=true.

Creamer, J. & Phillips, T. 1998. The ugliest show on earth: The use of animals in circuses. Animal Defenders International.

Dunayer, J. 2001. *Animal equality: Language and liberation*. Derwood, MD: Ryce Publishing, p1.

ECA. 2004. Animals in circuses" European Circus Association. http://www.european circus.info/ECA/.

Hancocks, D. 2001. *A different nature: The paradoxical world of zoos and their uncertain future*. Berkeley: University of California Press.

Hattersley, R. 2006. Beastly treatment. *The Guardian* (London). January 23, 2006.

IDA. Undated. Rodeo: Facts. In Defense of Animals. www.idausa.org/facts/rodeos.html.

Jamieson, D. 2003. Zoos revisited. In *Morality's progress: Essays on humans, other animals, and the rest of nature,* p. 177. New York: Oxford University Press.

Kiley-Worthington, M. 1990. Animals in circuses and zoos—Chiron's world? Basildon, U.K.: Little Eco-Farms Publishing.

Malamud, R. 1998. *Reading Zoos: Representations of animals and captivity.* New York: New York University Press, p. 85.

Mason, G. & Clubb, R. 2004. Guest Editorial. *International Zoo News.* 51(1): pp. 3–5.

MFA. 2003. An inside look at animal cruelty in Ohio rodeos. Mercy for Animals. http://www.mercyforanimals.org/rodeos.asp.

PRCA. 2008. About Us—History of the PRCA. Professional Rodeo Cowboys Association. www.prorodeo.com/prca.aspx?xu=1.

Rodeo Productions. 2008. Video on Media Page. New Zealand National Rodeo 2008. www.rodeogp.co.nz/mediapage.html.

Sample, I. 2003. Wide roaming animals fare worst in zoo enclosures. The Guardian (London). October 2, 2003.

Snyder, N.F.R., Derrickson, S.R., Beissinger, S.R., Wiley, J.W., Smith, T.B., Toone, W.D., Miller, B. 1996. Limitations of captive breeding in endangered species recovery. *Conservation Biology.* April 1996. 1(2): pp. 338–348.

Stern, A. 2008. Animals fare better in zoos as experts learn more. Reuters. May 30, 2008. http://uk.reuters.com/article/scienceNews/idUKN3044801120080530?rpc=401&feedType=RSS&feedName=scienceNews&rpc=401.

TVT. 2005. Expert opinion regarding rodeo events in the Federal Republic of Germany from a legal, ethological and ethical perspective. Tierärztliche Vereinigung für Tierschutz e.V. (Registered Association of Veterinarians for Animal Protection). April 25, 2005.

Viegas, J. 2007. Panda mating frenzy hits zoo. BBC News. May 4, 2007. http://news.bbc.co.uk/1/hi/sci/tech/6625789.stm.

WAZA. 2005. Building a future for wildlife—The World Zoo and Aquarium Conservation Strategy. World Association of Zoos and Aquariums.

Craig Redmond and Garry Sheen

ENVIRONMENTAL ETHICS

Anthropocentric environmental ethics bases concern for the nonhuman natural environment, including animals, on the benefits it provides humans. It treats only humans as of direct and intrinsic moral concern. Taking care of a pet or a park is done solely because it is useful to humans. Anthropocentrism is often defended by appeals to biblical passages that give humans "dominion over . . . every living thing that moves upon the earth" (Genesis 1:28). In contrast, nonanthropocentric environmental ethics bases protection of the environment on its intrinsic value. It conceives of nonhuman nature as important in ways that surpass its instrumental value to humans.

A sentiocentric environmental ethic holds that sentient creatures—those who can feel and perceive—are morally important in their own right. Some of the best-known defenders of animals accept this environmental ethic, including Peter Singer. Because it is likely that only vertebrate animals—mammals, birds, fish, amphibians, and reptiles—consciously feel and perceive, a sentiocentric environmental ethic treats nonvertebrate nature as solely of instrumental value for sentient creatures. Such an ethic protects trees and ecosystems, for example, not for their own sake, but because they provide habitat and other benefits for sentient creatures.

Sentiocentrism breaks down the boundaries of the traditional human-only moral club and is likely to have radical implications for animal agriculture, animal experimentation, hunting, and other human uses of animals. Nonetheless, from the perspective of a broader environmental ethic, sentiocentrism is but a small modification of the traditional, human-centered ethic. It extends moral concern beyond humans only to our closest cousins, the sentient animals, and it denies direct moral concern for 99 percent of living beings on the planet, as well as species and ecosystems. Sentiocentrists respond that it makes no sense to care directly about trees or ecosystems for their own sake because they don't matter to themselves, and experiencing and pursuing one's own good is what brings into the world the kind of value that we ought to directly morally consider.

Biocentric environmental ethics views all living beings as worthy of direct moral concern. Biocentrists contend that although plants and invertebrate animals do not have preferences, they nonetheless have benefits of their own that we should morally consider. Although a tree does not care if its roots are crushed by a bulldozer, crushed roots are still bad for the tree, and not just for the homeowner who wants its shade, or for the squirrel whose nest is there. Insentient living beings have a welfare of their own that should be part of direct environmental concern. Albert Schweitzer's reverence-for-life ethic is an example of biocentrism.

Ecocentric environmental ethics holds that entire species and ecosystems are morally important in their own right. Ecocentrists reject the idea that only individuals, for example, a particular animal or plant, are appropriate objects of direct moral concern. They believe that whole ecosystems and species are intrinsically valuable, not simply the individuals in them. Aldo Leopold's concern to preserve the integrity, stability, and beauty of the biotic community is an example of an ecocentric ethic. These broader environmental ethics view concern for animals as only a first step toward extending moral concern beyond humans to include the natural, nonhuman environment. This broadening of concern creates conflict. For example, hunters and fishermen can show great ecocentric concern for the perpetuation of species and ecosystems while placing little or no moral value on the lives and welfare of individual animals. Conversely, defenders of sentient animals can have great concern for the well-being of individual animals while placing little or no direct moral value on the protection of plants, the perpetuation of species, or the preservation of ecosystems.

These conflicts are not simply theoretical. Feral goats have been shot to protect rare plants. Conservation of endangered species like the California condor often involves captive breeding programs that harm individuals for the sake of the species. Preservation of ecosystems often calls for the elimination of exotics, as when lake trout introduced into Yellowstone Lake are poisoned to protect the integrity of the ecosystem. Restoration of ecosystems sometimes involves bringing back predators. This not only disrupts the lives of the predators, but puts responsibility for the suffering of their prey in the hands of humans.

Broader environmental ethics and animal ethics may also diverge on the alleviation of animal suffering in the wild. Some defenders of animals say that only human-induced suffering and death are bad things that should be prevented. It is human violation of animal rights that needs to be prevented, not natural suffering and death in the wild. However, if one believes that animal rights are logically analogous to human rights, then humans are responsible for failing to assist an animal in distress, just as we are culpable when we fail to assist a human in distress.

The worry that a consistent commitment to protect the lives and welfare of animals would involve massive human intervention into natural systems has led some to claim that defenders of animals cannot be environmentalists.

Further Reading

Callicott, J. Baird. 1989. Animal liberation: A triangular affair and animal liberation. In *In Defense of the Land Ethic*. Albany: State University of New York Press.

Callicott, J. Baird. 1989. Environmental ethics: Back together again. In *In defense of the land ethic*. Albany: State University of New York Press.

Cowen, Tyler. 2003. Policing nature. *Environmental Ethics* 25 (2).

Hettinger, Ned. 1994. Valuing predation in Rolston's environmental ethics: Bambi lovers versus tree huggers. *Environmental Ethics* 16(1) (Spring): 3–20.

Jamieson, Dale. 1998. Animal liberation is an environmental ethic. *Environmental Values,* 7 (1).

Raterman, Ty. 2008. An environmentalist's lament of predation. *Environmental* Ethics 30 (4).

Rolston, Holmes, III. 1994. *Conserving natural value*. New York: Columbia University Press.

Rolston, Holmes, III. 1998. *Environmental ethics*. Philadelphia: Temple University Press.

Sagoff, Mark. 1984. Animal liberation and environmental ethics: Bad marriage, quick divorce. *Osgoode Hall Law Journal* 22(2) (Summer): 297–307.

Varner, Gary. 1995. Can animal rights activists be environmentalists? In Donald Marietta and Lester Embree (eds.), *Environmental ethics and environmental activism*. Lanham, MD: Rowman and Littlefield.

Ned Hettinger

EQUAL CONSIDERATION

Equal consideration, whether for humans or animals, means in some way giving equal moral weight to the relevantly similar interests of different individuals. By itself this is very vague and abstract, yet it is extremely important. Aristocratic, feudalist, Nazi, and other elitist worldviews have often denied that human beings are subject to any sort of basic moral equality. Moreover, to extend equal consideration, in any reasonable interpretation of this idea, to animals would represent a major departure from common thinking and practice throughout the world.

At an abstract level, equal consideration for animals would rule out a general discounting of animals' interests, an across-the-board devaluing of their interests relative to ours. An example of such devaluing would be the judgment that a monkey's interest in avoiding pain is intrinsically less important than a human's interest in avoiding pain. At a practical level, equal consideration for animals would rule out the routine overriding of animals' interests in the name of human benefit. While equal consideration is in agreement with numerous ethical theories, it is not in agreement, if extended to animals, with any view that sees animals as essentially resources for human use and amusement.

Assuming that humans are entitled to equal consideration, then unequal consideration for animals is justified only if there is some morally relevant difference between humans and animals. Peter Singer has argued that there is no such difference between all humans and all animals, so that denying equal consideration to animals is speciesism.

Among leading philosophical arguments for a crucial moral difference between humans and animals are the following. Contract theories typically argue that only those who have the capacity to form contracts are entitled to full, equal consideration; such theories are often motivated by the belief that morality is constructed by humans primarily for human benefit. A somewhat related view is that only moral agents, that is, those who can have moral obligations, are entitled to equal consideration. In these views, only humans qualify as potential contractors and moral agents. A different approach appeals to social relations: How much moral consideration one is due depends on how closely or distantly moral agents are socially related to one. As bond-forming creatures, we human moral agents are much closer to other humans than to animals. Yet another argument appeals to the comparative value of human and animal lives. Equal consideration would require giving equal moral weight to the relevantly similar interests of humans and animals. According to the argument, a dog's life and a human's life are relevantly similar, that is, equally important to the dog and the human, respectively, so equal consideration implies that a dog's life is as morally valuable as a human's. A final argument appeals to the alleged authority of moral tradition. Because our moral tradition, the only source of moral authority, has always given animals' interests a subordinate place, there is no compelling reason to grant animals equal consideration.

The debate over equal consideration remains open because the issues are complex. Two points deserve mention. First, defenders of equal consideration generally deny that this principle means that human and animal lives are of equal value, but their supporting arguments have been incomplete at best. Second, defenders of unequal consideration for animals need to contend with the so-called problem of marginal cases. Any criterion that supposedly marks a relevant difference between humans and animals, for example, moral agency, will seemingly fail to apply to all humans, with the apparent suggestion that the exceptional humans are not due equal consideration.

Further Reading

Carruthers, Peter. 1992. *The animals issues: Moral theory in practice*. Cambridge: Cambridge University Press.

DeGrazia, David. 1995. *Taking animals seriously: Mental life and moral status*. Cambridge: Cambridge University Press.

Midgley, Mary. 1983. *Animals and why they matter*. Athens: University of Georgia Press.

Regan, Tom. 1983. *The case for animal rights*. Berkeley: University of California Press.

Singer, Peter. 1990. *Animal liberation*. New York: New York Review of Books.

David D. DeGrazia

EUTHANASIA

The major differences between veterinary medicine and human medicine are not biological, but ethical and economic; in no way is that more evident than in decisions and policies regarding euthanasia.

The term euthanasia comes from two Greek words: *eu* (good, well) and *thanatos* (death). Euthanasia is a central concern in human-animal relations, as several million animals are euthanized by people each year in animal shelters, veterinary clinics, and research laboratories. That number reaches the billions when food animals—to whom the word *slaughter* is more often applied than *euthanasia*—are added.

The definition of euthanasia differs in veterinary medicine and human medicine in important ways. In human medicine, the term is restricted to mercy killing, ending the life of a patient where death is a welcome relief from a life that has become too painful or no longer worth living. Not all forms of killing humans deserve the good death label of euthanasia; capital punishment, for example, no matter how painlessly performed, is not euthanasia.

Human euthanasia is controversial for many reasons. Critics of legalized human euthanasia and its close relative, assisted suicide, fear that seriously ill or old people could be coerced into having their lives ended. In that case, death would not be an act of mercy for the person being killed, but one of convenience or economics for the survivors.

Veterinarians are familiar with the euthanasia ideal of mercy for the suffering patient, as well as with the call to end animals' lives for such reasons as convenience and economics. Veterinarians often euthanize patients with serious or incurable diseases, in cases where death really does seem the animal's best option. However, veterinarians may also be called upon to end the lives of animals who are destructive in the home, or are inconvenient, or aggressive, or simply unwanted. Shelter workers are similarly required to end the lives of healthy but unwanted animals. In the middle, between the mercy killing of incurably suffering animals and the destruction of unwanted animals, are those animals who are suffering but not from untreatable conditions; these animals, too, may be put to death if their human decision-makers cannot or will not devote the time and money to their health needs.

How Animals Are Euthanized

Because the reasons for killing animals are so broad, the meaning of the word *euthanasia* in veterinary medicine is similarly broad. What makes euthanasia a good death, when speaking of animals, is not that it is better than continued life, but that the death is caused without pain or distress to the animal. It is method, not motive, that has traditionally defined animal euthanasia.

Human euthanasia comprises both active euthanasia (actions such as drug overdoses that kill patients) and passive euthanasia (withholding or stopping treatments, such as ventilators, that could sustain life). In veterinary medicine, withholding or withdrawing treatment is not typically referred to as euthanasia. Many veterinarians are distressed when animals' human guardians choose to let a suffering pet die slowly of disease when fast, painless, active medical euthanasia is an available option. Thus, passive euthanasia is not part of the veterinary ideal of euthanasia.

Not all methods of killing animals can be considered euthanasia, a truly good death. The American Veterinary Medical Association first published guidelines for animal euthanasia in 1963 and has updated them six times, most recently in 2007. Primary criteria for the evaluation of euthanasia techniques are the physical pain and psychological distress experienced by the animal. Other criteria include the emotional effect on humans who are present; the availability of appropriate drugs; and the compatibility with the subsequent examination or use of the animal's body and tissues. Strangely, the veterinary guidelines only cover methods of euthanasia, not issues of why, when, or whether specific animals should be euthanized. They offer no real guidance for veterinarians on how to advise clients whether or not to euthanize an animal.

The preferred method for euthanizing individual dogs or cats has not changed in

Racehorse Eight Belles is restrained on the track after breaking both front ankles at the 134th Kentucky Derby. She had to be euthanized moments later. (AP Photo/Brian Bohannon)

the 40 years in which the AVMA has published its guidance. Then, as now, a rapid injection of an anesthetic overdose, usually a barbiturate such as pentobarbital, is chosen, because it induces unconsciousness rapidly and painlessly. Only once the animal is peacefully anesthetized does the drug go on to stop the breathing and the heart. Sometimes the veterinarian recommends a tranquilizer several minutes before the anesthetic overdose, making the process even easier for animal and human. Often human caregivers choose to be present during the euthanasia of their companion animal and are relieved to see how a suffering animal can leave the world so peacefully.

The AVMA's guidelines have been updated so many times not because euthanasia of a loved, ill animal has changed, but because other circumstances are more challenging. What is the least painful way to euthanize very large animals such as zoo elephants or stranded whales? What is the best way to euthanize dangerous wildlife? How do we process the dozens of animals a busy animal shelter euthanizes every day? How do we euthanize laboratory rats and mice in a way that minimizes pain and distress while leaving their tissues suitable for study? The AVMA gathers a panel of veterinarians and scientists to review the available science and update recommendations for how to humanely kill these varied animals.

The 2007 edition of the AVMA's guidelines did not update any information relative to animal euthanasia at all. Rather, it clarified that AVMA guidelines were for nonhuman animals only, not for lethal injection of humans (a form of capital punishment in the United States). The AVMA guidelines strongly discourage the use of neuromuscular blocking agents that paralyze respiration, except in some defined emergency situations. Critics of human capital punishment by lethal injection have sought guidance from the AVMA guidelines, arguing that if the risk of pain and distress with these drugs is too great for animal use, it is too great for human use as well. Some forms of lethal injection that have been used on humans would not meet the AVMA's standards if performed on nonhumans.

Making the Euthanasia Decision

One of the hardest decisions for an animal's guardian is when and whether to euthanize an ill or aging animal. How can we know when it is the right time? This author believes there is no such thing as *the* right time, given the range of factors at play.

The euthanasia decision is only partly a medical decision, but it should certainly be made with a veterinarian's input. The veterinarian can do his or her best to provide a medical diagnosis of the animal's condition. But even a diagnosis of an incurable illness does not mean immediate euthanasia is warranted. A combination of good medical and nursing care may keep animals with certain terminal illnesses comfortable for months or years. Conversely, a diagnosis of some treatable injuries and illnesses may still result in the animal's euthanasia. This may be because of the cost of the treatments, since insurance coverage for payment of veterinary bills is not common, the time demands of some treatments, or the significant suffering that an animal would likely go through before starting to feel better.

Veterinarians can help animal guardians predict what the animal will experience with a particular illness. Not all heart diseases, for instance, are equal. Some heart disease may result in sudden death, some in decreased exercise tolerance, some in a distressful inability to breathe comfortably. From the animal's perspective, these are three very different heart conditions. Sudden death is sad, but the animal does not suffer in the months leading up to it. Decreased exercise tolerance means the animal will run and play less, but may be content to limit his or her activities without significant suffering. The inability to breathe comfortably, however, may be severely distressing for weeks or months on end. A veterinarian can help the guardian understand not just whether the condition is treatable, but how much suffering it causes.

As with heart disease, so with other life-limiting illnesses. Some cancers may be excruciatingly painful, while others are barely noticed until they are very advanced. Kidney disease can make animals feel extremely ill, but with dietary management and supplemental fluids, they may remain in relatively good health for several months.

There will be medical uncertainty. Veterinarians can give parameters for how the average case progresses, but not how an individual patient will. People want to know, "Is this animal suffering?" Like human patients, animals have better and worse days. Veterinarians can help the caregiver learn how to recognize the major signs of an animal's quality of life: interest in food, ability to eat and drink, the ability to move about or to sleep comfortably. None are particularly mysterious, but they require careful observation.

Rarely, however, is euthanasia solely a medical decision, which is why the decision rests with the animal's caregiver, not the veterinarian. The caregiver must decide how much time, energy, and money she or he can devote to end-of-life care for an animal. But even given infinite resources, she must assess when she considers the animal's life is somehow no longer worth living. This includes value assessments of how many good and bad days will tip the balance toward euthanasia. Moreover, a person's beliefs about the value of life and the possibility of an afterlife for an animal will affect the course chosen. One person may feel that a half an hour a day of apparent comfort and happiness means that the animal's life is still worth living. Another may believe that half an hour a day of serious sickness or pain makes that life intolerable. Most will believe somewhere in the middle.

Is there an animal equivalent of assisted suicide? It is impossible to know for sure what an animal is thinking, but it is clear that animals sometimes feel far too sick to eat or drink on their own, and that this can lead to their death. Most veterinarians will treat this anorexia as a clinical problem that can be managed and treated, just as fever, infection, and broken bones are treated; most do not treat this as the animal's attempt to end his or her own life.

Grieving

Pet guardians often grieve the euthanasia of a loved animal just as we grieve the death of our loved human friends and family. Social workers and therapists recognize this important response to animal death. They work to help people come to terms with this loss, rather than trivializing or ridiculing it. Some books on the topic are listed at the end of this article. In addition, following the lead of the University of California at Davis's veterinary college, various pet-loss support hotlines have been established, most of them associated with veterinary colleges.

Support during grief for the loss of an animal is important, as many people may find that their friends and family do not really understand. For many people, the love and companionship of their animal is a central part of their life, and the loss is devastating. This can be true for adults as well as children, but it may be ridiculed as immature or inappropriate by people who are less animal-focused.

Grief over the euthanasia of a companion animal is complicated by the animal guardian's knowledge that she or he made the conscious decision to end the animal's life. This decision is rarely easy, and many people will guiltily second-guess their decision in the following days and months. Not only must the decision-maker come to terms with the fact that she made a decision that may later feel wrong, but she must also decide how to discuss this with others, possibly including small children.

Loving pet guardians are not the only people who may feel grief and distress in connection with animal euthanasia. There are also professionals for whom killing animals is part of a day's work: veterinarians, veterinary technicians, research workers, and animal shelter workers. All participate in animal euthanasia, some as part of the decision-making, others powerless to make the decisions but required to perform the euthanasia procedure. Thus, euthanasia training for shelter workers includes not just technical training, but also seminars on dealing with the tragic irony that responsible animal care sometimes includes killing animals.

See also Laboratory Animal Use—Sacrifice

Further Reading

American Veterinary Medical Association. (2005). *How do I know it is time? Pet euthanasia* (brochure) and *How do I know it is time? Equine euthanasia* (brochure). American Veterinary Medical Association: Schaumburg, IL. Available online at http://www.avma.org/animal_health/brochures/default.asp.

American Veterinary Medical Association. (2007). *AVMA guidelines on euthanasia.* American Veterinary Medical Association: Schaumburg, IL. Available online at http://www.avma.org/issues/animal_welfare/euthanasia.pdf.

Association of Veterinarians for Animal Rights. (undated) *Position statement: Euthanasia of nonhuman animals.* Available online at http://www.avar.org/publications_position.asp#p14.

Carmack, B. J. (2003). *Grieving the death of a pet.* Minneapolis, MN: Augsburg Books.

Kay, W. J., Cohen, S. P., Fudin, C. E., Kutscher, A. H., Nieburg, H. A., Grey, R. E., et. al. (Eds.). (1988). *Euthanasia of the companion animal.* Philadelphia: The Charles Press.

Nakaya, S. F. (2005) *Kindred spirit, kindred care: Making health decisions on behalf of our animal companions.* Novato, CA: New World Library.

Larry Carbone

EVOLUTIONARY CONTINUITY

One hundred and fifty years after the publication of Charles Darwin's *On the Origin of Species,* humanity has yet to come to grips with the meaning of evolutionary continuity. Through a plethora of evidence, arguments, and examples, Darwin showed that all organisms are related by common descent. For example, zebras and horses evolved from a common ancestor, as did chimpanzees and humans, and wasps and ants. All six of these animal species also evolved from a common ancestor, only that ancestor existed and became extinct even further back in time. Species emerge like branches growing out of other branches on a single tree, all originating from the same root.

Before the Darwinian revolution, the dominant notion in Western culture was that animals were specially created and organized hierarchically according to a great chain of being. Mammals were positioned at the top of this hierarchy, with humans at the apex; then came birds, reptiles, and amphibians—that is, vertebrates, animals with a backbone like human beings. Invertebrates, which include insects, were placed at the bottom of the hierarchy. Instead of having a skeleton inside their bodies, like we do, insects wear their skeletons on the outside, a protective adaptation, and this is only one of the ways insects are different from us. Other ways they differ is that they are much smaller, they sense the world in totally unfamiliar ways (for example, bees see ultraviolet light), they communicate in ways we find hard to imagine (for example, by using chemicals), and they look almost alien to our eyes, causing many people to be afraid of them.

Invertebrates were positioned at the bottom of the hierarchical ladder for the arbitrary reason that the less an animal resembled human beings, the lower its place. Darwin, however, showed that the reason animals can appear unlike one another is not because they are lower or higher on some imagined scale, but because they have different adaptations. Because of their common descent, all living beings are related and interconnected, varying only in their manifest forms.

Through his discovery of evolution as a process of descent from common ancestors, with new species shaped through encountering novel conditions, Darwin destroyed the human conceit of the great chain of being. In its place he gave us a world in which there are no discontinuous leaps between species; all animals are bound together, and to all other organisms, by the single, very long story of life on Earth. Today we know that life has persisted on Earth for 3.8 billion years. Darwin went to great lengths to demonstrate this unbroken continuity at every level, not only in anatomy and physiology, but also in behavior and mental characteristics. His understanding of evolution has been supported and enriched through countless scientific discoveries since the late 19th century.

Despite the dismantling of the hierarchical chain of being, in our practices and ideas we continue to uphold a radical break between vertebrates and invertebrates. We resist the idea that invertebrates can feel pain, experience suffering, have intelligence, or lead lives that are meaningful to them. We underrate their critical importance in the health of ecosystems, and show little consideration for their intrinsic value as members of the biosphere. All these attitudes, be they conscious or unconscious are, from an evolutionary perspective, unfounded and anthropocentric.

Critically dissecting our attitudes towards all living beings is especially important today, as the Earth is in the midst of an anthropogenic mass extinction. Species, many of them invertebrates, are disappearing at unprecedented rates. By taking the fact of evolutionary continuity seriously, and embracing the oneness of all organisms, we may yet stem the losses of life caused by the destructive forces of human arrogance and ignorance.

See also Anthropocentrism; Dominionism

Further Reading
Bekoff, Marc. 2007. *The emotional lives of animals: A leading scientist explores animal joy, sorrow, and empathy—and why they matter.* Novato, CA: New World Library.
Bekoff, Marc, Allen, Colin, and Burghardt, Gordon (eds.). 2002. *The cognitive animal: Empirical and theoretical perspectives on animal cognition.* Cambridge: The MIT Press.
Crist, Eileen. 2000. *Images of animals: Anthropomorphism and animal mind.* Philadelphia: Temple University Press.
Darwin, Charles. 1964. *On the origin of species,* facsimile of the first edition (1859). Cambridge: Harvard University Press.
Darwin, Charles. 1985. *The expression of the emotions in man and animals* (1872) Chicago: Chicago University Press.
Darwin, Charles. 1985. *The formation of vegetable mould through the action of worms with observations on their habits* (1881) Chicago: Chicago University Press.
Fabre, Jean Henri. 1991. *The insect world of J. Henri Fabre* E. Teale, ed. Boston: Beacon Press.
Goodall, Jane. 2000. *Reason for hope: A spiritual journey.* New York: Warner Books.
Griffin, Donald. 2001. *Animal minds: Beyond cognition to consciousness,* second and revised edition (1992). Chicago: Chicago University Press.
Kellert, Stephen. 1996. *The Value of Life: Biological Diversity and Human Society.* Washington, DC: Island Press.
von Frisch, Karl. 1972. *Bees: Their vision, chemical senses, and language.* Ithaca, NY: Cornell University Press.
Wilson, E. O. 2002. *The future of life.* New York: Alfred A. Knopf.

Eileen Crist

EXOTIC SPECIES

Debates about animal rights have traditionally focused on the exploitation of animals for human food, clothing, transportation, medical research, and entertainment. Recently controversy has also arisen about the extermination of exotic animals to protect not only human interests, but also the interests of other animal or plant species.

For animal and plant species, the term exotic is used interchangeably with the terms nonnative, non-indigenous, alien, foreign, and immigrant. Related but not synonymous terms are *introduced* and *invasive*. Although exotic and the other terms are used widely in scientific, government, and popular publications, precise definitions remain elusive. There is general agreement, however, that the terms designate species whose spread beyond their historical native range has been assisted, either intentionally or unintentionally, by human activities. The term invasive is used particularly to designate species that significantly alter the environment.

Intentional Importation, Unintentional Dispersal

Humans traveling to new areas as colonists have transported animals for food, clothing, and labor. European colonists brought cattle, sheep, pigs, and horses to the Americas and Australia. Some of these animals escaped from human management and reproduced. Animals whose ancestors have a history of domestication, but who live apart from human management, are called feral animals. The wild horses of the American Southwest are both feral and exotic.

Humans have imported non-domesticated animals for economic gain, but then lost control of their movement. Gypsy moths, whose larvae have defoliated large areas of forest in the American East, were introduced from Europe into the Cape Cod area in 1868 in an attempt to promote an American silk industry.

Humans import wild, exotic species as household pets, and sometimes lose

or abandon them. Burmese pythons and Asian walking catfish, now thriving in the Everglades and competing with native species, were imported into Florida as pets.

Intentional Importation and Dispersal Humans have imported and dispersed non-domesticated animals for economic gain. In the late 1930s, Edward McIlhenney imported 13 nutrias from Argentina to Louisiana to establish a fur industry. After being released into a marsh, the nutrias reproduced at such an astounding rate that, within a few decades, their numbers had grown to an estimated 20 million. They have consumed enormous amounts of vegetation needed by native animals, and caused extensive soil erosion.

Human immigrants have introduced exotic species while attempting to recreate familiar environments. In 1890–91, Eugene Scheiffelin released about 100 English starlings in Central Park, New York, as part of a plan to bring to the United States all the bird species mentioned in the works of William Shakespeare. By 1940, starlings were found in California. Their population in North America now numbers about 200 million. Starlings encroach upon the nests of other birds and compete with them for food.

Humans have also introduced one exotic species in an attempt to eradicate or control another exotic species. The mongoose, a native of southeast Asia, was brought to the Hawaiian Islands in 1883 from Jamaica, where they had been imported in 1872, to destroy the population of exotic rats that were eating cultivated sugar cane.

Unintentional Importation and Dispersal Many animal species have been transported through the unintentional agency of humans. A host of exotic animals have hitched rides in the cargo containers of ships, trucks, trains, and airplanes. The rats that arrived in Hawaii were stowaways on boats. The zebra mussel, a thumbnail-sized mollusk native to the Caspian Sea, reached North America in the mid-1980s when the ballast water of a transatlantic freighter was discharged into Lake St. Clair, Michigan. The zebra mussels quickly colonized the Great Lakes and the Mississippi River basin, and their population in some areas may now be as high as 70,000 per square foot.

Human-Assisted Dispersal   All of the above species can be designated as imported, because they were transported by humans to an area in which they did not evolve. Some species, however, can be considered exotic, but not imported. These are species whose dispersal was assisted by disturbances to the environment caused by humans. The coyote evolved in the American Southwest, but now inhabits urban, as well as rural, areas as far east as the Atlantic coast. It has profited from humans' eradication of its competitors, such as the wolf, and from the availability of food which human habitation brings. Similarly, the cattle egret migrated on its own from Africa to South America in the 1870s and, by the mid-1940s, to North America. Its dispersal can be considered human-assisted because it benefited from the alterations that humans made to the American landscape, particularly the dedication of vast areas of the land to raising cattle.

The majority of exotic species do not survive if deprived of human care. Of the few that do thrive (perhaps as few as two percent), some are of little concern to humans. The opossum, a marsupial native to the American Southeast, was imported to the San Francisco Bay area of California, around 1890, to provide a new target for hunters. Because opossums do not harm humans or stress native species, their dispersal throughout California has been tolerated by humans. The cattle egret has also been easily accommodated in its new environment, albeit an environment that has been altered significantly by human activities. The egret has proved useful and therefore welcome to humans, because it feeds on insects attracted to human-managed cattle, another exotic species.

Some exotic species, however, are considered pests because of their impact on human industry, economics, health, safety, and recreation. Another concern is the stress they place on native species.

Industry and Economic and Health and Safety Problems   Rabbits introduced to Australia for hunting multiplied rapidly and began devouring both native plants and the exotic crops planted by farmers. The economic impact has been enormous in terms of crop destruction, loss of forage for livestock, and costs of largely unsuccessful attempts to control the rabbit population through poisons, viruses, warren demolition, and fences. Zebra mussels clog the water-intake pipes of factories and choke agricultural irrigation pipes, increasing the costs of raising human food. Masses of them in American waterways clog water-intake structures and reduce pumping capacity, threatening human water supplies and power generation.

Feral horses in the American Southwest graze on land that ranchers want to reserve for their livestock, and rats transported to Hawaii eat into the profits of plantation owners. Rats also concern humans because they can carry disease.

Recreation Problems Zebra mussels clog the engines of recreational boats. The round goby, a fish which, like the zebra mussel, is native to the Caspian Sea area and was introduced to the Great Lakes in the 1990s by the discharge of the ballast water of a transatlantic ship, is larger and more aggressive than most fish species native to the Great Lakes and has threatened species prized by sport fishermen.

Environmental Problems The few species that do successfully colonize areas new to them succeed because they are resilient, have high reproductive rates, are generalist feeders, that is, they eat a wide variety of foods, and have no predators in the new area, and because their food sources or competitors for food have not yet developed defenses against them. In a relatively short time, they can alter an environment extensively. The voracious Nile perch, imported into Lake Victoria in Africa in the mid-1950s as a food fish, is thought to be responsible for the extinction of about 100 species of native fish. Species of animals and plants that evolved on isolated islands are particularly vulnerable. Mammals introduced to New Zealand by Europeans, and also by Polynesian colonists over 1,000 years ago, caused the extinction of many native, ground-nesting flightless bird species that had evolved on the remote islands. Rabbits in Australia have destroyed several native plant species and caused soil erosion by denuding the land; they also endanger native animal species that cannot compete with them for food, or whose habitat has been destroyed by

their activities. The mongoose, which was introduced to the West Indies and the Hawaiian Islands for rat control, prefers to prey on native species of reptiles, amphibians, and birds.

Categorizing animals as exotic, nonnative, and invasive is a controversial matter. Although one criterion for designation as an exotic species is dispersal beyond one's native range or place of evolution, it is rarely possible to determine spatial or temporal boundaries for any species. Dispersal and colonization of new areas have always been naturally occurring phenomena. It is therefore appropriate to ask how long a species must inhabit an area before it is considered naturalized. Some scientists reserve the terms exotic, alien, and nonnative for species whose dispersal took place in the modern period of European exploration and migration, beginning about 1450. For the Americas, the dividing line is the arrival of Christopher Columbus in 1492. Species inhabiting these continents in the pre-Columbian period are considered native and indigenous; those that arrived after 1492 are nonnative. However, using European migrations as the line of demarcation between native and nonnative would mean that the species brought to the Hawaiian Islands by Polynesians from about 400 AD on should be considered native, a point which many biologists would dispute. Some scholars, therefore, focus on the element of human-facilitated dispersal as a key to distinguishing native from exotic. Human-facilitated dispersal is thought to be unnatural, in the sense that it has moved species much farther and more quickly than they would otherwise have moved, and has moved them across natural boundaries, particularly oceans and mountain ranges, which they would not otherwise have crossed. However,

calling human-assisted dispersal unnatural is problematic, because we consider
the migration of humans to be a natural
human behavior, and when they move,
humans take their biological possessions
with them. The transport of animals, even
across oceans, is thus a natural occurrence.
It is, moreover, difficult to reconcile that
humans, whether Europeans in the Americas or Polynesians in New Zealand, are
considered naturalized, but the biological
items intentionally transported by them
are categorized as alien or exotic. And
categorizing as alien and exotic animals
unintentionally transported by humans
is also problematic. If a rat, attracted by
human-cultivated food on a ship, is transported to a place where its species has not
before been, is this method of dispersal
logically less natural than if the rat, while
scavenging wild food, was carried on a
floating tree limb?

Some scholars argue that the terms
exotic, alien, nonnative, and invasive reveal an anthropocentric bias. They are not
applied to humans who have dispersed
across the planet with their domesticated
animals and crowded out native species. Moreover, nonnative species that
are judged to have a negative impact on
human economic, health, or recreational
interests are targeted for eradication,
while other nonnative species, considered
useful to humans or benign, are not. The
concern is thus not the exotic origin of a
species, but rather its perceived interference in how humans want to use an area
into which they have dispersed.

The use of terms such as alien and
invasive has an influence on the way
people think about these species. Invasive conjures up images of invading
armies. Militaristic metaphors abound
in contexts where humans are describing
species that they believe must be exterminated. In 1999, for example, Agriculture

Secretary Glickman declared an "all-out
battle" against the spread of alien species in the United States. Humans speak
of undertaking assaults on alien species
and waging war against invasive species.
Such metaphors prompt people to conclude that dispersal is a hostile act on the
part of the animals, when in reality the
animals are simply following their natural behavior in their efforts to survive. In
addition, framing the issue as a war then
seems to justify—and even encourage—
the harsh methods of extermination that
are employed. On the Channel Islands
of southern California, species imported
by European ranchers two hundred years
ago are now being killed by guns, traps,
poisons, and fires.

Proponents of such methods argue that
they need to eradicate resilient invasive
species as quickly as possible. Animal
protectionists, however, protest that the
methods are inhumane (even sharpshoot-
ers often leave injured animals to endure
lingering deaths) and indiscriminate,
because poisons, traps, and fires kill
non-targeted species as well. Even if
animal protectionists are persuaded that
eradication is justified, they advocate the
use of nonlethal methods of population
control such as sterilization. They raise
two moral issues: the infliction of pain
and distress, and the termination of life.
They believe that humans have a moral
obligation to refrain from doing harm or
causing death, and that each individual
animal has a right not to be harmed or
killed by a human. Nonetheless, among
people who protest eradication, there is
generally more sympathy for vertebrates
than invertebrates, and for mammals than
for fishes or reptiles.

Supporters of eradication contend
that it is both a natural behavior and a
moral obligation of humans to protect
the economic and health interests—the

very survival—of their own species and, consequently, to destroy creatures whose habits threaten those interests.

In recent decades, another reason for eradication has been advanced: to save other species, both animal and plant, from extinction. European and Asian species now dominate landscapes far from their original point of evolution. Proponents of eradication argue that humans have a moral obligation to preserve biodiversity; it is humans who are responsible for transporting exotic species across oceans and mountains, and who have the intellectual capacity to recognize the consequences of their actions. The moral intuition that there is value in biodiversity and, correspondingly, in landscapes that have not been altered by human activities, is a recent phenomenon, and it conflicts with the values of earlier generations of humans. Throughout their history as agriculturalists, humans have promoted the development of monocultures, that is, cultivated areas devoted to the production of one crop, such as wheat, rice, or cattle. In our efforts to alter the environment to suit our purposes, we have eliminated other species and considered that a landscape had value only if it served our needs.

Advocates of biodiversity, however, argue that species and landscapes have an intrinsic value that is independent of human needs. They maintain that exotic species degrade or harm the environment. Again, it is important to analyze the rhetoric of the statements. *Degrade* and *harm*, like *invasive*, are pejorative terms, intended to influence the way we think about a species. In truth, exotic species do not degrade or harm an environment; they change or alter it (more neutral words). If they cause the extinction of other species, the extinction is a permanent change, but the surviving organisms and relationships continue to evolve. Even staunch

conservationists now recognize that ecosystems are always in flux and that disturbance and change are persistent features of biocommunities. Nonetheless, it is undeniable that human-facilitated migrations of animals have altered ecosystems much more quickly and extensively than any nonhuman activity.

In traditional eradication programs, proponents and opponents disagree on whether human interests must always be given priority—whether, for example, the human interest in beef production justifies the elimination of wild horses from western American range lands, which they graze with cattle. Environmental restoration programs, however, focus on the interests of nonhuman species. Their proponents and opponents can therefore both rightfully claim to be protectors of animals, although their value systems differ. Proponents defend the harsh methods they employ to kill exotic species by maintaining that they place a high value on biodiversity and are trying to ensure the very survival of native species of animals and plants. Opponents respond that they assign the highest value to compassion and are concerned about the pain, distress, and death caused by humans to each individual animal. The development of humane methods of controlling animal populations, in particular methods of contraception and sterilization, would offer a resolution to the ethical issues raised by restoration practices.

See also Endangered Species Act; Endangered Species and Ethical Perspectives

Further Reading

Baskin, Y. 2002. *A plague of rats and rubbervines: The growing threat of species invasion.* Washington, DC: Island Press/ Shearwater Books.

Bright, C. 1998. *Life out of bounds. Bioinvasion in a borderless world.* New York: W.W. Norton.

Burdick, A. 2005, May. The truth about invasive species. *Discover, 26*(5), 33–41.

Cox, G. 1999. *Alien species in North America and Hawaii: Impacts on natural ecosystems.* Washington, DC: Island Press.

Glotfelty, C. 2000. Cold war, silent spring: The trope of war in modern environmentalism. In C. Waddell (Ed.), *And no birds sing: Rhetorical analyses of Rachel Carson's* Silent Spring, 157–73. Carbondale: Southern Illinois University Press.

Larson, B. 2005. The war of the roses: Demilitarizing invasion biology. *Frontiers in Ecology and the Environment, 3,* 495–500.

McGrath, S. 2005, March. Attack of the alien invaders. *National Geographic, 207*(3), 92–117.

Peretti, J. (1998). Nativism and nature: Rethinking biological invasion. *Environmental Values, 7,* 183–92.

Sagoff, M. 1999. What's wrong with invasive species? *Report from the Institute for Philosophy and Public Policy, 19,* 16–23.

Shelton, J. 2004. Killing animals that don't fit in: Moral dimensions of habitat restoration. *Between the Species* 4, http://cla.calpoly.edu/bts/index_04.htm.

Simberloff, D. 2003. Confronting invasive species: A form of xenophobia? *Biological Invasions, 5,* 179–92.

Woods, M., & Moriarty, P. 2001. Strangers in a strange land: The problem of exotic species. *Environmental Values, 10,* 163–91.

Zimmer, C. 2008. Friendly invaders. *New York Times*, Science section, September 9.

Jo-Ann Shelton

EXPERIMENTATION AND RESEARCH WITH ANIMALS

Despite over a century of animal rights and antivivisectionist protest, scientists, regulatory agencies, and others have remained convinced that experiments on animals yield important scientific and medical discoveries. At this time, animal experimentation is not only permitted by American laws, it is actively required. For example, before most drug studies can proceed to clinical trials in human patients, animal testing must first be performed. Animal use still seems to be increasing, despite its high cost, tight regulation, and the availability of cell cultures, advanced imaging procedures, and other technologies that can replace some animal studies.

A very wide variety and large number of animals serve in experiments. Great apes, such as chimpanzees, are used in small numbers in laboratories. *Drosophila* fruit flies, *Caenorhabditis* nematode worms, and other invertebrates are also common laboratory inhabitants. Also numerous are mice, rats, zebra fish, frogs, and others. Exact numbers of laboratory animals are impossible to come by in the United States. The U.S. Department of Agriculture publishes an annual report, including the numbers reported for the handful of species covered by the Animal Welfare Act. In 2006, they reported the use of 1,012,713 dogs, cats, primates, and other covered species (http://www.aphis.usda.gov/animal_welfare/downloads/awreports/awreport2006amend.pdf). This number excludes and is dwarfed by the vast numbers of mice, rats, fish, and frogs for which there is no required reporting. This author has estimated that some 80–100 million mice and rats are bred for use in laboratories annually. Comparing these numbers to vertebrate and invertebrate animals used for human food is difficult. This author has estimated that approximately one hundred mammals or birds are killed for food each year in the United States for every one laboratory mammal or bird, but this is a very, very rough estimate.

The variety of animal species used in the laboratory are derived from a number of factors. All things being equal, scientists are under an ethical and regulatory obligation to choose the least sentient species that will serve the scientific purpose. This is rarely the sole criterion in choosing an experimental animal, however. Fruit flies, for instance, might generally be considered better for genetic studies than chimpanzees or mice because they are thought to be less sentient, but their shorter life cycle, simpler genetic make-up, small size, and the ease with which they are kept in the lab are all also points in favor of choosing fruit flies for genetic studies. For genetic studies of uniquely mammalian traits, mice, rather than chimpanzees, are chosen to replace the fruit flies, not because they are less sentient than chimps, but because they are small and cheap, have short generation cycles, have a well-defined genome, can be easily genetically reengineered, are not an endangered species, and are less thoroughly regulated by the government. Despite the challenges and expense of working with them, however, the genetic closeness of chimpanzees and monkeys to humans, and their complex mental abilities, sometime make them a scientist's first-choice of study animal.

The overwhelming majority of animals used in laboratories are bred specifically for use in laboratories. Laboratory-bred animals, in general, are less likely to carry infections, are less likely to be distressed by life in the laboratory, and may be more genetically uniform. Most of the exceptions to this general rule raise ethical concerns. Laboratory-bred *Xenopus* or African clawed frogs are the most numerous frogs used in laboratories, but wild-caught frogs of other species are also used, and over-collection of species causes conservation problems.

Wild-caught nonhuman primates are used in some laboratories, raising serious concerns about conservation of species, as well as welfare concerns of capturing these highly social animals and removing them from their group. Overwhelmingly, the most controversial acquisition of laboratory animals is the use of so-called random source dogs and cats. The 1966 Laboratory Animal Welfare Act was passed largely to prevent theft of companion animals for sale to laboratories. People who work in laboratories, such as this author, believe companion-animal theft for laboratories to be rare; nevertheless, purchase of random source animals is still permitted in many states, and some of these animals can include former household animals that were rescued by or donated to animal shelters.

Animal Research: Critiques and Defenses

Animal research has long been controversial. Criticism comes in two main forms. First, there is the scientific claim that studies on animals are not only useless, but downright misleading. Critics claim that information gleaned from animal studies rarely applies well to humans, and that it is difficult to tell when animal studies would apply to humans and when they would not. They further warn that animal studies may result in falsely labeling a dangerous drug as safe or, conversely, that animal studies may lead a scientist to abandon a particular line of research because animal studies incorrectly show it to be useless. For example, if the study of penicillin had relied on guinea pigs, for whom it is often fatal, it would never have been developed. In other words, some critics say that animals are too different from people to serve as models of human health and biology.

The other criticism is more clearly moral: whether or not animal studies make scientific sense, it is wrong to inflict illness and pain on sentient animals solely for human benefit. Critics holding that view would say that even if a cure for a devastating disease could assuredly be found by harming a small number of animals, it would still be immoral to conduct that research. Most of these critics focus on traits that humans and animals share, such as some degree of consciousness, sentience, or the ability to feel pain and suffering as the basis for arguing that if it would be wrong to do something to a person, it would likewise be wrong to do that thing to an animal. In other words, animals are too much like people for us to justify using them in experiments.

These two arguments against animal research often work together. If moral critics believe it would take an extraordinary effort to justify some limited animal research but are convinced that the science of animal studies is weak, they will, of course, conclude that most animal research should be stopped.

Defenders of animal research tend to argue that animals are sufficiently different from people that it might be acceptable to use them in studies, as long as scientists are careful to do their best to limit their pain and distress in the laboratory. These research defenders mostly argue that animals may feel pain, but they do not have sufficient consciousness and self-awareness to be placed on a moral level with people. On this side of the argument there are also extreme views, such as the view that humans have no duties to animals whatsoever, or that, while animals are similar enough to humans to make research worth doing, they are too dissimilar to raise qualms about harming them.

The most consistent defense of animal research is that, in the eyes of most scientists, it works. Virtually every modern medical and surgical advance has involved some use of laboratory animals in its development. That claim is not necessarily the same as saying that there could not have been any other route to these advances which did not involve animals. That argument also, more importantly, does not rule out the day when medical advances will no longer require animal studies. Science would not stop if animal use, or even only harmful uses of animals, stopped, but it might be very different. Many projects might not be able to be accomplished. Others might be done in different ways. Still others might be largely unchanged.

At this point in time, the compromise position, as represented by laws such as the Animal Welfare Act, is that animal research is permitted by law, and even sometimes required by law, but only with systems of oversight to try to minimize the amount of pain and distress animals in laboratories experience. Scientists, animal care and use committees, regulatory agencies, and funding agencies all perform some sort of comparison of costs and benefits in using animals. The benefits are most often seen in terms of medical advances for human health. It is important to recognize that these are *potential* benefits, and that not all experiments will lead to cures for human diseases. If scientists knew in advance the results of their experiments, there would be no reason to do the research. Even experiments that yield hoped-for results must be seen in context, where any one finding is just part of the very large puzzle of how the body works. It would be incredibly rare to be able to say, "This is the crucial experiment that eradicated Disease X."

Just as the human benefits of animal research are potential and hard to predict, likewise there can be uncertainty in the cost of such research, cost measured in terms of animal suffering and death. Veterinarians and others must be able to predict the degree of pain and distress anticipated, make recommendations for ways to decrease the pain (by changing how the experiment is done or by adding more painkillers for the animals), and then make their best assessment of how the animals are actually faring.

See also all three alternatives

Further Reading

Carbone, L. 2004. *What animals want: Expertise and advocacy in laboratory animal welfare policy.* New York: Oxford University Press.

Greek, J. S., and Greek, C. R. 2004. *What will we do if we don't experiment on animals? Medical research for the twenty-first century.* Victoria, BC: Trafford Publishing.

Institute of Laboratory Animal Resources. 1996. *Guide for the care and use of laboratory animals.* Washington DC, National Academy Press.

Orlans, F. B. 1993. *In the name of science: Issues in responsible animal experimentation.* New York: Oxford University Press.

Orlans, F. B., Beauchamp, T. L., et al. 1998. *The human use of animals: Case studies in ethical choice.* New York: Oxford University Press.

Rowan, A. N. 1984. *Of mice, models, and men: A critical evaluation of animal research.* Albany: State University of New York Press.

Russell, W.M.S., & Burch, R. L. 1959. *The principles of humane experimental technique.* London: Methuen & Co.

Stevens, C. 1990. Laboratory animal welfare. *Animals and their legal rights.* Washington, DC: Animal Welfare Institute.

Larry Carbone

EXTINCTION AND ETHICAL PERSPECTIVES

Extinction is one of the most significant problems facing many wild animal species today. The English word extinct was originally applied to the extinguishing of a flame, and later to a human family or race that had died out and left no living representative; eventually the word was applied to species of animal or plant. The first example of this use given by the *Oxford English Dictionary* is from A. R. Wallace's *Island Life* (1880): "the most effective agent in the extinction of species is the pressure of other species." The extinction of animals, together with the appearance of new forms of life, has been occurring for millions of years. Following the publication of Charles Darwin's theory of evolution in the mid-19th century, the phenomenon has generally been seen as a response to environmental conditions and competition from more adaptable species. Extinction, then, can be perceived as the result of a species' developmental inadequacy and a natural or inevitable occurrence.

Before humans inhabited the Earth in great numbers, extinctions happened slowly, but in the last 100,000 years the rate of disappearance has accelerated, and it is believed that we are currently witnessing an extinction event. Scientists such as Richard Leakey and Roger Lewin forecast that up to 20 percent of all living animal populations will disappear within 30 years, and that human practices and actions are the major reason for this increase. In the face of this scenario, important ethical issues are raised regarding the obligations of humans toward animals. Many animal rights and welfare ethicists would agree that, as beings endowed with reason and emotion and the dominant species on earth, humans have duties to those who are powerless in the face of their actions. However, some believe that ethical issues arise not only when humans cause or contribute to the exploitation, suffering or death of another species, but also when an individual or species is threatened for any reason. Either way, when human activities result in the complete annihilation of another species of animal, it would seem to constitute an extreme form of unethical behavior.

In the 1980s, Paul and Anne Ehrlich outlined four arguments for the preservation of animal species: first, compassion: the right of animals to exist; second, the argument from aesthetics: the beauty, cultural and spiritual value or intrinsic interest of animals; third, the economic value of animals; and fourth, the argument from biodiversity. Biodiversity is a key concept in the discussion of extinction, because biologists consider it essential to maintain the Earth's variety of plants, animals, microorganisms, and ecosystems. This reinforces the Ehrlichs' other arguments. For example, in instrumental terms, diversity is the primary source of humanity's needs, such as food, medicines and industrial products and, as all life forms are interdependent, it provides a basis for the ability of every living thing to adapt to changing environments. The removal of one species, or even significant numbers of a localized population, can have radical effects on an entire ecosystem and shut down the processes by which diversity can be regenerated. Ecologist Aldo Leopold has stressed the transutilitarian value of wildlife by contending that humans and animals are part of a biotic community, while environmental philosophers and ethicists such as J. Baird Callicott and John Muir have

maintained that the conservation of biological diversity has intrinsic value. Ultimately, the utilitarian stance of ethicist Peter Singer, which is based on consideration of the aggregate benefit or harm of an action, is supported by the argument from biodiversity. But it would seem that Tom Regan's position, which leans more toward the rights of individual animals, is also upheld if maintaining diversity is in the long-term interests of each and every member of a particular species, as well as each species as an individual entity.

However, the actions that result in animal extinctions and the processes involved in their occurrence are varied and complex and raise a number of more specific ethical issues. Although the extermination of many species has been deliberate, some are accidental, and eradication programs are often considered necessary for economic progress, food production, or lifestyle improvement. Modern extinctions primarily arise from human-driven changes to the environment through habitat destruction, such as forestry practices, industrial development and other management of the land; urban sprawl; pollution of the air, water, or ground through the application, release, or concentration of chemicals that can cause a chain of disappearances; the transportation or introduction of new or invasive species; the harvesting of a species' food source; or the dispersal of animal populations. Some of these actions cause genetic transformations that occur over a considerable time, or a species gradually loses out in the competition for food with another species. Extinctions also occur as a result of the killing of species deemed pests, or through the overhunting of seemingly abundant species. To complicate the issue further, the moment when the last member of a species disappears may be difficult to determine. In many cases the range or numbers of individuals is uncertain or a species has not even been recorded. There may be closely related animals or subgroups that are hard to define; a small or widely dispersed breeding population, or one with low genetic diversity, may mean a species is effectively extinct before all members are gone. Species may be extinct in the wild, although individuals survive in captive situations. Ultimately, extinctions of tiny organisms or remote populations may go undetected and, occasionally, a species believed to be extinct is discovered in a remote location.

The results of human practices that cause extinction and highlight ethical issues can be seen in the histories of hundreds of species of animal that disappeared in the course of European colonial expansion. The North American passenger pigeon is believed to have once been the most abundant bird in the world, with massive flocks that blacked out the sun and nesting roosts that could span 100 miles. Yet in the space of 200 years this pigeon, sold commercially as game, was wiped out through a variety of killing methods. The term stool pigeon comes from the practice of using a decoy bird tied to a perch to attract others. Human actions resulted in shrinking the flocks, resulting in inbreeding and finally mortality from other predation, which affected the viability of the species. The dodo from the island of Mauritius was also a member of the pigeon family, but this larger bird was flightless and nested on the ground. Humans did not find the species particularly good to eat, yet from the late 1500s onward successive settlers on the previously uninhabited island caught and killed these relatively docile birds. Widespread destruction of habitat

on the island as well as the introduction of pigs, monkeys, and rats, destroyed the dodo's eggs and hastened the disappearance of the species in 1660.

Wolves have born the brunt of blame for attacks on livestock for centuries, and due to the implementation of bounties they were extinct in most European countries by the beginning of the 19th century. In America and many British colonies wolf-like animals were, and often still are, targeted in the same way. The Falkland Islands dog came into the water to greet sailors in the 18th century, but fur traders and sheep farmers killed large numbers until the last known individual died in 1876. In 1889 on the Japanese island of Hokkaido, the Ezo wolf disappeared after American advice to use strychnine-poisoned bait to reduce the species' numbers on horse and cattle ranches. In Japanese myth and legend, this animal was a benign creature and seen as a watchdog or guardian of travelers. On the island state of Tasmania, the striped, dog-like marsupial *thylacine,* called the Tasmanian tiger, disappeared in 1936 after successive private and government bounties were placed on the species because of failures in the sheep industry. Yet there is sparse evidence of stock predation by the *thylacine,* and recent research suggests that the species was adapted to kill much smaller native prey.

Much less obvious extinctions have occurred in these and other countries during the last 200 years through causes that can be traced to human practices. The effects of climate change can be seen in a decrease in populations of tiny animals such as frogs, insects, and organisms that live in coral reefs. The disappearance of animals often invisible to the human eye breaks links in the food chain and causes unpredictable effects on larger animals in an ecosystem. But animals can also be adaptable and resilient. Through their own agency, many species begin to eat different foods, to colonize new and unfamiliar human habitats such as urban areas, and to develop accommodating behaviors. Some species are better at adapting than others, and animal rights and welfare sympathizers stress the responsibility of humans to care for and protect vulnerable species. Difficulties arise when these views are incompatible with environmental philosophies and wildlife management practices. For instance, culling animal populations is sometimes considered necessary to preserve a habitat or save other species, particularly when animals invade or proliferate areas in which they are not native. Introduced species can devastate farms and forests, impede waterways, and affect human health. To complicate the issue, eradicating species such as these with poison baits or spraying can result in damage to non-target animals, while the introduction of a pest's traditional prey has also caused unforeseen problems for native wildlife. There is also the question of whether any species is such a threat to humans or other animals that its extermination is justified. In many parts of the world, mosquitoes pass disease from one animal species to another, including humans. The Asian tiger mosquito spreads encephalitis, yellow fever, and dengue fever, and the Anopheles mosquito transmits malaria. The latter kills millions of humans, with an estimated 515 million cases of infection per year. Some philosophers argue that if it is in the interests of species to evolve in response to environmental pressures, then allowing a species to die out if critically endangered is a right. They contend that human interference in the extinction process does not demonstrate respect for beings based on their intrinsic value, particularly if artificial means or human manipulation are used or if their is a human-centered motivation for the crisis.

Recently, advances in genetic technology have resulted in attempts at reversing the extinction of some animals through cloning. None of these projects have been successful, largely because of problems associated with degraded or fragmented DNA, but with accelerating progress in genetic technologies many of these difficulties may someday be overcome. However, there are ethical issues related to reviving extinct species. Animals suffer and die in attempts at cloning. A large number of genetically varied individuals needs to be produced to create a viable population. Complications often arise for the surrogate mother, and there are many health problems with the animals produced by these methods. They may endure confinement and suffering if used for display or future research purposes, and there are often problems associated with reintroduction into the wild, especially when habitat is increasingly degraded. With the inevitable publicity that surrounds cloned animals, few would be likely to be released into their natural environment to live undisturbed lives. Genetic technologies may be better employed in identifying areas of low genetic diversity in wild populations and then establishing insurance populations; in the establishment of banks of tissue samples, eggs, sperm and frozen embryos to ensure the preservation of gene pools; in noninvasive reproductive technologies such as animal husbandry used in zoos, reserves, or with semi-captive native species, thus providing the mechanisms for animals to respond to environmental change. A less invasive form of extinction reversal has been achieved through selective breeding both in captive situations, such as zoos, and in reserves. An animal that resembles the South African quagga, a sub-species of the Plains Zebra that has stripes only on the front of the body, has been rebred over the course of 15 years. However, this project has raised questions about what gives an animal its identity, its genetic makeup, history, behavior, and habitat. If quaggas can be returned to their original habitat in the South African Karoo, it is thought that the problems encountered will be balanced by raising awareness of extinctions and encouraging programs that protect species before they disappear.

See also Endangered Species Act; Endangered Species; Wild Animals

Further Reading

Armstrong, S. J., & Botzler, R. G. eds. 2003. *The animal ethics reader.* New York: Routledge.

Chessa, F. 2005. Endangered species and the right to die. *Environmental Ethics,* 27 (1), 23–41.

Ehrlich, P. & Ehrlich, A. 1982. *Extinction: The causes and consequences of the disappearance of species.* London: Victor Gollanz.

Fiester, A. 2005. Ethical issues in animal cloning. *Perspectives in Biology and Medicine,* 48, 328–43.

Flannery, T., & Schouten, P. 2001. *A gap in nature: Discovering the world's extinct animals.* Melbourne: Text Publishing.

Freeman, C. 2007. Imaging extinction: Disclosure and revision in photographs of the thylacine (Tasmanian tiger). *Society and Animals: Journal of Human-Animal Studies* 15, 241–256.

Iziko South African Museum, The Quagga Project South Africa. http://www.quaggaproject.org/. Retrieved April 5-May 7, 2008.

Knight, J. 1997. On the extinction of the Japanese wolf. *Asian Folkloric Studies,* 56, 129–159.

Leakey, R., & Lewin, R. 1996. *The sixth extinction: Biodiversity and its survival.* London: Weidenfeld and Nicholson.

Lee, K. 2001. Can cloning save endangered species? *Current Biology,* 11, R245–246.

Max, D. T. 2006. Can you revive an extinct animal? *New York Times Magazine,* http://www.nytimes.com/2006/01/01/magazine/01taxidermy.html?_r=2&8hpib&oref=slogin&oref=slogin. Retrieved April 28, 2008.

Ryder, O. 2002. Cloning advances and challenges for conservation. *Trends in Biotechnology* 20, 231–3.

Snow R. W., Guerra C. A., Noor, A. M., Myint H. Y., & Hay S. I. 2005. The global distribution of clinical episodes of Plasmodium falciparum malaria. *Nature* 434, 214–217. Varner, G.1995. Can animal rights activists be environmentalists? In *Environmental philosophy and environmental activism.* (Ed. by D. E. Marietta and L.Embree),169–202.NewYork:Rowman& Littlefield.

Carol Freeman

F

FACTORY FARMS

Americans seem to have an insatiable appetite for animal products. We each eat, on average, about 220 pounds of meat, (U.S. Department of Agriculture, 2008) 255 pounds of dairy products, (U.S. Department of Agriculture Economic Research Service, 2008) and 260 eggs every year, (U.S. Department of Agriculture, 2008) a huge increase from just a few decades ago.

As people have been eating more and more animals, they have also become distanced from farming operations. Less than two percent of Americans live on farms, (U.S. Department of Agriculture Cooperative State Research, Education, and Extension Service,2008) and most urbanites' only contact with farm animals occurs when they consume them.

Yet many care about animal welfare and are appalled when confronted with acts of cruelty. Most people accept that farm animals are individuals with their own needs and interests, just like their companion animals, and their pleasure and suffering is worthy of moral consideration. This awareness is fueling a growing opposition to factory farms.

What Is a Factory Farm?

Since the mid-20th century, animal agribusiness has mutated from small farms to factories, massive indoor facilities that can confine hundreds of thousands of animals in a single small location. Our storybook version of Old MacDonald's Farm has been replaced with industrial operations focused on maximizing the amount of product while minimizing costs and making the most profit from animals.

As a result, factory farms in the United States are now responsible for raising and killing nearly 10 billion animals each year (U.S. Department of Agriculture NASS, 2008; U.S. Department of Agriculture National Agricultural Statistics Service, 2008). And as our oceans' fish populations plummet, underwater factory farming has exploded, producing billions more aquatic animals annually.

There are thousands of factory farms across the country, and some experts believe they abuse animals on a scale and with an institutional ferocity unprecedented in human history. Critics also maintain that these farms poison the environment and mistreat their employees, who are often especially vulnerable due to poverty or immigration status.

Laws Affecting Farm Animals

There is no federal law regarding the treatment and welfare of animals on farms. Most states' cruelty codes exempt common agricultural practices, which means that if abuse is the industry standard, there is next to no protection in most states. So farm animals routinely endure cruelties

245

that would warrant felony charges if they were inflicted upon a dog or cat.

The Humane Methods of Slaughter Act applies to animals in their final moments, and it requires that companies render farm animals insensible to pain before slaughter. However, the U.S. Department of Agriculture is responsible for enforcing this law, and the agency excludes chickens, turkeys, and many other animals from its protection. Since these are the vast majority of animals we kill for food, the USDA renders the HMSA nearly meaningless.

Worst Welfare Problems with Factory Farms

Slaughter Even at USDA-inspected plants, rampant slaughter abuse can be the norm. In January 2008, an undercover Humane Society of the United States investigation documented shocking cruelty (Weiss, 2008) at a Southern California slaughterhouse. Footage caught workers torturing downer dairy cows—animals too sick or injured to walk—by dragging them with forklifts, jabbing them in the eyes, electrically shocking them, and simulating drowning by forcing high-pressure jets of water into their nostrils in vain attempts to get them to stand and march to their own death. The USDA had multiple inspectors on the premises and had even awarded the plant with a Supplier of the Year distinction, as it was a top supplier of the National School Lunch Program.

As bad as things are for cows and pigs during slaughter, chickens have it even worse, and they comprise the vast majority of animals killed for food (U.S. Department of Agriculture NASS, 2008);. Workers unload them from transport crates and shackle them upside down while they are fully conscious. They are stunned by being moved through electrified water, and then their throats are slit. Many of them miss the blade and are still fully conscious when they're immersed and finally drown in tanks of scalding-hot water used to loosen their feathers.

Cages and Crates Critics believe that one of the worst abuses in raising animals for food involves cramming hundreds of millions of egg-laying hens, breeding sows, and veal calves into cages that are so small they can hardly move for months on end.

Factory egg farms confine about 280 million laying hens inside battery cages (United Egg Producers, 2008), tiny and often filthy enclosures, where they cannot even spread their wings or walk, much less nest, perch, or dust bathe. Multiple birds are stuffed into a file drawer-sized cage, where each hen has less space than a sheet of letter-sized paper in which to spend her life of up to a year and a half. Undercover investigations (Miller and Ghiotto, 2008) have revealed hens impaled on cage bars, trapped without food and water access, packed into cages with dead and rotting birds, and suffering from a litany of painful health problems.

About four million breeding sows, used to produce meat pigs, endure similarly intensive confinement. Breeding facilities restrict them for pregnancy after pregnancy to individual two-foot wide metal gestation crates that prevent them from turning around or walking. For up to four years, they live on concrete floors above suffocating manure pits.

Like breeding pigs, about 750,000 calves raised for veal each year are packed into narrow individual crates barely larger than their bodies. From the time they're one or two days old until they reach

Pigs at Doug Ruth's factory farm operation Downing Missouri. Ruth built his feeding operation in Scotland County—just 100 feet from the county border—because the regulations in neighboring Schuyler County would make it almost impossible for him to build there. (AP Photo/Al Maglio)

market weight at about five months, they cannot turn around, lie down comfortably, or meaningfully interact with their mothers or other calves.

Rapid Growth More than nine billion chickens are killed each year (U.S. Department of Agriculture NASS, 2008), their numbers dwarfing all other farm animals combined. Bred to reach market weight in an ever-shorter amount of time, chickens are ready for slaughter at only 45 days of age. Since they gain weight so unnaturally fast, these animals endure often painful, sometimes fatal metabolic and skeletal disorders that cause a tremendous amount of suffering.

Dairy As the above slaughter investigation revealed, dairy cows' final moments can be horrific. Slaughter isn't the only

problem with dairy production; many cows at the plant involved in the scandal were already so ill or crippled they couldn't walk by the time they arrived at the slaughterhouse.

Factory farms typically confine the nine million U.S. dairy cows (U.S. Department of Agriculture NASS, 2008) slaughtered annually inside concrete-floored sheds for about four years. Like all mammals, cows don't lactate unless they've given birth, so dairies remove their newborn calves and often sell them to veal producers or beef cattle ranchers.

Dairies often inject cows with bovine growth hormone to further increase milk yield, which can cause or exacerbate a number of health problems. Many of these operations mutilate cows by cutting off their horns and tails.

Underwater Factory Farms

Aquaculture now accounts for about one-third of all aquatic animal production, and it involves problems similar to those found in industrialized farm animal production. Underwater factory farms confine animals in restrictive, unnatural enclosures, where they can develop painful injuries and problems such as lesions, infections, deformities, parasitic infestations, and more. Producers may starve fish for several days before slaughter—a process that can take up to 15 minutes. Often fish are completely conscious when their gills are slit and they bleed to death.

The Environmental Cost of Meat Production

The evidence is strong that animal agribusiness is among the most serious causes of environmental destruction. A United Nations Food and Agriculture Organization report concludes that raising animals for food is a greater contributor to global warming than automobiles, with the author stating, "Livestock are one of the most significant contributors to today's most serious environmental problems" (Food and Agriculture Organization of the United Nations, 2006).

Each year, these animals produce about 500 million tons of manure, which can pollute our soil, air, and water. And factory farms are resource-intensive. These massive operations are responsible for a large share of domestic water use, and nearly three-fourths of our grain harvest is used to feed farm animals. Meat production also uses a significant amount of raw materials and fossil fuels.

Water and air pollution can also threaten the health of workers and nearby residents. The American Public Health Association has called for a moratorium on factory farms (American Public Health Association, 2003).

The Human Cost of Factory Farms

Pollution threats aren't the only social justice problem that animal agribusiness creates. Worker safety is often abysmal at factory farms and slaughterhouses, which together employ more than half a million workers. many of whom are especially vulnerable because they're poor, unaware of their rights, and sometimes unable to speak English.

Slaughterhouse line speeds have increased, and workers spend long days doing dangerous work with sharp knives and equipment. They're sometimes untrained or undertrained, and they can develop crippling repetitive strain injuries and be cut, stabbed, dismembered, or worse. Slaughterhouse and factory farm workers can be exposed to a number of illnesses by inhaling blood, feces, dirt, pesticides, and other particulates. And of course, bacteria and viruses can enter the food supply.

When workers do develop illnesses or injuries, management may intimidate them to prevent them from reporting the problem. High turnover means that workers often don't accrue sick time or obtain insurance coverage.

In the midst of all the problems cited by critics of large industrial farms, however, there is some good news.

Hope for the Future

Because of the work of animal advocates across the country, Americans have approved legislation in a few states to prevent some of the worst abuses.

On November 5, 2008, California voters approved Proposition 2: The Prevention of Farm Animal Cruelty Act, in a landslide. This landmark citizen ballot initiative criminalized, with a phase-out period, the confinement of animals in battery cages, gestation crates, and veal crates. Despite an agribusiness-funded campaign opposing the modest requirement to provide animals with enough space to stand up, lie down, turn around, and extend their limbs, more than 63 percent of the California electorate voted in favor of Prop 2, in the nation's top agriculture state, no less.

California is in good company. Since 2002, Florida and Oregon have passed laws against gestation crates, while Arizona and Colorado have banned both gestation crates and veal crates. Although these improvements won't prevent every problem with using animals for food, they're a step toward ending the worst confinement abuses, and they've sparked major changes at the corporate level.

Over the past several years, retailers and restaurants, including Safeway, Whole Foods, Burger King, Wolfgang Puck, and even animal producers such as Smithfield Foods, have begun to move away from supporting the use of crates and cages on factory farms.

They're also catering to the growing number of Americans who demand vegetarian and vegan foods that are more humanely produced, sustainable, and socially responsible. Plant-based meat, dairy, and egg alternatives are exploding in popularity and are readily available at nearly every supermarket.

Gourmet restaurants are increasingly featuring vegetarian and vegan options as haute cuisine, and exclusively vegetarian eateries are now commonplace. Even fast-food chains that used to be vegan wastelands offer menu choices, and many restaurants happily accommodate vegan customers. Finally, vegan and vegetarian cookbooks have flooded the bookshelves, proving that plant-based cooking is accessible, easy, and delicious.

See also Food Animals, Ethics and Methods of Raising Animals

Further Reading
American Public Health Association. 2003. Precautionary moratorium on new concentrated animal feed operations. Policy # 20037.

Eisnitz, G. A. 2006. *Slaughterhouse: The shocking story of greed, neglect and inhumane treatment inside the U.S. meat industry.* Buffalo, NY: Prometheus Books.

Food and Agriculture Organization of the United Nations. 2006. Press release: Livestock a major threat to environment. November 29, 2006. http://www.fao.org/newsroom/en/news/2006/1000448/index.html. Accessed December 23, 2008.

Greger, M. 2006. *Bird flu: A virus of our own hatching.* New York: Lantern Books.

Miller, J., and Ghiotto, G. 2008. Video shows alleged mistreatment of chickens at egg ranch. *Riverside Press-Enterprise.* October 14, 2008.

Masson, J. 2003. *The pig who sang to the moon.* New York: Ballantine Books.

Schlosser, E. 2001. *Fast food nation: The dark side of the All-American meal.* Boston: Houghton Mifflin.

Scully, M. 2002. *Dominion: The power of many, the suffering of animals, and the call to mercy,* New York, St. Martin's Press.

Singer, P., and Mason, J. 2007. *The ethics of what we eat: Why our food choices matter.* New York: Rodale.

United Egg Producers. 2008. UEP animal husbandry guidelines for U.S. egg laying flocks, 2008 edition. http://www.uepcertified.com/media/pdf/UEP-Animal-Welfare-Guidelines.pdf. Accessed December 23, 2008, p.1. See also U.S. Department of Agriculture, NASS, 2008. Chickens and eggs 2007 summary. http://usda.mannlib.cornell.edu/usda/current/ChickEgg/ChickEgg-02-28-2008.pdf, p.4. Accessed December 23, 2008.

U.S. Department of Agriculture. 2008. USDA agricultural projections to 2017. http://www.ers.usda.gov/Publications/OCE081/OCE20081d.pdf, pp. 52, 55. Accessed December 18, 2008.

U.S. Department of Agriculture Cooperative State Research, Education, and Extension Service. 2008. http://www.csrees.usda.gov/qlinks/extension.html. Accessed December 18, 2008.

U.S. Department of Agriculture Economic Research Service. 2008. Food availability (per capita) data system. http://www.ers.usda.gov/data/foodconsumption/FoodAvailIndex.htm. Accessed December 18, 2008.

U.S. Department of Agriculture NASS. 2008. Poultry slaughter: 2007 annual summary. http://usda.mannlib.cornell.edu/usda/current/PoulSlauSu/PoulSlauSu-02-28-2008.pdf, p. 1. Accessed December 18, 2008.

U.S. Department of Agriculture NASS. Milk Cows: Inventory by year, US. 2008. http://www.nass.usda.gov/Charts_and_Maps/Milk_Production_and_Milk_Cows/milkcows.asp. Accessed December 23, 2008.

U.S. Department of Agriculture National Agricultural Statistics Service. 2008. Livestock slaughter: 2007 annual summary. http://usda.mannlib.cornell.edu/usda/current/LiveSlauSu/LiveSlauSu-03-07-2008_revision.pdf, p. 3. Accessed December 18, 2008.

Weiss, R. 2008. Video reveals violations of laws, abuse of cows at slaughterhouse. *The Washington Post.* January 30, 2008, A04.

Williams, E., and DeMello, M. 2007. *Why animals matter: The case for animal protection.* Amherst, NY: Prometheus Books.

Erin E. Williams

FACTORY FARMS AND EMERGING INFECTIOUS DISEASES

The first major period of disease since the beginning of human evolution likely started approximately 10,000 years ago with the domestication of farm animals (Armelagos et al., 2005). Human measles, for example, which has killed roughly 200 million people over the last 150 years, likely arose from a rinderpest-like virus of sheep and goats (Weiss, 2001). Smallpox may have resulted from camel domestication (Gubser et al., 2004), and whooping cough may have jumped to us from sheep or pigs (Weiss, 2001). Human influenza may have arisen only about 4,500 years ago with the domestication of waterfowl (Shortridge, 2003), and leprosy may have originated in water buffalo (McMichael, 2001). Rhinovirus, the cause of the human cold, may have come from cattle (Rodrigo & Dopazo, 1995). Indeed, before domestication, the common cold may have been common only to them.

Over the last few decades, there has been a dramatic resurgence in emerging infectious diseases, approximately three-quarters of which are thought to have come from the animal kingdom. The World Health Organization coined the term zoonoses, from the Greek *zoion* for "animal" and *nosos* for "disease," to describe this phenomenon (Mantovani, 2001). This trend of increasing zoonotic disease emergence is expected to continue (WHO/FAO/OIE, 2004), and the U.S. Institute of Medicine suggests that without appropriate policies and actions, the future could bring a "catastrophic storm of microbial threats" (Smolinski et al., 2003).

Animals have been domesticated for thousands of years. What new changes are taking place at the human/animal interface that may be responsible for this resurgence of zoonotic disease in recent decades?

In 2004, a joint consultation was convened by the World Health Organization, the Food and Agriculture Organization of the United Nations, and the World

Organization for Animal Health, to elucidate the major drivers of zoonotic disease emergence (WHO/FAO/OIE, 2004). A common theme of primary risk factors for both the emergence and spread of zoonoses was the expansion and intensification of animal agriculture associated with the increasing demand for animal protein.

Strep Suis

In 2005, China, the world's largest producer of pork (RaboBank International, 2003), suffered an unprecedented outbreak in scope and lethality of *Streptococcus suis,* a newly emerging zoonotic pig pathogen (Gosline, 2005). *Strep suis* is a common cause of meningitis in intensively farmed pigs worldwide and presents most often as meningitis in humans as well (Huang et al., 2005), particularly those who butcher infected pigs or handle infected pork products (Gosline, 2005). Due to involvement of the auditory nerves connecting the inner ears to the brain, half of human survivors of the disease become deaf (Altman, 2005).

The World Health Organization reported that it had never seen such a virulent strain (Nolan, 2005) and blamed intensive confinement conditions as a predisposing factor in its sudden emergence, given the stress-induced suppression of the pigs' immune systems (World Health Organization, 2005). The U.S. Department of Agriculture explains that these bacteria can exist as a harmless component of a pig's normal bacterial flora, but stress due to factors such as crowding and poor ventilation can drop the animal's defenses long enough for the bacteria to become invasive and cause disease (U.S. Department of Agriculture, Veterinary Services, Center for Emerging

Issues, 2005). China's Assistant Minister of Commerce admitted that the disease was "found to have direct links with the foul environment for raising pigs" (China View, 2005). The disease can spread through respiratory droplets or directly via contact with contaminated blood on improperly sterilized castration scalpels, tooth-cutting pliers, or tail-docking knives (Du, 2005). China boasts an estimated 14,000 confined animal feeding operations (CAFOs) (Nierenberg, 2005), colloquially known as factory farms, which tend to have stocking densities conducive to the emergence and spread of disease (Arends et al., 1984).

Nipah Virus

The 2005 *Strep suis* outbreak followed years after the emergence of the Nipah virus on an intensive industrial pig farm in Malaysia. Nipah turned out to be one of the deadliest of human pathogens, killing 40 percent of those infected, a toll that propelled it onto the United States' list of potential bioterrorism agents (Fritsch, 2003). This virus is also noted for its intriguing ability to cause relapsing brain infections in some survivors (Wong et al., 2002) many months after initial exposure (Wong et al., 2001). Even more concerning, a 2004 resurgence of Nipah virus in Bangladesh showed a case fatality rate on par with Ebola, 75 percent, and showed evidence of human-to-human transmission (Harcourt, 2004). The Nipah virus, like all contagious respiratory diseases, is a density-dependent pathogen (U.S. Central Intelligence Agency 2006). "Without these large, intensively managed pig farms in Malaysia," the director of the Consortium for Conservation Medicine said, "it would have been extremely difficult for the virus to emerge" (Nierenberg, 2005).

Bovine Spongiform Encephalopathy

Global public health experts have identified specific dubious practices used in modern animal husbandry beyond the inherent overstocking, stress, and unhygienic conditions that have directly or indirectly launched deadly new diseases (Phua & Lee, 2005). One such misguided practice is the continued feeding of slaughter plant waste, blood, and excrement to farm animals to save on feed costs (Stapp, 2004).

A leading theory on the origin of BSE, also known as mad cow disease, is that cattle, which are naturally herbivores, became infected by eating diseased sheep (Kimberlin, 1992). In today's corporate agribusiness, protein concentrates, or meat and bone meal, euphemistic descriptions of "trimmings that originate on the killing floor, inedible parts and organs, cleaned entrails, fetuses" (Ensminger, 1990) are fed to dairy cows to increase milk production (Flaherty, 1993), as well as to most other farm animals (*Economist*, 1990). According to the World Health Organization, nearly 10 million metric tons of slaughter plant waste is fed to farm animals every year (WHO/OIE, 1999). The recycling of the remains of infected cattle into cattle feed was likely what led to the British mad cow epidemic's explosive spread (Collee, 1993) to nearly two dozen countries around the world in the subsequent 20 years (U.S. Department of Agriculture and Animal and Plant Health Inspection Service, 2005). Dairy producers can use corn or soybeans as a protein feed supplement, but slaughter plant by-products can be cheaper (Albert, 2000).

Multidrug-Resistant Bacteria

Another risky industrial practice is the mass feeding of antibiotics to farm animals. The Union of Concerned Scientists estimate that up to 70 percent of antimicrobials used in the United States are utilized as feed additives for chickens, pigs, and cattle for non-therapeutic purposes (Mellon, 2001). Indeed, the use of growth-promoting antibiotics in industrial animal agriculture may be responsible for the majority of the increases in antibiotic-resistant human bacterial illness (Tollefson et al., 1999), the emergence of which is increasingly being recognized as a public health problem of global significance (Moore et al., 2006).

Alarmingly high rates of methicillin-resistant *Staphylococcus aureus* (MRSA) detected in farm animals and retail meat in Europe, for example, have led to increased scrutiny of the agricultural use of antibiotics. The Dutch Agriculture, Nature, and Food Standards Minister, Cees Veerman, was recently reported as saying that "the high usage of antibiotics in livestock farming is the most important factor in the development of antibiotic resistance, a consequence of which is the spread of resistant microorganisms (MRSA included) in animal populations" (Soil Association, 2007). The 2008 discovery of MRSA in the majority of pigs tested in Iowa and Illinois suggests that the potential public health risk attributed to farm animal-associated MRSA may be a global phenomenon (Goldburg, 2008).

Avian Influenza

The dozens of emerging zoonotic disease threats must be put into context. SARS, which emerged from the live animal meat markets of Asia (Lee & Krilov, 2005), infected thousands of humans and killed hundreds. Nipah infected hundreds and killed scores. *Strep suis* infected scores and killed dozens.

AIDS, which arose from the slaughter and consumption of chimpanzees (Hahn et al., 2000), has infected millions, but there is only one virus known that can infect billions—influenza.

Influenza, the "last great plague of man" (Kaplan & Webster, 1977), is the only known pathogen capable of truly global catastrophe (Silverstein, 1981). Unlike other devastating infections like malaria, which is confined equatorially, or HIV, which is only fluid-borne, influenza is considered by the Centers for Disease Control and Prevention's Keiji Fukuda to be the only pathogen carrying the potential to "infect a huge percentage of the world's population inside the space of a year" (Davies, 1999). In its 4,500 years of infecting humans since the first domestication of wild birds, influenza has always been one of the most contagious pathogens (Taylor, 2005). Only since 1997, with the emergence of the highly pathogenic strain H5N1, has it also emerged as one of the deadliest.

H5N1 has so far only killed a few hundred people (World Health Organization, 2008). In a world in which millions die of diseases like malaria, tuberculosis, and AIDS, why is there so much concern about bird flu?

The risk of a widespread influenza pandemic is dire and real because it has happened before. An influenza pandemic in 1918 became the deadliest plague in human history, killing up to 100 million people around the world (Johnson & Mueller, 2002), and that 1918 flu virus was likely a bird flu virus (Belshe, 2005) that made more than one-quarter of all Americans ill and killed more people in 25 weeks than AIDS has killed in 25 years (Barry, 2004). Despite the harrowing effects of that influenza nearly a century ago, the case mortality rate in 1918was less than five percent (Frist, 2005). H5N1,

in comparison, has so far officially killed half of its human victims (World Health Organization, 2008).

Free-ranging flocks and wild birds have been blamed for the recent emergence of H5N1, but people have kept chickens in their backyards for thousands of years, and birds have been migrating for millions. What has changed in recent years that led us to this current crisis? According to Robert Webster, the "godfather of flu research," it is because

> farming practices have changed. Previously, we had backyard poultry . . . Now we put millions of chickens into a chicken factory next door to a pig factory, and this virus has the opportunity to get into one of these chicken factories and make billions and billions of these mutations continuously. And so what we've changed is the way we raise animals . . . That's what's changed. (Council on Foreign Relations, 2005).

The United Nations specifically calls on governments to fight what they call factory-farming: "Governments, local authorities, and international agencies need to take a greatly increased role in combating the role of factory farming [which, combined with live bird markets] provide ideal conditions for the virus to spread and mutate into a more dangerous form" (United Nations, 2005).

Michael Osterholm, the director of the U.S. Center for Infectious Disease Research and Policy and an associate director within the U.S. Department of Homeland Security, tried to describe what an H5N1 pandemic could look like in one of the leading U.S. public policy journals, *Foreign Affairs*. Osterholm suggests that policy makers consider the devastation of the 2004 tsunami in South Asia: "Duplicate it in every major urban centre and rural community around the planet simultaneously, add in the paralyzing fear and panic of contagion, and we begin to get some sense of the potential of pandemic influenza" (Kennedy, 2005).

"An influenza pandemic of even moderate impact," Osterholm continued, "will result in the biggest single human disaster ever—far greater than AIDS, 9/11, all wars in the 20th century and the recent tsunami combined. It has the potential to redirect world history as the Black Death redirected European history in the 14th century" (Kennedy, 2005).

It is hoped that the direction world history will take is away from raising birds by the billions under intensive confinement, so as to potentially lower the risk of our ever being in this same precarious situation in the future.

According to a spokesperson for the World Health Organization, "The bottom line is that humans have to think about how they treat their animals, how they farm them, and how they market them—basically the whole relationship between the animal kingdom and the human kingdom is coming under stress" (Torrey & Yolken, 2005). Along with human culpability, though, comes hope. If changes in human behavior can cause new plagues, changes in human behavior may prevent them in the future.

Further Reading

Albert, D. 2000. EU meat meal industry wants handout to survive ban. *Reuters World Report*, December 5.

Altman, L. K. 2005. Pig disease in China worries UN. *New York Times*, August 5. iht.com/bin/print_ipub.php?file=/articles/2005/08/05/news/pig.php.

Arends, J. P., Hartwig, N., Rudolphy, M., and Zanen, H. C. 1984. Carrier rate of Streptococcus suis capsular type 2 in palatine tonsils

of slaughtered pigs. *Journal of Clinical Microbiology* 20(5):945–947.

Armelagos, G. J., Barnes, K. C., and Lin, J. 1996. Disease in human evolution: the re-emergence of infectious disease in the third epidemiological transition. *National Museum of Natural History Bulletin for Teachers* 18(3).

Barry, J. M. 2004. Viruses of mass destruction. *Fortune*, November 1.

Belshe, R. B. 2005. The origins of pandemic influenza—lessons from the 1918 virus. *New England Journal of Medicine* 353(21): 2209–11.

China View. 2005. China drafts, revises laws to safeguard animal welfare. November 4. news.xinhuanet.com/english/2005-11/04/content_3729580.htm.

Collee, G. 1993. BSE stocktaking 1993. *Lancet* 342(8874): 790–3. www.cyber-dyne.com/~tom/essay_collee.html.

Council on Foreign Relations. 2005. Session 1: Avian flu—where do we stand? Conference on the Global Threat of Pandemic Influenza, November 16. cfr.org/publication/9230/council_on_foreign_relations_conference_on_the_global_threat_of_pandemic_influenza_session_1.html.

Davies, P. 1999. The plague in waiting. *Guardian*, August 7. guardian.co.uk/birdflu/story/0,,1131473,00.html.

Du, W. 2005. Streptococcus suis, (S. suis) pork production and safety. Ontario Ministry of Agriculture, Food and Rural Affairs.

Economist. 1990. Mad, bad and dangerous to eat? February, 89–90.

Ensminger, M. E. 1990. *Feeds and nutrition.* Clovis, CA: Ensminger Publishing Co.

Flaherty, M. 1993. Mad Cow disease dispute U.W. conference poses frightening questions. *Wisconsin State Journal*, September 26, 1C.

Frist, B. 2005. Manhattan project for the 21st Century. Harvard Medical School Health Care Policy Seidman Lecture, June 1. frist.senate.gov/_files/060105manhattan.pdf.

Fritsch, P. 2003. Containing the outbreak: Scientists search for human hand behind outbreak of jungle virus. *Wall Street Journal*, June 19.

Goldburg, R., Roach, S., Wallinga, D., and Mellon, M. 2008. The risks of pigging out on antibiotics. *Science* 321(5894): 1294.

Gosline, A. 2005. Mysterious disease outbreak in China baffles WHO. Newscientist.com. July. www.newscientist.com/article.ns?id=dn7740.

Gubser, C., Hue, S., Kellam, P., and Smith, G. L. 2004. Poxvirus genomes: A phylogenetic analysis. *J. Gen. Virol.* 85: 105–117.

Hahn, B. H., Shaw, G. M., De Cock, K. M., and Sharp, P. M. 2000. AIDS as a zoonosis: Scientific and public health implications. *Science* 287: 607–14.

Harcourt, B. H., Lowe, L., Tamin, A., Liu, X., Bankamp, B., Bowden, N., et al. 2004. Genetic characterization of Nipah virus, Bangladesh, 2004. Centers for Disease Control and Prevention, Emerging Infectious Diseases 11(10). www.cdc.gov/ncidod/EID/vol11no10/05-0513.htm. www.cdc.gov/ncidod/EID/vol11no10/05-0513.htm.

Huang, Y. T., Teng, L. J., Ho, S. W., and Hsueh, P. R. 2005. Streptococcus suis infection. *Journal of Microbiology, Immunology and Infection* 38: 306–13. jmii.org/content/abstracts/v38n5p306.php.

Johnson, N.P.A.S., and, Mueller, J. 2002. Updating the accounts: global mortality of the 1918–1920 'Spanish' influenza pandemic. *Bulletin of the History of Medicine* 76: 105–15.

Kaplan, M. M., and Webster, R. G. 1977. The epidemiology of influenza. *Scientific American* 237: 88–106.

Kennedy, M. 2005. Bird flu could kill millions: global pandemic warning from WHO. 'We're not crying wolf. There is a wolf. We just don't know when it's coming'. *Gazette* (Montreal), March 9, A1.

Kimberlin, R. H. 1992. Human spongiform encephalopathies and BSE. *Medical Laboratory Sciences* 49: 216–17.

Lee, P. J., and Krilov, L. R. 2005. When animal viruses attack: SARS and avian influenza. *Pediatric Annals* 34(1): 43–52.

Mantovani, A. 2001. Notes on the development of the concept of zoonoses. WHO Mediterranean Zoonoses Control Centre Information Circular 51. www.mzcp-zoonoses.gr/pdf.en/circ_51.pdf.

McMichael, T. 2001. *Human frontiers, environments and disease*. Cambridge, UK: Cambridge University Press.

Mellon, M. G., Benbrook, C., and Benbrook, K. L. 2001. *Hogging it! Estimates of antimicrobial abuse in livestock*. Cambridge, MA: Union of Concerned Scientists.

Moore, J. E., Barton, M. D. Blair, I. S. Corcoran, D., Dooley, J. S., Fanning, S. et al. 2006. The epidemiology of antibiotic resistance in Campylobacter. *Microb. Infect.* 8: 1955–1966.

Nierenberg, D. 2005. Happier meals: Rethinking the global meat industry. Worldwatch Paper 171, September. www.worldwatch.org/pubs/paper/171/.

Nolan, T. 2005. 40 people die from pig-borne bacteria. AM radio transcript. www.abc.net.au/am/content/2005/s1441324.htm.

Phua, K., and Lee, L. K. 2005. Meeting the challenges of epidemic infectious disease outbreaks: an agenda for research. *Journal of Public Health Policy* 26: 122–32.

RaboBank International 2003. China's meat industry overview. Food and Agribusiness Research. May. www.rabobank.com/Images/rabobank_publication_china_meat_2003_tcm25-139.pdf.

Rodrigo, M. J., and Dopazo, J.. 1995. Evolutionary analysis of the Picornavirus family. *J. Mol. Evol.* 40: 362–371.

Shortridge, K. F. 2003. Severe acute respiratory syndrome and influenza. Am. J. Resp. Crit. Care Med. 168: 1416–1420.

Silverstein, A. M. 1981. *Pure politics and impure science, the swine flu affair*. Baltimore: Johns Hopkins University Press.

Smolinksi, M. S., Hamburg, M. A., and Lederberg, J. eds. 2003. *Microbial threats to health: Emergence, detection and response*. Washington, DC: National Academies Press.

Soil Association. 2007. MRSA in farm animals and meat. http://www.soilassociation.org/Web/SA/saweb.nsf/89d058cc4dbeb16d80256a73005a2866/5cae3a9c3b4da4b880257305002daadf/$FILE/MRSA%20report.pdf. Accessed October 29, 2008.

Stapp, K. 2004. Scientists warn of fast-spreading global viruses. *IPS-Inter Press Service*, February 23.

Taylor, M. 2005. Is there a plague on the way? *Farm Journal*, March 10. www.agweb.com/get_article.asp?pageid=116037.

Tollefson, L., Fedorka-Cray, P. J., and Angulo, F. J. 1999. Public health aspects of antibiotic resistance monitoring in the USA. *Acta. Vet. Scand. Suppl.* 92: 67–75.

Torrey, E. F., and Yolken, R. H. 2005. *Beasts of the earth: Animals, humans, and disease*. Piscataway, NJ: Rutgers University Press.

U.S. Central Intelligence Agency. 2006. Malaysia. CIA World Fact Book. March 29. cia.gov/cia/publications/factbook/geos/my.html.

U.S. Department of Agriculture and Animal and Plant Health Inspection Service. 2005. List of USDA-Recognized Animal Health Status of Countries/Areas Regarding Specific Livestock or Poultry Diseases, April, 12. oars.aphis.usda.gov/NCIE/country.html.

U.S. Department of Agriculture, Veterinary Services, Center for Emerging Issues. 2005. Streptococcus suis outbreak, swine and human, China: Emerging disease notice. www.aphis.usda.gov/vs/ceah/cei/taf/emergingdiseasenotice_files/strep_suis_china.htm.

United Nations. 2005. UN task forces battle misconceptions of avian flu, mount Indonesian campaign. UN News Centre, October 24. un.org/apps/news/story.asp?NewsID=16342&Cr=bird&Cr1=flu.

Weiss, R. A. 2001. Animal origins of human infectious disease, The Leeuwenhoek Lecture. *Philos. Trans. R. Soc. Lond. B. Biol. Sci.* 356: 957–977.

Wong, K. T., Shieh, W. J., Zaki, S. R., and Tan, C. T. 2002. Nipah virus infection, an emerging paramyxoviral zoonosis. *Springer Seminars in Immunopathology* 24:215–28.

Wong, S. C., Ooi, M. H., Wong, M.N.L., Tio, P. H., Solomon, T., and Cardosa, M. J. 2001. Late presentation of Nipah virus encephalitis and kinetics of the humoral immune response. *Journal of Neurology, Neurosurgery & Psychiatry* 71: 552–4.

World Health Organization and Office International des Epizooties (WHO/OIE). 1999. WHO Consultation on Public Health and Animal Transmissible Spongiform Encephalopathies: Epidemiology, Risk and Research Requirements. December 1–31.

World Health Organization, Food and Agriculture Organization of the United Nations, and World Organization for Animal Health (WHO/FAO/OIE). 2004. Report of the WHO/FAO/OIE joint consultation on emerging zoonotic diseases. whqlibdoc.who.int/hq/2004/WHO_CDS_CPE_ZFK_2004.9.pdf.

World Health Organization. 2005. Streptococcus suis fact sheet. www.wpro.who.int/media_centre/fact_sheets/fs_20050802.htm.

World Health Organization. 2008. Cumulative number of confirmed human cases of avian influenza A/(H5N1). September 10.

Michael Greger

FIELD STUDIES AND ETHICS

While there are many obvious ethical concerns that need to be addressed in studies of captive animals, there are also ethical issues associated with the study of wild animals. Nonetheless, it is important to stress that field studies of many animals contribute information on the complexity and richness of animal lives that has been, and is, very useful to those interested in animal rights and animal welfare. Students of behavior want to be able to identify individuals, assign gender, know how old animals are, follow them as they move about, and possibly record various physiological measurements including heart rate and body temperature. Animals living under field conditions are generally more difficult to study than individuals living under more confined conditions, and various methods are often used to make them more accessible to study. These include activities such as handling, trapping using various sorts of mechanical devices that might include using live animals as bait, marking individuals using colored tags or bands, and fitting individuals with various sorts of devices that transmit physiological and behavioral information telemetrically, such as radio collars, other instruments that are placed on an animal, or devices that are implanted in the animal.

Trapping is often used to restrain animals while they are marked or fitted with tags that can be used to identify them as individuals, or equipped with radio-telemetric devices that allow researchers to follow them or to record physiological measurements. However, the trapping and handling of wild animals is not the only way in which their lives can be affected, for just being there and watching or filming them can influence their lives. What seem to be minor intrusions can really be major intrusions. Here are some examples:

1. Magpies, which are not habituated to human presence, spend so much time avoiding humans that this takes time away from essential activities such as feeding

2. Adélie penguins exposed to aircraft and directly to humans showed profound changes in behavior including deviation from a direct course back to a nest and increased nest abandonment. Overall effects due to exposure to aircraft that prevented foraging penguins from returning their nests included a decrease of 15 percent in the number of birds in a colony and an active nest mortality of eight percent. There are also large increases in penguins' heart rates. Trumpeter swans do not show such adverse effects to aircraft. However, the noise and visible presence of stopped vehicles produces changes in incubation behavior by trumpeter females that could result in decreased productivity due to increases in the mortality of eggs and hatchlings

3. The foraging behavior of Little penguins (average mass of 1,100 grams) is influenced by their carrying a small device (about 60 grams) that measures the speed and depth of their dives. The small attachments result in decreased foraging efficiency. However, when female spotted hyenas wear radio collars weighing less that two percent of their body weight, there seems to be little effect on their behavior. Changes in behavior such as these are called the instrument effect.

4. Mate choice in zebra finches is influenced by the color of the leg band used to mark individuals, and there may be all sorts of other influences that have not been documented. Females with black rings and males with red rings had higher reproductive success than birds fitted with other colors. Blue and green rings were found to be especially unattractive on both females and males

5. The weight of radio collars can influence dominance relationships in adult female meadow voles. When voles wore a collar that was greater than 10 percent of their live body mass, there was a significant loss of dominance

6. Helicopter surveys of mountain sheep that are conducted to learn more about these mammals disturb them as well as other animals, and greatly influence how they use their habitat, and increase their susceptibility to predation as well as nutritional stress

While there are many problems that are encountered both in laboratory and field research, the consequences for wild animals may be different from and greater than those experienced by captive animals, whose lives are already changed by the conditions under which they live. This is so for different types of experiments that do not have to involve trapping, handling, or marking individuals. Consider experimental procedures that include visiting the home ranges, territories, or dens of animals, manipulating food supply, changing the size and composition of groups by removing or adding individuals, playing back vocalizations, depositing scents or odors, distorting body features, using dummies, and manipulating the gene pool. All of these manipulations can change the behavior of individuals, including their movement patterns, how they utilize space, the amount of time they devote to various activities including hunting, antipredatory behavior, and various types of social interactions including caregiving, social play, and dominance interactions. These changes can also influence the behavior of groups as a whole, including group hunting or foraging patterns, caregiving behavior, and dominance relationships, and also influence non-targeted individuals. Lastly, there are individual differences in responses to human intrusion.

Field workers are becoming more sensitive to how their presence and methods of study influence the animals they are studying. In a study evaluating long-term capture and handling effects on bears, wildlife researcher and veterinarian Marc Cattet and his colleagues discovered that we really do not know much about bears, and we could be gathering spurious data in the absence of this knowledge. Bears captured for research are more prone to injuries and death. One bear suffered from such "a severe case of capture myopathy—a kind of muscle meltdown

some captured animals suffer when they overexert themselves trying to escape— that its chest, bicep and pectoral muscles were pure white and as brittle as chalk." Blood analyses of 127 grizzlies caught in Alberta between 1999 and 2005 revealed a significant number of those animals showed signs of serious stress for alarmingly long periods of time after they were processed and released back in the wild and about two-thirds of the animals caught in leg-hold traps suffered muscle injuries.

Animal activist and carnivore expert Camilla Fox has shown that there are extensive negative effects of trapping many different species that significantly compromise their wellbeing and thus their behavior, and produce misleading results. Consider what she wrote about trapping aquatic animals in the *Encyclopedia of Animal Behavior*:

> Leghold and submarine traps act by restraining the animals underwater until they drown. Most semi-aquatic animals, including mink, muskrat, and beaver, are adapted to diving by means of special oxygen conservation mechanisms. The experience of drowning in a trap must be extremely terrifying: animals have displayed intense and violent struggling and were found to take up to four minutes for mink nine minutes for muskrat, and ten to thirteen minutes for beaver to die. Mink have been shown to struggle frantically prior to loss of consciousness, an indication of extreme trauma.

> Because most animals trapped in aquatic sets struggle for more than three minutes before losing consciousness,

wildlife biologists have concluded that these methods did not meet basic trap standards and could therefore not be considered humane. Fox concluded, "For an activity that affects millions of wild animals each year, it is astounding that so little is known about the full impact of trapping on individual animals, wildlife populations and ecosystem health."

While Cattet and other researchers are not ready to give up wildlife research it is heartening that he concludes that we can do much more:

"I think that a number of things can be done to perhaps minimize restraint times and capture-related injuries," Cattet said. "We could use motion activated video cameras at trap sites that would allow researchers to assess animals' reactions to capture. I think that what this study underscores is that the status quo is not the answer. It also underscores the reality that it is not only bears that suffer. There's every reason to believe that other animals are suffering too when they are captured and released."

I have personally experienced the good use of noninvasive field research. When I visited elephant expert Iain Douglas-Hamilton and his coworkers, who have been studying elephants in Samburu National Reserve in Northern Kenya, I had the pleasure of collecting elephant dung with George Wittemyer. Samples of dung are collected, then sent off for genetic analyses that help George and his colleagues further understand the elephants at Samburu. By analyzing fecal hormones, information can also be gathered on stress levels. It is known that stress hormones increase when a matriarch is killed, and are higher in areas where there are high levels of poaching.

Research ecologist Robert Long and his colleagues recently published a book titled *Noninvasive Survey Methods for Carnivores* that will surely help the animals and be a win-win for all involved in field research. John Brusher and Jennifer Schull have developed nonlethal methods for determining the age of fish using the characteristics of dorsal spines. Many researchers realize that they don't have to kill animals to study them, and we can look forward to the development of more and more noninvasive techniques for studying a wide variety of animals. Admittedly, it's a difficult situation, because we need to do research to learn more about the animals we want to understand and protect. But we can always do it more ethically and humanely and be sure that the information we collect truly reflects the behavior of the animals, and that we don't harm them while we pursue this knowledge.

While we often cannot know about various aspects of the behavior of animals before we arrive in the field, our presence does influence what animals do when we enter into their worlds. What appear to be relatively small changes at the individual level can have wide-ranging effects in both the short and long term. On-the-spot decisions often need to be made, and knowledge of what these changes will mean to the lives of the animals involved deserve serious attention. A guiding principle should be that the wild animals we are privileged to study should be respected, and when we are unsure about how our activities will influence their lives, we should err on the side of the animals and not engage in these practices until we know the consequences of our acts. By being careful about what we do in field work, we will also collect more reliable data that can be used in future studies.

Further Reading

Aitken, G. 2004. *A new approach to conservation: The importance of the individual through wildlife rehabilitation.* Burlington, VT: Ashgate Publishing Limited.

Bears captured for research more prone to injuries, death: http://www.canada.com/topics/news/national/story.html?id=7be8722e-083c-42ce-af35-6ecaa7f3ee36.

Bekoff, M., and Jamieson, D. 1996. Ethics and the study of carnivores. In Gittleman, J. L. (ed.), *Carnivore Behavior, Ecology, and Evolution,* Volume 2, 15–45. Ithaca, NY: Cornell University Press.

Bekoff, M. 2000. Field studies and animal models: The possibility of misleading inferences. In M. Balls, A.-M. van Zeller and M. E. Halder (eds.), *Progress in the reduction, refinement and replacement of animal experimentation,* 1553–1559. New York: Elsevier.

Bekoff, M. 2001. Human-carnivore interactions: Adopting proactive strategies for complex problems. In J. L. Gittleman, S. M. Funk, D. W. Macdonald, and R. K. Wayne (eds.), *Carnivore conservation,* 34–89. London and New York: Cambridge University Press.

Bekoff, M. 2002. The importance of ethics in conservation biology: Let's be ethicists not ostriches. *Endangered Species Update* 18, 23–26.

Bekoff, M. 2006. *Animal passions and beastly virtues: Reflections on redecorating nature.* Philadelphia: Temple University Press.

Brusher, J. H., and Schull, J. 2008. Non-lethal age determination for juvenile goliath grouper *Epinephelus itajara* from southwest Florida. http://www.int-res.com/prepress/n00126.html.

Cooper, N. S., and Carling, R. C. J. (eds.) 1996. *Ecologists and ethical judgments.* New York: Chapman and Hall.

Farnsworth, E. J., and Rosovsky, J. 1993. The ethics of ecological field experimentation. *Conservation Biology* 7: 463–472.

Festa-Bianchet, M., and Apollonia, M. (eds.). 2003. *Animal behavior and wildlife conservation.* Washington, DC: Island Press.

Jamieson, D., and Bekoff, M. Ethics and the study of animal cognition. In M. Bekoff and D. Jamieson (eds.), *Readings in Animal Cognition,* 359–371. Cambridge, MA: MIT Press.

Kirkwood, J. K., Sainsbury, A. W., and Bennett, P. M. 1994. The welfare of free-living wild animals. Methods of assessment. *Animal Welfare* 3: 257–273.

Laurenson, M. K., and Caro, T. M. 1994. Monitoring the effects of non-trivial handling in free-living cheetahs. *Animal Behavior* 47: 547–557.

Long, R. A. MacKay, P., Zielinski, W. J., and Ray, J. C. 2008. *Noninvasive survey methods for carnivores.* Washington, DC: Island Press.

Rollin, B. E. 2006. *Science and ethics.* London and New York: Cambridge University Press.

Marc Bekoff

FIELD STUDIES: ANIMAL IMMOBILIZATION

Immobilization, in the context of animal ethics, is the forced restriction of movement of all or part of an animal's body, either by physical or chemical means. It is used to impose management of some kind, for human and/or animal benefit. Immobilization is a common practice in many animal management procedures. Here we'll examine the impact of immobilization on animal welfare, outline the ethics of use in different situations, and consider ways of improving standards in these areas.

Physical immobilization methods usually involve traps to restrain the whole animal (e.g., pitfall traps, cage traps, box traps, crush cages, plastic tubes, restraint boards, restraint chairs), or part of the animal (e.g., snares, leg-hold traps, chutes, head-holding devices) or just use of direct handling restraints.

Chemical immobilization is achieved using drugs, which have a range of intended effects, from those which produce a widespread muscular paralysis while the animal is fully conscious, to those which produce unconsciousness with anesthesia (lack of sensation, e.g., of pain).

Immobilization Is a Welfare Issue

Immobilization of an undomesticated or anxious animal may cause considerable stress. When animals are immobilized, they may undergo some or all of a series of acute stressors including pursuit, restraint, pain, fear and anxiety, all of which are capable of inducing harmful responses and pathological changes. Repeated stressors, such as are imposed on some laboratory and wild animals, are likely to result in very poor welfare outcomes.

Animals in physical traps experience stress similar to that of being caught by a predator, but their struggle to escape may continue until released from the trap. Traps may be remote from the human who set them, and a trapped animal may be left unattended for long periods. Physical injury is also a risk. For example, steel-jaw leg-hold traps, widely condemned as inhumane, cause high levels of fractures and tissue necrosis in target and nontarget species. A good account of capture and physical restraint techniques for zoo and wild animals is given by Todd Shury (2007), and a general veterinary account by Sheldon et al. (2006).

With chemical immobilization there are different welfare issues. Immobilizing drugs have the potential to disturb normal regulatory systems, particularly respiratory and thermoregulation, which in turn can lead to negative outcomes such as respiratory depression, overheating (hyperthermia), lowered blood pH (acidosis), and oxygen deficit (hypoxemia). These in turn can lead to neurological or myocardial problems and multi-organ failure. A chase by ground or air to dart an animal can lead to extreme muscular activity and hyperthermia, as well as a potentially fatal outcome, capture myopathy syndrome, which can lead to death in minutes to weeks after the inciting event. Drugs may behave differently in combination, and in individual animals, depending on their physiological status. Dosages often have to be estimated for animals of unknown weight, and where drugs are remotely delivered by unpredictable darts to a moving target animal, delivery of the correct dosage is very difficult to control. These scenarios would present a nightmare for a human anesthetist, as would the resulting morbidity and mortality rates, but both can be routine in situations where wild or untamed animals are immobilized.

While these stressor situations are much less common under controlled conditions, for example, in the immobilization of laboratory or companion animals, there are welfare issues for each animal being immobilized.

Immobilization Is Also an Ethical Issue

Perhaps the majority of us think of animal immobilization in the context of veterinary procedures conducted on companion animals, exhibit or zoo animals, or valuable sports animals, for example horses. Here, under controlled circumstances and with primary emphasis on the welfare of the animal, immobilization standards are usually high and improved technologies rapidly adopted.

Ethical concerns around the immobilization of farm animals are very different, with the prime concern being the economics of production. Cattle, sheep and pigs are routinely immobilized for management procedures such as castration, dehorning and Caesarian section. Immobilization techniques range from humane to highly unethical and stressful techniques such as electro-immobilization (EI). Many immobilization procedures for mutilation, such as castration, tail docking, beak trimming, teeth-clipping etc., are carried out on young animals using physical restraint without anesthesia. All evidence shows that these cause unnecessary pain and distress. The organization Compassion in World Farming gives more information at www.ciwf.org.

The sheer numbers of immobilizations undertaken prior to slaughter, primarily for the meat and byproducts industries, outweigh those in all other categories combined. In 2005, in the United States alone, 10 billion land animals were immobilized and then slaughtered for the food/byproducts industry (U.S. Department of Agriculture, 2006). Welfare standards for chickens and turkeys, which comprise more than 95 percent of all animals slaughtered in the United States each year, are the poorest. They are unprotected by existing legislation in either the United States or Britain. Electric immobilization is the standard method of preparation for slaughter, and causes a wide range of animal welfare, economic, and worker-safety problems. More information can be obtained from People for the Ethical Treatment of Animals at http://www.peta.org. Temple Grandin, a professor at Colorado State University, has done much work to improve the standards of immobilization for other meat animals (http://www.grandin.com/references/humane.slaughter.html) although, as she has pointed out, standards that are applied still depend to a large extent

on the personal ethics of the slaughterhouse manager rather than legislation. In particular, the use of religious slaughter, involving immobilization by physical restraint of the animal prior to blood-letting, has also been the subject of much ethical debate. VIVA, the Vegetarians International Voice for Animals has published an online account of this controversy (http://www.viva.org.uk/campaigns/ritual_slaughter/goingforthekill01.htm), and the UK government agency DEFRA has online information relating to their stance on this issue http://www.defra.gov.uk/animalh/welfare/farmed/slaughter.htm#religiousslaughter.

Laboratory animals are routinely immobilized for various procedures in research. Just over 3.2 million scientific procedures on laboratory animals were started in the UK in 2007, the majority of which entail some restraint or immobilization. Around 39 percent of all procedures used some form of anesthesia (UK Government Home Office, 2007). When laboratory animals are subject to repeated immobilization, they begin to learn the preparatory stimuli, which entails increased stress. This is particularly serious in highly intelligent animals such as primates, who respond badly to repeated physical immobilization. Many researchers now question the validity of data gathered using stressful techniques, because they undoubtedly affect the normal physiology and behavior of the animal (Baldwin, 2007), and their emotional welfare (Bekoff, 2007).

Wildlife researchers may need to immobilize wild animals to mark them for later identification, to provide veterinary treatment, or to relocate them from dangerous or overpopulated areas. Marking may involve mutilation, such as ear-notching, digit or tooth removal, etc., tagging and banding, or external or internal radio-transmitter attachment. In the last 20 years, the immobilization of wild animals for the fitting of tags and markers has increased dramatically, to the point where this is the starting point for many monitoring studies.

Wildlife immobilization increasingly employs chemical means. The immobilization of large or potentially dangerous wild animals may pose huge challenges, with risks for both operators and target animals. Drug choices and combinations must be of proven safety for each species and calculated for the weight, age, physiological and reproductive status, body condition, and presence of young or companions with the target animal. If the onset of anesthesia effect is slow, this increases the risk of physical injury such as lacerations, limb injuries, head trauma, etc. It isn't surprising that capture- and immobilization-related mortalities in wild animals are more frequent and more serious than in domestic animals. Arnemo & Caulkett (2007) detail useful precautions which can be taken to help reduce the effects of stressors.

Evidence for the negative effects of immobilization for marking is beginning to emerge in several areas of wildlife research (Murray & Fuller, 2000). It is no longer the case that survival of a wild animal through the process of immobilization implies the safety of that procedure. Longer-term views of capture and handling are beginning to reveal problems. Cattet et al. (2008) showed negative effects of immobilization on ranging behavior and body condition in grizzly and black bears in Canada, and similar effects have been suggested for polar bears (Dyck et al., 2007). Immobilization may also negatively impact the fertility of target species. Alibhai & Jewell (2001)

reported a negative effect of repeated immobilization for radio-collar fitting and maintenance on the fertility of female black rhinoceros. While these findings and others often give rise to heated debate among wildlife researchers, most domestic animal veterinarians would not expect their patients to sustain a pregnancy, or perhaps even survive, under similar circumstances. Some authorities (e.g., the government of New South Wales, Australia) have now begun to issue ethical guidelines for wildlife research: http://www.agric.nsw.gov.au/reader/wildlife-research/arrp-radio-tracking.htm

The physical trapping of animals for research or killing is an area in which the quality of immobilization is of ethical and welfare importance. A good account of trapping and marking terrestrial vertebrates for research is given by Roger Powell and Gilbert Proulx (2003). Some of the more responsible hunting and trapping authorities issue ethical guidelines, for example, in the United States, by the Pennsylvania Game Commission: http://www.pgc.state.pa.us/pgc/cwp/view.asp?a=514&q=168724.

First, the need for immobilization can be reduced. Many of the conditions described above are consumer-driven, and could be avoided if demand was reduced. In wildlife research, the ethics of some practices requiring prior immobilization, e.g., radio-telemetry, can be questioned when there is a high failure rate of collars and/or transmitters (Alibhai & Jewell, 2001), and an accepted, but also poorly documented, potential for injury (see illustration). Training laboratory animals can avoid the need for immobilization in some circumstances; nonhuman primates can be trained to present themselves for routine blood-sampling without restraint (Reinhardt, 1995).

Second, current techniques can be replaced with those which provide better welfare. The UK National Centre for the Replacement, Refinement and Reduction of animal in research (NC3RS), (http://www.nc3rs.org.uk/news.asp?id=924) has begun this process. Better husbandry and management conditions in farming, and the adoption of noninvasive techniques for wildlife monitoring, including camera-trapping and biometric techniques such as footprint identification (Alibhai et al., 2008) and coat-pattern identification (Burghardt et al., 2008), can be considered.

Third, research can be prioritized into reduction or replacement. The Dr. Hadwen Trust for the replacement of animals in medical research does excellent work in this field: (http://www.drhadwentrust.org.uk/).

Lastly, standards of immobilization can be regulated by developing and monitoring protocols and legislation as a foundation for change. Much unnecessary stress in immobilization is imposed by economic time constraints on the competitiveness of commercial practitioners. Legislation and consumer-awareness campaigns could greatly improve conditions for animals undergoing the stressful process of immobilization.

Further Reading

Alibhai, S. K., & Jewell, Z. C. 2001. Hot under the collar: The failure of radio-collars on black rhino (*Diceros bicornis*). *Oryx 35* (4). 284–288.

Alibhai, S. K., Jewell, Z. C., & Towindo, S. S. 2001. The effects of immobilisation on fertility in female black rhino (*Diceros bicornis*). *J. Zool.* 253: 333–345.

Alibhai, S. K., Jewell, Z. C., & Law P. R. 2008. Identifying white rhino (*Ceratotherium simum*) by a footprint identification technique, at the individual and species levels. *Endangered Species Research* 4: 219–225.

Arnemo, J. M., & Caulkett, N. 2007. Stress. In G. West, D. Heard and D. Caulkett, eds., *Zoo animal immobilization and anesthesia*, 103–109. Iowa: Iowa State University Press.

Baldwin, A., and Bekoff, M. 2007. Too stressed to work. *New Scientist.* 2606, 24.

Bekoff, M. 2007. *The emotional lives of animals.* Novato: CA: New World Library.

Burghardt, T., Barham, P. J., Campbell, N., Cuthill, I. C., Sherley, R. B., Leshoro, T. M. 2007. A Fully Automated Computer Vision System for the Biometric Identification of African Penguins (*Spheniscus demersus*) on Robben Island. In Eric J Woehler (ed.), *6th International Penguin Conference (IPC07)*, 19. Hobart, Tasmania, Australia.

Cattet, M., Boulanger, J., Stenhouse, G., Powell, R. A., & Reynolds-Hogland, M. J. 2008. An evaluation of long-term capture effects in Ursids: Implications for wildlife welfare and research. *J. of Mammalogy*, 89 (4): 973–990.

CIWF. Animal Welfare Aspects of Good Agricultural Practice: http://www.ciwf.org.uk/resources/education/good_agricultural_practice/default.aspx.

Dyck, M. G., Soon, W., Baydack, R. K., Legates, D. R., Baliunas, S., Ball, T. F., & Hancock, L. O. 2007. Polar bears of western Hudson Bay and climate change: are warming spring air temperatures the 'ultimate' survival control factor? *Ecol. Complexity* 4, 73–84. doi:10.1016/j.ecocom.2007.03.002.

Moberg, G. P., & Mench, J. A. 2007. *The biology of animal stress: Basic implications for animal welfare.* Wallingford, UK: CABI publishing.

Murray, D. L. & Fuller, T. K. 2000. A Critical Review of the Effects of Marking on the Biology of Vertebrates. In L. Boitani & T. K. Fuller, eds. *Research techniques in animal ecology: Controversies and consequences*, 14–64 New York: Columbia University Press.

Powell, R. A., Proulx, G. 2003. Trapping and marking terrestrial mammals for research: integrating ethics, performance criteria, techniques and common sense. *ILAR J;* 44:259–276.

Sheldon, C. C., Topel, J., & Sonsthagen, T. F. 2006. *Animal restraint for veterinary professionals,* 1st ed. St. Louis, MO: Mosby.

Shury, T. 2007. Capture and physical restraint of zoo and wild animals. In G. West, D. Heard and D. Caulkett, eds. *Zoo animal immobilization and anesthesia*, 131–144. Iowa: Iowa State University Press.

Reinhardt, V., Liss, C., & Stevens, C. 1995. Restraint methods of laboratory non-human primates: a critical review. *Animal Welfare* 4: 221–238.

UK Government Home Office 2007. *Animals (Scientific Procedures) Inspectorate, Annual Report 2007*. London: Home Office publications.

U.S. Department of Agriculture National Agricultural Statistics Service. 2006. Poultry slaughter: 2005 annual summary. usda.mann lib.cornell.edu/usda/current/PoulSlauSu/ PoulSlauSu-02-28-2006.pdf.

Zoe Jewell
Sky Alibhai

FIELD STUDIES: ETHICS OF COMMUNICATION RESEARCH WITH WILD ANIMALS

A Personal Essay

For many years I have been using music in an attempt to communicate with the orcas that reside off the north coast of Vancouver Island. Trying to meet another species halfway tends to make one perceive them differently than a researcher who views them as the cool subject of objective observation. For one example, the whales swim past our cove several times a day in their matrilineal pods consisting of a grandmother, offspring and mates. The younger orcas, juveniles, as biologists refer to them, vocalize with us whenever they wish. It seems appropriate that our own human family groups conduct communication research with their family groups. So, for many years, our research group has had a similar percentage of children involved in the activity as the orcas have in the water.

We chose the orcas for our experiment in interspecies communication because, in contrast to almost all other dolphin species, orcas vocalize nearly all the time in a frequency range within the confines of human hearing. They also vocalize loudly, and we sometimes hear them fifteen minutes before we see them swimming our way. These whales cruise close to shore. Biologists refer to them as residents, which simply means they live here, and the whales we played with yesterday are the same whales we hear today.

These residents signal one another in two modes: the frequency modulated whistle and the pulsed click train. Frequency modulated means melodic. The pulsed click train is rhythmical. In other words, the orcas use musical concepts to communicate among their own kind. To hear these orcas calling back and forth to one another, and then interact with them, I have assembled a sound system with underwater recording and transmitting capabilities built inside a comfortable boat which is anchored just offshore. A single switch powers up a keyboard, microphones, and an electric guitar, all of which are run through an amplifier and output to underwater speakers. This sound system is basically a telephone line to the whales. If we like the conversations we hear, we record them for posterity.

If it's little children using our orca telephone, the whale's innate loudness and edgy abruptness breeds both excitement and fear. Some parents who come aboard assume that these large dolphins will naturally be drawn to children, invoking a naive view of this charged border between species as a Peaceable Kingdom where innocence is celebrated and hard work unnecessary. Whales are compassionate and wise. They love us. They love our children even more.

But when the orcas fail to respond, these parents wonder what might be wrong with their ideal. Maybe the studio isn't child friendly? They turn up the thermostat. Hide the synthesizer. Lead the kids in a rendition of "Row, row, row, your boat." It makes no difference. The whales' rubbery, bone-jarring screams remain child-unfriendly and aloof.

Playing music with orcas is better understood as an expression of conceptual art than a variation of an Edward Hicks painting. To keep going at this work, musicians must revel in counterintuitive phrasing, dissonance, and nearly unbearable stretches of silence. The slightest hints of synchronized rhythm become the measure of our correspondence. Those who persevere for more than an hour, more than a week until, finally, we visit this whale habitat every summer for two decades, celebrate a radical paradigm that insists animals are sentient beings both capable and amenable to an aesthetic interaction. Most people feel no such motivation. Most musicians find the sonic rewards too few and far between, and the intellectual rewards too unmusical.

We hear the orcas vocalizing through the speakers, like a cross between an elephant and an soprano sax. They are still a mile up the strait. I turn on the switch and let anyone play what they like. To limit the experience seems prejudicial and pompous. We have uncovered no evidence that a whale responds better to Bach played by a virtuoso than to some determined girl singing "Come little orca, sha-lalalalala-la-la." We've tried it both ways. Sometimes one gets a response, sometimes the other.

My rationale to permit children and musicians access to the sound system is sometimes judged unprofessional by objectivists who insist we attach scientific rigor to this long-term study. They insist we control our transmissions to a specific few notes, or better yet, focus pure tones from a sine wave generator, and monitor on an spectrogram. It all seems worthy. I would gladly fit any valid experiment into our schedule if someone would just administer it, and also agree not to interfere with the music-making regimen. Therein lies the problem. Scientific control is like virginity. You either have it or you don't.

That our work prospers without control is the reason our research attracts musicians, not cognitive scientists and behavioral biologists. We are laypeople whose relationship with the whales is more an affair of the heart, the ear, and the gut, than of the mind and the spreadsheet.

Those of us who have observed many different people play with the whales over several years have reached an admittedly unverifiable conclusion about the orca's response. While the orcas display no special interest in virtuosity—for example, a soloist rendering Mozart with precision—they seem highly attuned to soloists and ensembles who play with soulfulness. These whales are attracted to music-makers who are having a good time. Musicians refer to this as getting into the groove. The mechanics of rhythm, harmony, and timing take on a substance greater than the sum of its parts. What affects the players likewise affects the audience, turning the sensuous experience communal. The groove is, apparently, capable of mitigating the species barrier as easily as it cuts through the performer/audience barrier.

At Orcananda, we impose a few rules to guide interspecies etiquette. First, we conduct our musical experiment only after dark. One does not presume to play and record underwater music with orcas during daylight without contending with considerable noise pollution from boat motors rumbling and whining along the freeway of the strait. Second, we never chase the whales. We play our music from a boat anchored at the same spot year after year. If the whales choose not to come to us, the interaction cannot happen. Third, our objective is interspecies communication, so we never transmit recorded music into the water. Although a whale may certainly respond to a recording, a recording cannot respond to a whale. Fourth, we never retransmit whale sounds, reflecting an orca call back into the water. Such technology offers nothing vital to the communal ground we nurture.

Over the years, musicians have discovered techniques to facilitate interspecies music-making. Foremost is the routine of adding rhythmical silent spaces to an improvisation as an invitation for a whale to fill in the hole. If the orca vocalizes only in the allotted space, it may be a response. However, congruency is not always what it seems. For instance, a player may hear an orca call a phrase, and respond by repeating the same notes. Back and forth it goes.

Except that the whale would have made the same sounds even if the musician hadn't played anything. This simultaneity of response is of the same ilk as Paul Winter's affable studio compositions that include animal calls as overdubbed elements.

Pointing this out to a musician can lead to dispute. "What do you mean I wasn't communicating? I heard it!"

One may well ask why players confuse orca Karaoke with real-time communication. The mistake is mostly a function of a charged playing environment. Our studio is a boat in a wilderness cove. The sessions occur late at night, often with rain pounding on the roof. The candlelight we favor to conserve electricity casts an eerie glow over the proceedings, contorting shadows. When the wind comes up, the boat rocks, sometimes enough to knock a musician from one wall to the other. The underwater speakers resound with colossal gurgles, oddball kerplunks, the obscure croaking of bottom fish. The total effect is disorienting. Certain water sounds can prompt listeners to examine their clothing for signs of wetness.

From faraway, orca whistles resound through the speakers like horns playing a bebop refrain. Certain calls rise above the fray, slithering and soaring with the abandon of a Charlie Parker solo. Other calls balance this boldness; they fold in upon themselves like a flower closing its petals. A musician plays a few tentative notes. The whales turn silent for a minute. When we hear them vocalize again, it is much louder, a sure sign they have moved closer. If they come close enough, the orcas echolocate the boat. At two hundred feet, the clicks remind us of a woodpecker knocking on a tree. At twenty feet, they sound like a machine gun firing at the boat cabin.

Now the orcas are whistling at such a volume that their calls explode into the darkened room. The sensation is not so much that the orcas are close by but, rather, that one of them has inhaled the boat. When they vocalize at the volume of a loud rock band, every sound an orca makes, and some it doesn't, suggests linkage. When a skilled musician mimics their calls with aplomb, no one is left unaffected. By the time the whales take their exit, everyone feels spat out, exhausted . . . and witness to a bona fide encounter. At that moment, the question of whether the dialogue was genuine or counterfeit seems moot, a sorry attempt to superimpose an analytical frame over a profoundly emotional and spiritual experience.

One might imagine it takes a little practice to tell the difference. It takes more than that. These respondents really are whales, a truth that confounds a player even as it hints of a secret knowledge. Although I have devoted twenty summers of my life to exploring music with orcas, I did not learn the difference between interaction and simultaneity by paying attention to the sessions on the boat. I learned it, instead, by studying the recordings in the comfort of my home studio. The knowledge came to me in a rush, like glimpsing a face hidden within the textures of a surrealistic painting. The moment I heard the difference, I heard it ever after. Unfortunately, the distinction defies a literal explanation.

Though describing the truth of interaction may be difficult, the techniques that foster communication are straightforward. A sense of courtesy is fundamental. Start off playing quietly. Treat the music as an invitation. Visualize the moment as a sanctuary filled with music. Feel what it means to get on whale time. If the orcas start to leave, give them up immediately. Don't *try* to communicate; it's a contradiction in terms that impedes bonding. Keep aware that beautiful music is a species-specific presumption. The sounds a musician casts into the water may be interpreted by an orca as an intrusion or, even worse, as the acoustic analogue to poisoned meat set out for coyotes. Some orcas in these waters possess bullet scars; reminders of violence perpetrated by fishermen who perceived the salmon-eating whales as a threat to their livelihood. Fortunately, the advent of whale-watching has put an end to such wanton gunfire.

I have discovered a simple technique to test my thesis of interaction versus simultaneity. The notes D-C-D describe one orca phrase heard in these waters. Playing the riff a whole tone higher opens a door of opportunity. About once in every ten tries, a whale will rise to the occasion by mirroring the alteration: E-D-E. About once in every five hundred tries, a whale has treated my tonal variation as the start of a pattern, responding another whole tone up: F#-E-F#.

I also discovered that it was not the orcas playing with me, but two specific whales that often gravitated to our boat. One was a young male, the other was his mother, named Nickola by local biologists. Nickola was generally regarded to be the most outgoing whale in the strait. Over several years the male, A6, developed into an inspired soloist, inventing melodies that occasionally attained a fluidity reminiscent of a jazz solo. There were nights the two whales remained to vocalize with us long after the rest of their pod had departed the immediate area.

The question has been posed whether this music with orcas is interspecies communication or just avant-garde music. In fact the latter derives from the former. The best examples of communication express recognizable harmony, rhythm, and melody, and are, therefore, the most *musical*. Music also evokes unquantifiable concepts such as emotion and community. As a recorded medium, music demonstrates a capability to engage the listener as intensively as the players. To deflect the covert criticism of the just music label, I have learned to hand the critic a recording of orca sessions with the comment, "It's music. So if there is communication, you're going to hear it. Right?" To this bold statement I would add one caveat. Whatever the verdict may be, there is nothing avant-garde about it. Indigenous people have been talking and singing with animals since before history.

Jim Nollman

FIELD STUDIES: NONINVASIVE WILDLIFE RESEARCH

The status of global wildlife populations is of grave concern to conservation biologists, with members of the order Carnivora (e.g., wolves, bears) at particular risk in many parts of the world. Terrestrial carnivores typically require large areas of habitat to meet their needs for food, mating, and dispersal, and are vulnerable to persecution when they are forced to live in close proximity with people. Thus, as habitat loss and fragmentation continue to increase against a backdrop of global climate change, a growing number of carnivore populations are in urgent need of protection.

Given the above scenario, it is more critical than ever for wildlife researchers to acquire information about the distribution, habitat use, and general ecology of carnivores. Unfortunately, the elusive and wide-ranging nature of these species, which also tend to exist in low densities, makes them notoriously

This entry was adapted from *The charged border: Where whales and humans meet* (New York: Henry Holt, 1999), by Jim Nollman.

challenging to study. Although radio telemetry and other traditional, live-capture based methods can produce valuable data pertaining to carnivore movement, survival, and related measures, such methods are labor-intensive, costly, and potentially hazardous for the animals of concern.

In recent years, a new suite of noninvasive survey techniques has become available to carnivore biologists. These techniques do not require the handling or even the direct observation of wildlife, but rather allow for the remote collection of biological samples (e.g., hair, feces—hereafter called *scat*) and other information (e.g., photographs, tracks). Beyond the advantage of requiring no physical contact with study animals, these methods are extremely effective if used appropriately. For example, survey devices can be deployed across large remote areas, and can be left in place for days or weeks without requiring researchers to return to them. This attribute can make noninvasive methods more affordable and efficient than alternative methods for collecting certain types of data. Further, the ability to use these methods across expansive terrain increases the number of animals that can potentially be surveyed. Finally, some noninvasive methods (e.g., scat detection dogs, tracking) permit animals to be studied without luring them with bait or other attractants. This can help to reduce some of the biases that may result when wildlife are drawn to locations that they might not otherwise visit, or when a subset of individuals in the population (e.g., males, young animals) is less likely to respond to attractants.

At their most basic level, noninvasive field methods probably date back to primitive humans, who no doubt engaged in tracking wildlife for food and other resources. Indeed, tracking and the interpretation of animal signs have long been fundamental tools for the field naturalist, and early published accounts of wildlife tracks and trails (e.g., Murie, 1954) served as an important foundation for wildlife biologists. But it wasn't until the mid-1990s that noninvasive survey methods for carnivores began to explode (e.g., Zielinski and Kucera, 1995), with newly emerging photographic and laboratory technologies helping to fuel the revolution.

Today's genetic techniques allow for the identification of an animal's species, sex, and genotype from noninvasively collected hair and scat samples containing adequate amounts of high-quality DNA. Coupled with modern statistical and computer modeling methods, genetic sampling potentially allows researchers to make inferences about the distribution and abundance of species across extensive survey areas and, if surveys are conducted repeatedly over time, to monitor changes in the status of populations. Scat samples in particular can also yield detailed information about the diet, health, and reproductive status of the source individual. Not surprisingly, hair and scat samples are in high demand by carnivore researchers, who continue to develop innovative techniques for their collection in the field.

Described below are several of the noninvasive survey methods currently being used by students, biologists, and other researchers who seek to better understand the population status and habitat needs of terrestrial carnivores. These methods and their various applications are discussed at length in Long et al., 2008, as well as in a growing body of peer-reviewed literature.

Tracking and Track Stations

Modern field biologists use tracking techniques very similar to those of our ancestors to determine which animals are present, where they have traveled, how many are in the area, and what types of habitats they are using. Following track trails in snow, mud, or other natural substrates can also be a good way to locate recent kills, scat, or hair samples. In some cases, researchers create special track stations to collect tracks from certain species at targeted locations. Such surveys require the special preparation of a tracking surface, such as sand or soil, in which animals leave foot impressions as they pass through. Tracks can also be collected on baited track plates, thin plates of wood or metal coated with soot, chalk, or other media. Surveys for mid-sized carnivores (e.g., fishers) often enclose track plates in box-like cubbies to protect them from the elements. Track plates advantageously provide permanent records of tracks, which can be removed from the field and studied in a laboratory setting.

Remote Wildlife Photography

Film cameras with motion-sensitive triggers have been used for decades to record photos of wildlife visiting trails, bait sites, or natural features (e.g., watering holes). Remote cameras, which are generally attached to trees or posts at field sites, can be used to document the presence of even very rare species, and provide permanent, visually compelling records of the animals of interest. Recently, advances in digital camera technology and camera designs have made this survey method much more reliable and effective. Digital technology permits cameras to collect thousands of images before reaching capacity, as opposed to the maximum of 36 photos that can be captured with a film camera. This increased capacity translates into researchers needing to visit digital camera stations far less frequently than film camera stations, thus significantly reducing labor costs. In addition, rapid-fire settings can be used to take multiple photos with very short (e.g., one-second) delays between images, resulting in pseudo-video that can be valuable for studying animal behavior.

Hair Sampling

Despite their relatively minute size, hairs from wildlife contain an amazing amount of information. For example, DNA extracted from the tiny root of an animal's hair can potentially be used to genotype or genetically fingerprint the individual so that it can be distinguished from the remainder of the population. Wildlife researchers have devised numerous creative methods for collecting hair samples. Some methods use attractants (e.g., rotten fish) to entice animals to slide under a strand of barbed wire or sticky tape, which captures small samples of hair much the way a comb does human hair. Other approaches take advantage of natural behaviors, such as when bears rub on trees and leave hair samples behind. Meanwhile, mid-sized carnivores (e.g., pine martens) can be lured into small cubbies containing bait, where they inadvertently rub hair onto small brushes affixed to the side of the enclosure while they're enjoying a free meal. Once hair samples are collected, they must be handled carefully to ensure that their DNA remains intact.

A pine marten visits a tree cubby device designed to snag a hair sample when the marten climbs inside to get the bait. (Western Transportation Institute)

A scat detection dog awaits her reward for locating a marten scat. (Western Transportation Institute)

Scat Detection Dogs

Domesticated dogs, like their wild ancestors, have highly sensitive noses. Conservation biologists have learned how to harness this sensitivity to find carnivore scats in forests and other natural settings (e.g., Long et al., 2007). With training methods similar to those used for narcotics and search-and-rescue dogs, scat detection dogs are taught to associate an enticing toy, a rubber ball on a string, for example, with scat from a particular species. Dogs can search for scats over huge areas, and are generally far more effective than human searchers at finding samples. Given the physical demands of this occupation, the best canine candidates are large, agile working breeds that have ample drive and energy. They must also be very object-focused, as their reward-toy serves as an ongoing incentive

for seeking out scat. In addition to locating scats in the field, detection dogs have been trained to detect a variety of carnivore-related odors, including carcasses and the scent of burrowing animals such as black-footed ferrets (Reindl-Thompson et al., 2006).

The Future of Noninvasive Carnivore Research

The noninvasive methods described above enable wildlife researchers to closely examine the lives of secretive species that are typically unseen by people. Given the many threats that carnivores face in our crowded world, the ability to assess and monitor wild populations is crucial if we are to ensure a future for this remarkable group of animals. Although the responsible capturing and collaring of animals will continue to be necessary in some situations, a rapidly expanding toolbox of noninvasive alternatives is now available to field biologists. These alternatives present an exciting opportunity to enhance our knowledge about carnivores while minimally disturbing them.

Further Reading
Long, R. A., Donovan, T. M., MacKay, P., Zielinski, W. J., and Buzas, J.S.. 2007. Effectiveness of scat detection dogs for detecting forest carnivores. *Journal of Wildlife Management* 71:2007–2017.
Long, R. A., MacKay, P., Zielinski, W. J., and Ray, J. C. 2008. *Noninvasive survey methods for carnivores*. Washington, DC: Island Press.
Murie, O. J. 1954. *A field guide to animal tracks*. Peterson Field Guide Series. Boston: Houghton Mifflin.
Reindl-Thompson, S. A., Shivik, J. A., Whitelaw, A., Hurt, A., Higgins. K. F. 2006. Efficacy of scent dogs in detecting black-footed ferrets at a reintroduction site in South Dakota. *Wildlife Society Bulletin* 34:1435–1439.

Zielinski, W. J., and Kucera, T. E. 1995. *American marten, fisher, lynx, and wolverine: survey methods for their detection.* USDA Forest Service, Pacific Southwest Research Station General Technical Report PSW-GTR-157, Albany, CA.

Paula MacKay and Robert A. Long

FISH

In contrast to mammals and birds, little consideration has traditionally been given to the welfare of fish. Increasing evidence indicate that fish are sentient beings, capable of suffering. Many ethicists consider sentience the key capacity for an animal to enter the moral circle, that is, to be given moral concern for their own sake. But also, according to other theories of ethics which do not focus on sentience, it may be argued that humans must care for and respect the individual fish.

Fish and Human Interaction

Human actions affect the lives of enormous numbers of individual fish through commercial fisheries as well as aquaculture. For leisure purposes, some people love the thrill of angling sport, or enjoy the beauty of ornamental fish kept in aquaria. In most of these cases, human interests may come at the expenses of the fish. In all of them, it seems relevant to discuss the moral status of fish.

The Moral Circle

Today most people in Western societies agree that animals such as mammals and birds deserve moral consideration, at least to a certain extent. That is, when humans plan to do something that will affect the welfare or interests of these animals, they must consider such effects. Fish are seldom included in these moral concerns. However, there are signs that the moral circle (see Figure 1) is now expanding to also include fish.

Throughout history, a philosophical discussion has been going as to whether or not animals ought to be moral objects, that is, worthy of moral consideration for their own sake. Early arguments against giving animals such consideration usually focused on differences between animals and humans, such as the fact that we belong to different species, that animals are not rational beings, don't have the ability to reason or not even a language, or that they can't take on moral responsibility. That is, you can't make a deal with a cat not to claw you if you promise the same. Arguments can also be religious (e.g., "this is the will of God"). The same kinds of arguments are still around today. In the late 18th century, the British utilitarian philosopher Jeremy Bentham successfully promoted the idea that only the capacity of suffering should decide a being's moral status. Thus, Bentham's argument made way for animals to enter the moral circle. One of the most well-known animal ethicists of today, Peter Singer, has further developed Bentham's utilitarian arguments that humans have the responsibility to evaluate the burdens and benefits of all sentient individuals, irrespective of species, affected by a course of action. The morally right action, then, is the one which in sum yields the best consequences for all sentient beings involved.

Today sentience has to a large extent come to mark the limit where moral concern begins. This includes several other ethical theories, for example those promoting the idea of animal rights. Tom Regan, the best-known animal rights ethicist, argue that not only humans but also

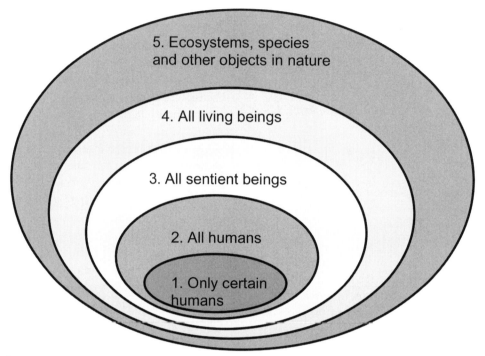

FIGURE 1. The moral circle. To be a member of the moral community or included in the moral circle is to be a being whose interests are given serious moral consideration for their own sake. The moral circle has expanded over time. During humankind's early history, only the family group or clan was included in the moral circle, while the expansion of the circle to eventually include all of humanity, for example, resulted in the UN Declaration of Human Rights. Peter Singer, among others, has argued for expanding the moral circle to include all sentient beings.

animals that are subjects of a life have inherent value and thus moral rights. The concept of subject of a life includes beings with a complex mental life, including perception, desire, belief, memory, intention, and a sense of the future—in other words, sentient beings. The basic right Regan wants to ascribe to these beings is derived from Immanuel Kant: the right never to be treated merely as a means to the ends of others. Another philosopher, Bernhard Rollin, includes sentient animals exclusively in the kind of rights-based ethics he advances. He argues that animals should be treated so that they may express or fulfill their evolutionarily imprinted natural behaviors or lives (i.e., their *telos*).

The Intangible Sentience

If fish are to enter the moral circle, sentience is the key, according to many ethicists. Sentience has also become the basis for legislative protection of animals in most countries. But what is sentience? Sentience is the ability to subjectively feel or perceive pain, for example. The International Association for the Study of Pain defines pain as "an unpleasant sensory and emotional experience associated with actual or potential tissue damage."

Pain is thus distinguished from nociception, which is merely the activity in nerves from the nociceptors (pain receptors) to the central nervous system elicited by a noxious stimulus. Nociception does not include the subjective, conscious perception of this stimulation as pain. Consciousness is another troublesome term in this discussion. Obviously, to feel pain an animal must be conscious, in the sense of being aware of its surroundings, not anesthetized or sleeping. The discussion concerns whether awareness or a higher order consciousness, a perception of self, is needed in order to experience feelings. Interestingly, no specific anatomic location for consciousness has ever been identified in the brain, and self-consciousness is so far only scientifically demonstrated in humans above a certain age, in other primates, and possibly in dolphins, elephants, and magpies. Thus, if proven self-consciousness is made a premise for sentience, all other kinds of animals are nonsentient, something which, for example, most dog owners probably would firmly reject.

Subjective experiences such as emotions and feelings are not possible to scientifically prove directly, thus they must be assessed indirectly. Several criteria are used to document pain in animals. The first condition that must be in place is that the animal possess a nociceptive system capable of transmitting signals to a sufficiently developed central nervous system, where the stimulus can be interpreted and perceived as pain by the individual. Secondly, the animal should show physiological and behavioral signs indicative of aversion when exposed to a potentially painful stimulus. These signs should disappear or diminish when a painkiller is administered. Reactions to pain may include physiological stress responses (e.g. increased heart rate, respiration, and stress hormone levels in the blood) and behavioral signs (e.g. vocalization, limping, retraction, etc.). Thirdly, the animal should learn to avoid the noxious stimulus. Self-medication, that is, when an animal in chronic pain selects food with an analgesic effect if given the opportunity, is considered a strong indicator of pain.

Are Fish Sentient Beings?

Science still lacks fundamental knowledge about the sensory apparatus of fish species and how fish perceive sensory inputs from their environment. However, there is growing evidence indicating a capacity for pain perception in at least some fish species, leading many researchers to conclude that affective states of pain, fear, and stress are likely to be experienced in fish. The evidence includes neuroanatomical similarities between fish and other vertebrates in regards to nociceptors, nerve fiber types, and neurophysiology. Most neuropeptides, neurotransmittors, and opioid receptors involved in nociception and pain modulation in mammals are also found in fish. However, the fish brain is organized very differently from the mammalian brain. The most striking difference is the lack of a cerebral cortex, a difference fish share with birds. The physiologist James Rose points to this difference when he concludes that it is implausible that fish can experience pain or other feelings. In humans, the cerebral cortex is essential for cognitive functions and is believed to play a central role in human pain perception. However, the well-known brain researcher Paul McLean has shown that a wide range of human emotions can be evoked by stimulation of those parts of the brain that are common to all verte-brates. Thus, the cortex modifies human feelings, but does not create them. Birds, in particular, but also some fish species, show advanced behaviors supporting the theory that other parts of avian and fish brains have cognitive functions which in mammals are dealt with by the cortex.

Fish display behaviors indicating pain in situations that would be painful to mammals, and this pain behavior is re-duced when analgesics are administered. Fish also learn to avoid noxious stimuli, for example, to avoid particular baits after being hooked. There is a debate among scientists about whether such behavioral reactions to nociceptive stimuli are just reflexes, like when a hand is retracted from a hot item. Obviously, some are. Nevertheless, the conscious experience of a situation is important for the learning process. It will not only minimize future risk by teaching the animal to be more careful next time, but allows it to benefit from previous experiences by modifying its behavior in new circumstances. Stud-ies have shown that fish learn to avoid aversive situations in ways that cannot readily be explained as simple reflexes.

Other Arguments for Extending Moral Consideration to Fish

The fact that fish are farmed and thus dependant on human care may automati-cally bring along a moral duty to provide for their needs. The influential philoso-pher Mary Midgley argues from an ethics of care that our moral responsibilities are derivative of our relationships with oth-ers. Our moral communities also include the animals in our care. This sense of community or connectedness comes from our shared evolutionary backgrounds and close human-animal relationships, and

serves as the basis of our ethical obliga-tions to animals. The domestication of fish and keeping them for farming pur-poses thus entails having a moral respon-sibility to care for their needs. Others have extended the idea to argue that the human-animal relationship in farming could be formulated according to the idea of a tentative contract. This can enjoin us to share the wealth created in aquaculture with all those sentient beings contribut-ing to it and care for the welfare of the individual animal, protecting the fish from exploitation, just as human workers should not be exploited.

There are also other ethical arguments that do not focus on sentience but still insist that fish should be handled with care and respect. Some of these belong to biocentric ethics, which holds that we should extend moral consideration to all living beings. Albert Schweitzer's prin-ciple of reverence for life is one example of biocentric ethics, which regards every living entity as intrinsically valuable and something that should be respected. Also, within the deep ecology movement, a principle of species egalitarianism is for-warded, but here the common denomi-nator is not life *per se* but the fact that all living beings are equally part of the Earth's biosphere. The symbiosis among humans and animals may be perceived on both a mystical and practical level, and it urges humans to handle animals with great respect.

Where to Go from Here?

One may conclude that there are many good ethical arguments why we should think twice before we expose fish to painful procedures or handle them dis-respectfully, perhaps regardless of their contingent sentience. However, some form of sentience seems likely in at least some fish species. The fact that there are some 30,000 teleost species, comprising an extremely diverse group of fish, makes it difficult to draw general conclusions about fish capacities. In this situation it could be wise to apply the precautionary principle, based on a simple risk analysis. The estimation of risk is usually based on the expected value of the conditional probability of the event occurring, multiplied by the consequence of the event, given that it has occurred. The event in this case is that fish are sentient. The consequence is the suffering of an enormous number of individual fish. The risk can be great, even though probability may be low. A reasonable risk management strategy in this case would be to implement animal welfare in fish farming, even though there still is scientific uncertainty regarding fish capacities.

The next challenge is how to provide these finned animals with a good life, since our knowledge regarding what fish require for their welfare is so far very limited. That must be the subject of further extensive research.

See also Fishing as Sport

Further Reading

Avian Brain Nomenclature Consortium. 2005. Avian brains and a new understanding of vertebrates' brain evolution. *Nature Reviews Neuroscience*, 6, 151–159.

Browman, H. I., & Skiftesvik, A. B., eds. 2007. Welfare of aquatic organisms. *Diseases of aquatic organisms*, special issue 2007, 75, no 2. Free internet access at http://www.int-res.com/abstracts/dao/v75/n2/

Chandroo, K. P., Moccia, R. D., & Duncan, I.J.H. 2004. Can fish suffer?—Perspectives on sentience, pain, fear and stress. *Applied Animal Behaviour Science*, 86, 225–250.

Chandroo, K. P., Yue, S., & Moccia, M. D. 2004. An evaluation of current perspectives on consciousness and pain in fish. *Fish and Fisheries*, 5: 281–295.

Huntingford, F.A., Adams, C., Braithwaite, V.A., Kadri, S., Pottinger, T.G., Sandøe, P. et al. 2006. Current issues in fish welfare. Review paper. *Journal of Fish Biology*, 68, 332–372.

International Association for the Study of Pain. Definition of pain. http://www.iasp-pain.org/AM/Template.cfm?Section=Home&template=/CM/HTMLDisplay.cfm&ContentID=6648#Pain

MacLean, P. 1990. *The triune brain in evolution*. New York: Plenum Press.

Midgley, M. 1983. *Animals and why they matter*. Athens: University of Georgia Press.

Regan, T. 1983. *The case for animal rights*. Berkeley: University of California Press.

Rollin, B. E. 1995. *Farm animal welfare*. Ames, IA: Iowa State University Press.

Rose, J. D. 2002. The neurobehavioral nature of fishes and the question of awareness and pain. *Reviews in Fisheries Science*, 10, 1–38.

Schweitzer, A. 1936. The Ethics of Reverence for Life. *Christendom*, 1, 225–239.

Singer, P. 1981. *The expanding circle: Ethics and sociobiology*. New York: Farrar, Straus & Giroux.

Singer, P. 1990. *Animal liberation*. 2nd ed. New York: New York Review of Books.

Cecilie M. Mejdell
Vonne Lund

FISHING AS SPORT

Many humans interact with fish on a regular basis, although for most people this is not an intimate relationship. Fish are cold blooded, slimy, and inhabit an alien waterworld in which humans travel with difficulty. Despite that, fish are a mainstay of human diets. In the Western world, dieticians and health gurus tell us that if we want to lead long, happy lives, we need to eat more fish rich in heart-friendly Omega-3 and Omega-6 fatty acids. Fish also drive the symbolism and life rhythms of entire cultures, such as those for many of North America's Pacific Coast First Nations, whose year revolves around the Pacific salmon. Despite this, most humans interact with fish only at the seafood counter at the local supermarket, where our piscine friends generally arrive filleted and skinned from industrial commercial fisheries, or aquaculture operations.

It wasn't always this way. For millennia the primary interaction people had with fish was to enter the fish's world and devise ways to catch them. The modern, technology-driven fishing fleets of today are a far cry from the one-on-one struggle that for most of human history dominated the capturing of fish. At some point in our evolution, people consciously or unconsciously came to the realization that the process of fishing was pleasant, even spiritual. Out of this was born the pastime of sport fishing, the quest of an individual angler armed with a fishing rod to capture a fish.

The earliest sport-fishing record we have, at least in the English language, is that of Dame Juliana Berners, *The Treatyse of Fysshynge with an Angle.* Dame Juliana was reportedly a nun and prioress of an abbey in Hertfordshire, England, but there is dispute over whether or not she actually existed. Some believe that the name is a pseudonym for the true author, who wished to remain anonymous. The book, however, definitely exists, is written in the English language style of the 15th century, and appeared in 1496. The *Treatyse's* primary purpose was to inspire people to go sport fishing, but it was also the start and inspiration for the voluminous English language angling literature which continues to pour forth to this day.

Consistent with her supposedly being a prioress, Dame Juliana starts her treatise by quoting the parables of Solomon, noting in particular that a healthy, happy,

righteous life flowed from a beauty of spirit ("a good spyrite maketh a flouring age that is a fayre age and a longe."). She believed that to achieve that beauty a person needed to pursue activities that nurtured the spirit (". . . a mery occupacion which may rejoice his harte, and in which his spirites may haue a mery delyte."). Not for her the contemporary popular pastimes among the noble-born of hunting, hawking or fowling, which were "laborious and greuous (grievous)" occupations and did not get people out of bed early enough to be ". . . holy, helthy & happy." Angling was the ticket, and in her how-to book she takes prospective anglers with simplicity and great accuracy through the equipment and techniques needed, on a species-by-species basis, for catching fish with a fishing pole. She even includes a description of the first reported artificial flies, and the materials needed to tie them.

Sport anglers today are more or less divided into two major groups: those with hardware and those devoted to fly fishing. Hardware fishermen use a variety of artificial metal lures, and/or baits to try and entice a fish to get caught. The equipment is primarily designed to securely hook and retain a fish, with the intent to take it home and eat it.

The fly fisherman typically approaches the sport differently. Fly fishing is full of social hierarchies, elaborate rituals, and techniques that have to be perfected in order to become a respectable fly fisherman. For example, aficionados believe that they must master fly-tying, which requires artistic capacities, manual dexterity, and a house full of esoteric materials like jungle cock feathers and fur strands that can be woven into the dress of an effective artificial fly. They must equip themselves from head to toe; proper dress includes waders, a fishing vest stuffed with tools, and a good hat. Finally, they need a good fly rod and reel, and through patience and hard work develop the motions that cast a nearly weightless fly accurately to the places in the water where the fish are waiting. All of this develops within the true partisan a particular identify as a fly angler, camaraderie with other like minded individuals, and at times a particular bent to the psyche. The sum of these characteristics was cogently captured by writer Fen Montaigne in his description of Atlantic salmon anglers: "In the angling world, there is no snob like an Atlantic salmon snob. And while being mindful not to tar all Atlantic-salmon fishermen with the same brush, the truth is this: many devotees of the 'sport of kings' are insufferable, elitist, tweedy, name-dropping bores" (p. 41).

Fly fishing goes on in unlikely places, under unlikely circumstances, with unlikely species. Atlantic salmon anglers were among the first wave of Westerners to enter Russia when the Soviet Union dissolved. They were seeking the undisturbed rivers of the Kola Peninsula, and in these turbulent times some of them found themselves being escorted back out of the country at gunpoint. Fly fishing sport camps for peacock bass have been established in the Amazon River basin, and at least one of them has been overrun by guerillas, with the anglers escaping into the jungle. Saltwater fly fishermen prize bonefish, and some are now even pioneering techniques for catching sharks!

Sport fishing is big business. In North America, people spend millions of days and billions of dollars each year on fishing trips. These expenditures create valuable employment in rural areas for guides and small businesses like hotels and restaurants, and play to the traditional nature-oriented skills of people in these regions such as boat handling and river navigation. Since people take care of the things that they value, the economic benefits of sport fishing provide a powerful incentive to conserve fish populations and maintain clean water. Despite this, there are many instances where too many sport anglers are chasing too few fish, which can have severe impacts on fish populations (Arlinghaus & Cooke, 2005).

Recent surveys of recreational anglers consistently show that the thing they value most is not catching a fish. Rather, it is the joy of being in the natural world and the gentle pace of life on the water. They are seeking to massage their spirits, which is exactly why Dame Juliana recommended the activity over 500 years ago.

Some anglers so prize the fishing experience and the conservation of fish populations that they can no longer bring themselves to kill a fish that they have caught. This has given rise to the practice of live release, also known as catch-and-release. Simply put, live release means that you treat a fish gently as you reel it up next to your boat or into a net, that you remove the hook as quickly as possible, preferably without taking the fish out of the water, to minimize stress, and you then let it swim back into the wild. Many studies have shown that many species of fish treated this way will survive, reproduce, and even be caught again by anglers a second time or more. However, while live release has proved to be a successful and valuable conservation tool, it has not been without controversy.

Humans have to eat, and most societies accept the capture of fish for consumption as an ethical and necessary human behavior. However, there has been a growing movement that has questioned the ethics of angling in general, and live-release fishing in particular, irrespective of the conservation and water quality benefits that the presence of a sport fishery can bring. With live-release fishing, many believe it is cruel to capture fish by impaling it on a metal hook, forcibly coercing it up to wherever the angler happens to be positioned, and then releasing it back to the wild to try and do the same again. Caught-and-released fish are often injured.

A key component of the cruelty argument revolves around fish awareness and whether or not they feel pain. The available scientific evidence is conflicting and contradictory. Some hold that the brains and neural systems of fish are not sufficiently developed to experience pain and awareness (Rose, 2002). However, recent experiments have generated results that consistently showed fish detecting and non-reflexively attempting to avoid noxious stimuli and pain (Sneddon, 2003, Sneddon et al., 2003). Scientific work is ongoing in this important field (for a recent review, see Arlinghaus et al., 2007), and there is a great deal at stake.

See also Fish

Further Reading

Arlinghaus, R., & Cooke, S. J. 2005. Global impact of recreational fisheries. *Science*, 307, 1561–1562.

Arlinghaus, R., Cooke, S. J., Lyman, J., Policansky, D., Schwab, A., Suski, C. et al. 2007. Understanding the complexity of catch-and-release in recreational fishing: an integrative synthesis of global knowledge from historical, ethical, social and biological perspectives. *Reviews in Fisheries Science*, 15, 75–167.

Berners, Dame Juliana. 1496. *The treatyse of fysshynge with an angle*. Available at http//darkwing.uoregon.edu/~rbear/berners/berners.html. Published online by Risa Stephanie Bear.

Economic and Policy Analysis Directorate 2003. 2000 Survey of recreational fishing in Canada. Economic and commercial analysis report No. 165, Canadian Department of Fisheries and Oceans.

Montaigne, F. 1999. *Hooked: Fly fishing in Russia*. London: Phoenix.

Muoneke, M. I. 1994. Hooking mortality: a review for recreational fisheries. *Reviews in Fisheries Science*, 2, 123–256.

Rose, J. D. 2002. The neurobehavioral nature of fishes and the question of awareness and pain. *Reviews in Fisheries Science*, 10, 1–38.

Sneddon, L. U. 2003. The evidence for pain in fish: the use of morphine as an analgesic. *Applied Animal Behaviour Science*, 83, 153–162.

Sneddon, L. U., Braithwaite, V. A., & Gentle, M. J. 2003. Do fishes have nociceptors? Evidence for the evolution of a vertebrate sensory system. Proceedings of the Royal Society London B, 270:, 1115–1121.

Whoriskey, F. G., Prusov, S., & Crabbe, S. 2000. Evaluation of the effects of catch-and-release angling on the Atlantic salmon (*Salmo salar*) of the Ponoi River, Kola Peninsula, Russian Federation. *Ecology of Freshwater Fish* 9: 118–125.

Fred Whoriskey

FOOD ANIMALS: ETHICS AND METHODS OF RAISING ANIMALS

A fundamental ethical choice concerns whether it is morally acceptable for humans to use nonhuman animals at all, for any purpose, including food. Once a choice has been made, by an individual or a society, to raise certain animals for food, ethical issues center around what considerations humans owe these animals, both as species and individuals, in life and in death. Answers may be influenced by the ethical frameworks within which issues are examined (e.g., utilitarian, contractualist, and so forth), the relative weights given by decision-makers to human and animal interests, which may rest on personal value systems, and whether or not decision-makers perceive animals' interests to be consistent with or opposed to humans' interests.

Some Ethical Considerations Regarding Species

Which species are suitable or appropriate for domestication, that is, can be expected to have a reasonably good life if under human control?

Should humans modify animals raised for food genetically in ways that threaten their ability to survive as a species without human intervention. for example, broad-breasted turkeys which can only reproduce by artificial insemination, or double-muscled breeds of cattle that require repeated caesarians to deliver their calves?

At what point are genetic changes likely to become irreversible, and should humans stop selecting for changes before they reach that point, in case individuals or society deem at a future time that these changes have been taken too far?

Should animals be genetically selected or modified to survive in production environments that humans have been unwilling to improve in ways that would meet their species-specific needs? Should producers be required to improve production environments and management techniques that result in a high occurrence of poor welfare indicators, instead of or before further selecting animals to meet human interests? And should breeding efforts be directed toward increasing viability of modern food-producing animals rather than on further increases in production and growth?

Once an ethical choice has resulted in individuals or species becoming unable to survive on their own, should future research and selection be directed toward reversing the unsustainable condition, thereby eliminating the condition that creates pain and distress? Should humans try to genetically select for reduced calf sizes in double-muscled cattle or research the most effective pain killers to administer during and after deliveries? The first restores to future animals a natural ability necessary for the breed to sustain itself and relieves future individuals of the distress of difficult births. The second, if attempted independently of the first, reinforces the condition requiring human intervention, but attempts to mitigate the impact on each individual animal. A decision could be made to attempt both: aim for a long-term, sustainable solution for the breed, while providing relief from suffering for individuals undergoing procedures now. Alternatively, a more radical decision could be made to stop breeding double-muscled cattle altogether and allow the breeds to die out.

Some Ethical Considerations Regarding Individuals How far should humans go to ensure that animals raised for food are spared pain, fear, distress, boredom and suffering, that is, eliminate or reduce the occurrence of negative consequences of human management?

How far should humans go to ensure that animals raised for food have positive life experiences, that is, permit them to satisfy innate needs such as mothering or enable freedom of movement and choice of social companions or ensure that experiences provide functional feedback? That is, from the animal's point of view, its actions with respect to its environment have the intended outcome. For example, chickens' dust bathing behavior results in cleaner feathers, which would only happen in a proper dust bath and not on a wire cage floor. Nest building behavior results in creating a secure and comfortable place for the sow to bear her young, which can only happen in a natural or artificially-enriched production environment and not in a farrowing crate..

Decisions about Methods and Systems of Raising Animals for Food Decisions might be taken from the perspective that *humans* are better off spiritually, economically, physically, morally, or ecologically if animals raised for food are provided with positive life experiences, genetic resources are managed so that animals are healthy and self-sustaining, and death comes to them swiftly and without fear or pain or arguments could be made that animals ought to be afforded these things because animals themselves have direct moral status. However, without the power to command humans to respect their rights or their moral status, the possession of rights may have as little practical advantage to animals as the possession of human rights appears to have to oppressed peoples. Legislatures in major farm states in the United States, at industry urging, have enacted legislation exempting animals raised for food from protections afforded other animals under state anticruelty statutes (Wolfson, 1996). Hence, at least in those states, without legal mechanisms in place to protect animal rights, arguments from this position may have insufficient practical value to make a difference in animals' lives.

A growing number of people, including many farmers, appears to accept that animals have a moral status in which their interests count directly in the assessment of actions that affect them, but do not count for as much as humans' interests (Wilson, 2006). They accept that animals

are raised for food for humans, but also desire that animals have lives worth living and come to their deaths without fear or pain. In a lecture to veterinary students, Waldau (2005) notes that "what is at issue for many people today . . . is not necessarily the value of traditional practices, but, rather, the ethical dimensions of certain modern practices and methods chosen because they create economic efficiencies."

Since the publication of *Animal Machines: The New Factory Farming Industry* by Ruth Harrison in 1964, much attention has been devoted to the conditions to which animals raised for food are subjected. *Animal Machines* directly influenced the development of a new scientific discipline, animal welfare science, which in the intervening years has produced a vast literature on the biological and behavioral needs of animals raised for food. This body of research has gone a long way toward illuminating conditions that can afford such animals a life worth living. Harrison herself advocated an ethic of fair play as the only way humans can repay animals for the sacrifices humans ask of them.

Several sets of criteria have been put forth regarding the adequacy of farming systems for meeting welfare needs of animals raised for food. The most well-known of these is the Five Freedoms enumerated by the Farm Animal Welfare Council (FAWC) in the United Kingdom:

1. Freedom from hunger, thirst, and malnutrition

2. Freedom from physical and thermal discomfort

3. Freedom from pain, injury, and disease (including parasitical infections)

An injured goat bleeds after a horn is ripped off during transit. (Farm Sanctuary)

4. Freedom to express normal behavior

5. Freedom from fear and distress, including predators

In an essay on health and wellbeing of companion animals, Dr. Michael W. Fox has enumerated Five Principles for Animal Health and well-being:

1. Right understanding and relationship

2. Right breeding/genetics

3. Right nutrition

4. Right environment

5. Right holistic veterinary care

The principles apply equally well to animals raised for food.

Fraser et al. (1997) note three overlapping ethical concerns expressed by the public for the welfare of animals raised for food. These are:

1. Animals should lead natural lives through the development and use of their natural adaptations and capabilities

2. Animals should feel well by being free from prolonged and intense fear, pain, and other negative states and by experiencing normal pleasures, and

3. Animals should function well in the sense of satisfactory health, growth and normal functioning of physiological and behavioral systems

And Hurnik (1988) conceptualized animal wellbeing as

> a condition of physical and psychological harmony between the organism and its surroundings. [In this conceptualization,] harmony [is] based on an acceptance of [a] basic moral principle that every sentient, living organism subjected to full, direct human control, should have an opportunity to experience an environment for which its own genotype is predisposed, in order to develop into a physically and psychologically healthy organism.

Ethical Performance of Different Systems of Raising Animals for Food

Some may argue that human interest in cheap food outweighs animals' interests in having lives worth living. A growing comprehension of the environmental consequences, human health risks, and net economic costs of industrialized animal production lead others to question whether human and animal interests are as far apart as has been believed, and whether society is really better off when food is cheap. These are human interests. This chapter considers only the ethical performance of systems in meeting animal needs and interests.

Using the above criteria, one can examine systems of raising animals as to their ability to deliver to animals a life worth living. Modern animal production systems range widely along a spectrum from most exploitive to most supportive of animal interests, for example:

System 1: Conventional, industrially-oriented systems characterized by high capital investments, high volumes of production, and a high degree of control over or restrictions on animal biology and behaviors

System 2: Modest changes to conventional systems such as banning battery cages and gestation crates, but leaving the basic animal genetics and industrial approach in place

System 3: Confinement systems taken a step further with behaviorally-appropriate space allowances for freedom of movement and positive social interactions, high levels of environmental enrichment with natural materials such as deep straw bedding for occupation, fiber fill, and comfort, limited access to the outdoors, opportunities for mothers of most species to care for their young

System 4: Free-range systems where normal behaviors are not restricted, breeding programs emphasize the ability of individuals to sustain themselves and their breed, and appropriate shelter for weather extremes and protection from predators and supplemental environmental enrichment are provided

In practice, management and husbandry range widely from extreme abuse, as demonstrated by undercover videos taken inside industrial facilities and slaughter plants, to high levels of care and consideration of animal needs. Here systems are compared assuming that husbandry reflects the highest level of competence for operators in all systems, so that only the systems themselves are being examined. Possible results are shown in Table 1.

TABLE 1. Comparison of systems using 4 ethical models

Ethical criteria	System 1. Industrial	System 2. Industrial Cage-, crate-free	System 3. Enriched confinement	System 4. Free-range
Five Freedoms (FAWC)				
1. Freedom from hunger, thirst, malnutrition	–	–	+	+
2. Freedom from thermal and physical discomfort	–	–	+	+
3. Freedom from pain, injury and disease	–	–	+/–	+/–
4. Freedom to express normal behavior	–	–	+/–	+
5. Freedom from fear and distress	–	–	+	+
Five Principles (Fox)				
1. Right understanding and relationship	–	–	+	+
2. Right breeding/genetics	–	–	+/–	+
3. Right nutrition	–	–	+	+
4. Right environment	–	–	+/–	+
5. Right holistic veterinary care	–	–	+	+
Fraser, et al.				
1. Use of natural adaptations & capabilities	–	–	+/–	+
2. Free from negative emotional states. Feeling normal pleasures.	–	–	+/–	+
3. Satisfactory health, growth, normal functioning of physiological and behavioral systems	–	–	+/–	+
Hurnik	–	–	+	+

The basic biological and behavioral characteristics of an animal comprise his or her initial endowment of tools and resources for coping with the environment. When these basic characteristics are compromised, for example by selective breeding for single traits such as fast growth or milk yield, the animal's ability to adapt to his or her environment may be compromised. Industrial systems clearly fail to meet animals' basic needs and interests. However, despite the slight increase in freedom afforded to cage- and crate-free animals, if all else in the production system remains unchanged, these animals may still experience a poor level of welfare. Only System 4, the free-range system, where birds and animals are exposed directly to variations in climate and terrain, requires more robust genetics for the birds or animals to thrive in their environments. Similarly, System 3, while allowing greater freedom of expression and opportunities for occupation and comfort than Systems 1 and 2, still prevents the animal from exercising certain normal patterns of behavior, such as grazing, an important feeding behavior for herbivores and for pigs as well.
See also: Factory Farms.

Further Reading

Farm Animal Welfare Council, U.K. Five freedoms. http://www.fawc.org.uk/freedoms.htm, accessed October 1, 2008.

Fox, M. W. (no date). Companion animals: Responsibilities, care, and rights: A synopsis,. http://tedeboy.tripod.com/drmichaelwfox/id11.html, accessed October 1, 2008.

Fraser, D., Weary, D. M, Pajor, E. A., & Milligan, B. N. (1997). A scientific conception of animal welfare that reflects ethical concerns. *Animal Welfare, 6(3)*: 187–205.

Halverson, M. (2002). *Animal health and wellbeing.* Technical Working Paper for the State of Minnesota Generic Environmental Impact Statement on Animal Agriculture. St. Paul, MN: Environmental Quality Board, http://www.eqb.state.mn.us/geis/TWP_Animal Health.pdf.

Harrison, R. (1964). *Animal machines: The new factory farming industry.* London: Vincent Stuart.

Hurnik, J. F. (1993). Ethics and animal agriculture. *Journal of Agricultural and Environmental Ethics, 6,* Special Supplement 1.

Singer, P. (1990). *Animal liberation.* New York: Avon Books.

Teutsch, G. M. (1987.) Intensive farm animal management seen from an ethical standpoint, in von Loeper, E., Martin, G., Muller, J., Nabholz, A., van Putten, G., Sambraus, H.H., et al., *Ethical, ethological and legal aspects of intensive farm management.* Basel: Birkhäuser Verlag.

Wilson, S. (2006). Animals and ethics. *The internet encyclopedia of philosophy.* http://www.iep.utm.edu/a/anim-eth.htm, accessed July 15, 2008.

Wolfson, D. (1996). *Beyond the law: Agribusiness and the systemic abuse of animals raised for food or food production.* New York: Archimedian Press.

Marlene Halverson

G

THE GENDER GAP AND POLICIES TOWARD ANIMALS

Differences in the attitudes and behavior of women and men towards animals have long been observed. Women comprise the majority of activists, members, and donors in the animal protection movement. In study after study, women generally express more favorable attitudes than men towards animals and animal protection policies.

According to Kellert and Berry (1987), gender is "among the most important demographic factors in determining attitudes about animals in our society." Women are more likely than men to support animal welfare positions and to express concerns about the moral treatment of animals (Jerolmack, 2003; Hills, 1995; Herzog and Galvin, 1997; Peek et al., 1996). Women are less likely to support animal use. While women and men share similar levels of concern about conservation, women are more supportive of strengthening the Endangered Species Act (Czech et al., 2001). Women also are more likely to oppose lethal wildlife management (Korval et al., 2004; Teel et al., 2002). In his comprehensive review of thirty-one human-animal interaction studies, Herzog (2007) found women consistently more sympathetic than men to animals, although the effect sizes varied.

In the political arena, the term gender gap is used to describe the differences between male and female attitudes and voting patterns. Since women register to vote in higher numbers and have a higher rate of turnout, the gender gap can be the margin of difference in close political races (Smeal, 1984).

While the gender gap is most often associated with divergent party preferences, candidate choices, and positions on war, social welfare, and women's rights, the magnitude of gender-based attitudinal differences on animal-related issues is comparable to and in some cases exceeds these more traditional gender gaps. The gender gap, often in double digits, has been a constant factor in animal protection victories in state-level ballot measure campaigns in which the public votes directly on policy measures.

Animal protection organizations have increasingly turned to ballot measures when legislative and administrative channels have been blocked (Pacelle, 2001). In response, opponents of animal protection have placed measures on the ballot to reverse pro-animal gains. Since 1990, animal protection organizations have prevailed in 28 out of 41 ballot measure contests (Humane Society of the United States, 2008). Animal protection ballot measure victories in California and Oregon have included bans on sport hunting of mountain lions and the use of baiting, hounds, and body gripping traps for bear and other furbearing species in Colorado, Arizona, Massachusetts, Washington, and Oregon. Animal advocates also have won measures prohibiting sow gestation crates in Arizona, Florida, and California, veal crates in Arizona and California, and battery cages for chickens in California. Arizona and Missouri have recently banned cockfighting. Slaughter of horses and sale of horsemeat for human consumption is banned in California, and greyhound racing has been banned in Massachusetts.

Pre- and post-campaign public opinion polls and ballot measures provide insights into the size of the gender gap on animal policy issues and its political consequences. Polls conducted in 10 ballot measure campaigns between 1995 and 2008 suggest that women voters favored the animal protection position by gender gaps ranging from 7 to 25 points (see Table 2).

In some contests, women voters have been decisive in animal protection victories. In these cases, supermajorities of women voters provided the margin of victory for animal protection measures in the face of opposition from the majority of male voters. For example, in Michigan, voters rejected the attempts of the legislature to repeal a ban on dove hunting by 69 percent to 31 percent. Pre-election polls in Michigan showed less than half of male voters supporting the ban, in contrast to almost three-fourths of female voters.

Animal policy and traditional gender gap issues share a common basis in women's greater levels of compassion and opposition to the use of force. Women are more likely to support social welfare programs for the needy and disadvantaged and oppose discrimination on the basis of sex, race, and sexual orientation (Center for American Women and Politics, 2008; Smeal, 2004). Similarly, women

are more likely to attribute mental capacity to animals and to regard animals as sentient beings, which influences their attitudes towards animals (Herzog and Galvin, 1997; Hills, 1995).

Women's strong opposition to hunting and trapping converges with women's negative attitudes towards weapons and the use of force. Women demonstrate far greater support for gun control and opposition to military intervention (Howell and Day, 2000). The gender gap on force issues has been found to be even greater than gender differences on compassion issues (Shapiro and Mahajan, 1986).

Causes of the Gender Gap

Social scientists posit cultural, structural, and ideological explanations for the gender gap. Often grounded in the work of Carol Gilligan (1982), cultural explanations of the gender gap maintain that differential socialization of boys and girls produces value differences which, in turn, contribute to distinctive political attitudes and behaviors (Howell and Day, 2000). Females are socialized to be more oriented toward caring, nurturance, cooperation, interpersonal relationships and responsibility. Males are socialized to be more oriented towards rules and rights and to be more competitive.

Feminist animal care theory is in part based on the assumption that women's greater concern for relationships is reflected in feelings of connection to nature and other living beings. Lauber et al. (2001) found that women contextualized their positions on deer management issues, considering more criteria than men. Women expressed concerns about whether management techniques would result in the suffering or death of deer, have unintended effects on pets or

TABLE 2   Gender Gaps in Pre-Election Polls of Likely Voters in Animal Protection Ballot Measure Contests

State/Year	Ballot Measure Question	Pollster/Sample Size	Gender Gap	Animal Protection Election Outcome
CA 2008	Prevention of farm animal cruelty	Survey USA 661 sample	14 pts.	Victory 63%-47%
MI 2006	Dove hunting	Lake Snell Perry Mermin 500 sample	25 pts.	Victory 69-31%
AZ 2006	Sow gestation and veal crates	Lake Snell Perry Mermin 200 sample	15 pts.	Victory 62-38%
ME 2004	Bear baiting, hounding, and trapping	Decision Research 400 sample	11 pts.	Defeat 48-52%
OK 2002	Cockfighting	Decision Research 500 sample	19 pts.	Victory 56-44%
WA 2000	Steel traps and poisons	Decision Research 600 sample	8 pts.	Victory 55-45%
OR 2000	Steel traps and poisons	Decision Research 600 sample	12 pts.	Defeat 41-59%
OH 1998	Morning dove hunting	Decision Research 800 sample	21 pts.	Defeat 41-59%
MA 1996	Body-gripping taps, hound hunting of bears and bobcats, and wildlife board	Decision Research 500 sample	14 pts.	Victory 64-36%
WA 1996	Bear baiting and hound hunting of bears, cougars, bobcats, and lynx	Decision Research 600 sample	7 pts.	Victory 63-37%

Polling data made available courtesy of Humane Society of the United States, with the exception of the 2008 SurveyUSA poll of likely California voters. The polls have sampling error rates from $+/-$ 3.5–4.5%.

nontarget wildlife, or involve weapons. Studies of attitudes toward the environment also attribute women's greater support for the environment to awareness of the consequences of human actions and concern for nonhuman beings (Zelezny, 2000).

A related cultural explanation for the gender gap has been women's experience of motherhood. This approach argues that women's responsibility for children translates into an ethic of caring and nonviolence. However, a number of studies have questioned maternal thinking as the basis for the gender gap, since gendered differences in attitudes predate motherhood and are found to be especially strong among those who have never had children. The absence of children in households has been associated with greater concern for animals (Kendall et al., 2003).

Others tie the gender gap to feminism and women's structural position in a patriarchal society (Peek et al., 1997). Feminist identity correlates with a sense of egalitarianism, liberal ideology, a modern view of sex roles, and expression of sympathy for the disadvantaged. Donovan and Adams (2007) have articulated the connection between sexism and speciesism. Peek et al. (1996) argue that women's experiences with oppression make women more disposed to support animal rights, although egalitarian attitudes tend to account for differences among women on animal rights rather than between women and men. Additional structural explanations for the gender gap include women's increasing personal and economic autonomy and women's closer relationship to the state as beneficiaries and public employees.

Obstacles to Measuring the Gender Gap

Despite the consistency and prominence of the gender gap, examination of gender differences in attitudes towards animal policy has been hindered by several methodological and data collection challenges. First, with several notable exceptions, few studies of animal attitudes have been conducted with national, random sample surveys. Most attitudinal surveys have relied on convenience samples. Second, while wildlife researchers have used random sample surveys to a greater extent than other animal-related attitudinal studies, wildlife surveys most often use sampling frames such as telephone lists that significantly under-represent women. As a result, findings from these studies are based on samples that are disproportionately male, in some cases as high as 80 percent (Czech et al. 2001). Third, to date, media-sponsored exit polls have not included animal-related ballot measures or issues in their Election Day surveys.

The intersection of gender and animal protection interests can be seen in the substantial gender gap in attitudes towards animals and animal policy issues. The gender gap in voting behavior has profound implications for the success of animal protection measures in the political arena.

Further Reading

Anderson, K. 1999. The gender gap and experience with the welfare state, *PS*: 17–19; Center for American Women and Politics, Women's Vote Watch (2008). Accessed at http://www.cawp.rutgers.edu/fast_facts/elections/wvwatch/index.php.

Conover, P. 1988. Feminists and the gender gap. *The Journal of Politics* 5, 4: 985–1010.

Czech, B., Devers, P. K., and Krausman, P. R. 2001. The relationship of gender to species conservation attitudes. *Wildlife Society Bulletin* 29: 1: 187–194

Donovan, J. and C. Adams, Eds., *The feminist care tradition in animal ethics* (New York: Columbia University Press, 2007)

Gilligan, C. 1982. *In a different voice: Psychological theory and women's development.* Cambridge: Harvard University Press.

Herzog, H. 2007. Gender differences in human-animal interactions: A Review. *Anthrozoos* 20:1: 7–21.

Herzog, H. A. and Galvin, S. 1997. Common sense and the mental lives of animals: An empirical approach. In R. W. Mitchell, N. S. Thompson, and H. L. Miles, *Anthropomorphism, anecdotes, and animals.* Albany: State University of New York Press.

Hills, A. M. 1995. Empathy and belief in the mental experience of animals. *Anthrozoos* 8:3: 132–142.

Howell, S. and Day, C. 2000. Complexities of the gender gap. *Journal of Politics* 62:3: 858–874.

Humane Society of the United States. Post-1990 initiative and referendum summary—Animal issues, (2008): Accessed at http://www.hsus.org/legislation_laws/ballot_initiatives/past_ballot_initiatives/.

Jerolmack, C. 2003. Tracing the profile of animal rights supporters: A preliminary investigation. *Society and Animals* 11: 3.

Kendall, H., Lobao, L., and Sharp, J. 2006. Public concern with animal well-being: Place, social structural, location and individual experience. *Rural Sociology* 71:3: 399–428.

Kellert, S. and Berry, J. 1997. Attitudes, Knowledge and Behaviors toward Wildlife as Affected by Gender. *Wildlife Society Bulletin* 15: 363–371.

Korval, M. and Mertig, A. 2004. Attitudes of the Michigan public and wildlife agency personnel toward lethal wildlife management. *Wildlife Society Bulletin*, 32:1: 232–243.

Lauber, T.B., Anthony, M., and Knuth, B. 2001. Gender and ethical judgments about suburban deer management. *Society & Natural Resources* 14:571–583.

Pacelle, W. 2001. The animal protection movement: A modern day model use of the initiative process. In M.D. Waters, ed., *The battle over citizen lawmaking*. Durham: Carolina Academic Press.

Peek, C., Bell, N. and Dunham, C. 1996. Gender, gender ideology, and animal advocacy. *Gender & Society* 10: 4: 464–478.

Peek, C., Dunham, B. Chorn, and Dietz, B. 1997. Gender, relational role orientation, and affinity for animal rights. *Sex Roles* 37: 11/12: 905–920.

Shapiro, R. and Mahajan, H. 1986. Gender differences in policy preferences: A summary of trends from the 1960s to the 1980s. *Public Opinion Quarterly* 50: 1: 42–61.

Smeal, E. 1984. *Why and how women will elect the next president.* New York: Harper and Row.

Teel, T., Krannich, R., and Schmidt, R. 2002. Utah stakeholders attitudes toward selected cougar and black bear management practices. *Wildlife Society Bulletin*, 30:1: 2–15.

Zelezny, L., Chua, P., and Aldrich, C. 2000. Elaborating on gender differences in environmentalism. *Journal of Social Issues* 56:3: 443–457.

Jennifer Jackman

GENETIC ENGINEERING

Although humans have always genetically engineered domesticated animals to suit their uses of these animals, the only tool available to accomplish this in the past was to breed animals selected specifically for this purpose. This in turn required many generations of gradual change in order to produce significant changes in the animals, and also limited manipulation of genes to those that could be introduced by normal reproduction. Since the late 1970s, however, the technology for inserting all manner of genes into an animal's genome, including radically foreign genes (for example, genes from human beings), has progressively developed in sophistication. This opens up a vast range of possibilities for manipulating animals' genetic makeup and thus their phenotypic traits. In 1989, the U.S. Patent Office announced that it had issued the first animal patent for a mouse that was genetically engineered to be highly susceptible to developing tumors, a trait rendering the animal extremely valuable for cancer research.

Genetic engineering and the potential for patenting the resulting animals have evoked strong negative criticism, largely from theologians and animal advocates. Theologians express concern that genetic engineering does not show proper respect for the gift of life and implies that humans are playing God. Although such religiously based criticisms are perhaps meaningful within the context of a religious tradition, it is difficult to extract from them any ethical content that can be used to illuminate the issue of genetic engineering of animals in the context of social ethics. Animal advocates, on the other hand, express the concern that genetic engineering and animal patenting will result in increased animal suffering.

It is certainly not necessarily the case that genetic engineering of animals must inevitably result in increased suffering for animals. Genetic engineering can, in principle, significantly reduce animal suffering by, for example, increasing animals' resistance to disease. This has already been accomplished in chickens which have been genetically engineered to resist some cancers. Furthermore, genetic engineering could be employed to correct suffering created by traditional breeding, as in the case of the more than 400 genetic diseases in purebred dogs that have been introduced into these animals by breeding them to fit aesthetic standards. Third, genetic engineering could be used to make animals more suited to the harsh environments in which we raise them, for example, hens kept in battery cages, though both common sense and common decency suggest that it makes more sense to change the environment to fit the animals than vice versa.

But animal advocates are correct in their concern that if current tendencies in animal use continue unchanged, they will favor genetic engineering being used in ways whose result, albeit unintended, will increase animal suffering. Consider animal agriculture. Traditional pre-mid-20th-century agriculture was based on animal husbandry, that is, caring for animals, respecting their biological natures, and placing them into environments for which they would be optimally suited; the producer did well if and only if the animals did well. Animal suffering worked as much against the farmer's interests as against the animal's interests, and thus animal welfare was closely connected with animal productivity. However, the advent of high-technology agriculture allowed farmers to put animals into environments that did not suit them biologically (e.g., battery cages), yet in which they could still be productive.

One major and legitimate concern is that genetic engineering not be used as yet another tool to augment productivity at the expense of animal welfare. Thus, for example, in the early 1980s, pigs were genetically engineered to produce leaner meat, faster growth, and greater feed efficiency. While this was accomplished, the negative effects of this genetic engineering were unexpected and striking, with the animals suffering from kidney and liver problems, diabetes, lameness, gastric ulcers, joint disease, synovitis, heart disease, pneumonia, and other problems.

To prevent the use of genetic engineering as a tool enabling us to further erode animal welfare for the sake of efficiency, productivity, and profit, Bernard Rollin proposed the principle of conservation of welfare as a check on commercial use of genetic engineering of animals,: Genetically engineered animals should be no worse off than the parent stock would be if they were not so engineered. Such a principle should serve to forestall new suffering based in genetic engineering for profit.

The second major source of suffering growing out of genetic engineering of animals comes from our increasing ability to create transgenic animal models for human genetic disease. Genetic engineering gives researchers the capability to genetically create animals who suffer from human genetic diseases. This means that vast numbers of defective animals will be created to research these human diseases. In many if not most cases of genetic disease, there is no way to control the painful symptoms, and reducing the animals' suffering through early euthanasia is excluded, since researchers wish to study the long-term development of the disease. Thus this sort of genetic engineering creates a major problem of animal suffering. Thus far, neither the research community nor society in general has addressed this issue, despite society's 1985 expression in federal law of its ethical commitment to limit animal suffering in biomedical research.

Further Reading

Duvick, Donald N., ed. 1991. *National agricultural biotechnology at the crossroads: Biological, social, and institutional concerns*, NABC Report 3. Ithaca, NY: NABC.

Fox, Michael W. 1992. *Superpigs and wondercorn*. New York: Lyons and Burford.

Pursel, Vernon, et al.. 1989. Genetic engineering of livestock. *Science* 244: 1281–1288.

Rifkin, J., and Kegan Paul,1985. *Declaration of a heretic*. Boston: Routledge.

Rollin, Bernard E. 1995. *The Frankenstein syndrome: Ethical and social issues in the genetic engineering of animals*. New York: Cambridge University Press.

Bernard E. Rollin

GENETIC ENGINEERING AND FARMED ANIMAL CLONING

The farming of animals for human medical and commercial purposes is being intensified through two new biotechnologies. One is genetic engineering, which involves either the splicing of alien genes into target animal embryos to create transgenic animals, or the deletion of certain genes to create genetically modified knockout animals. The other is cloning, which entails taking cells from the desired type of animal, which may be transgenic or a knockout, or from a conventionally-bred genotype possessing such qualities as rapid growth or high milk or wool yield, and inserting the nuclei of these cells into the emptied ova from donor animals of the same species. Once activated by electrical fusion of the nucleus to the egg wall, these embryo-developing ova are inserted into surrogate mothers to be gestated.

Cloning conventionally-bred and genetically engineered animals is now well underway in several countries. Transgenic farm animals are being cloned to create flocks and herds for gene pharming; many carrying human genes that make them produce various novel proteins in their milk, such as antithrombin 111 and alpha-trypsin, that the drug industry seeks to profit by. The animals are called mammary bioreactors. Commercial aims are directed toward developing animals that have leaner and more meat and healthful fats for human consumption; have greater disease resistance, fertility, and fecundity; produce more wool or milk with higher protein, even hypoallergenic and infant milk high in human lactoferrin; and that produce environmentally less harmful wastes containing lower levels of phosphorus. Pigs with transgenes from spinach, jelly fish, and a species marine worm have been cloned. The spinach gene lowers saturated fats and increase linoleic acid levels in body fat. The jellyfish gene make the pigs fluorescent, thus serving as a genetic marker, and the nematode worm gene converts omega 6 fatty acids into more consumer-beneficial omega 3 fatty acids Genetically altered pigs are also being created to serve as organ donors for humans, to produce human blood substitutes, and to produce monoclonal and polyclonal antibodies. Models of human diseases have also been created in transgenic animals, like Denmark's cloned pigs, which have genes for Alzheimer's disease.

Advocates for the creation of genetically engineered and cloned animals claim that this new biotechnology is simply an extension of the process of

Seven-month-old Dolly, a genetically cloned sheep, at the Roslin Institute in 1997. (AP Photo/ Paul Clements)

human-directed natural selection for desired genetic traits that began thousands of years ago when animals were first domesticated. Some of these production traits, coupled with how these animals are husbanded in crowded factory farms, are now recognized as causing a host of animal health, welfare, public health, and economic problems. Critics contend that the creation of transgenic and knockout animals, as well as cloning, are biologically aberrant if not abhorrent technologies that the life science industry and others cannot, from any sound scientific or bioethical basis, claim to be simply an extension of natural selective breeding. Clones are not identical to the original foundation-prototype, because of epigenetic environmental influences and different maternal mitochondrial DNA.

In 2008, the FDA announced that the meat and milk from cloned cattle and pigs is as safe to eat as food from more conventionally bred animals. But greater genetic uniformity can mean significant economic losses from diseases that become contagious when there is a fatal combination of genetic susceptibility and uniformity. The loss of genetic diversity in a livestock population increasingly displaced and replaced by homozygous clones is a bioethical and potential financial issue that governments and regulatory agencies have not fully addressed.

Health and Welfare Concerns

The incorporation of other species' genes into farm animals, such as the human growth hormone gene into pigs, can have so-called multiple deleterious pleiotropic effects. These unforeseen consequences on transgenic animals' development and physiology include abnormal and excessive bone growth (acromegaly), arthritis, skin and eye problems, peptic ulcers, pneumonia, pericarditis and diarrhea (implying an impaired immune system), as well as decreased male libido and disruption of estrus cycles. Inserted/spliced genes may be overexpressed, meaning overactive, and produce excessive amounts of certain proteins such as growth hormone, or create an insertional mutation problem, disrupting the functions of other genes and organ systems. These Russian roulette-like adverse consequences of genetic engineering can result in serious health problems later in life, if they do not cause fetal deformities and pre- or early postnatal death. Many transgenic creations are either stillborn or are reabsorbed by the mother, or soon after birth they die from internal organ failure or circulatory, or immune system collapse. This is especially so with cloned animals, with the success rate being extremely low in terms of survivability. For example, a U.S. Department of Agriculture research experiment to create cows resistant to mastitis had a success rate of 1.5 percent, with only eight calves being born from 330 transgenic cloned ova and gestated to term as live calves. Three of these died before maturity.

Cloning can result in abnormally large fetuses, which can mean suffering and death for the mothers. Abnormal placentas, deformed stillborn fetuses, and live offspring with defective lungs, hearts, brains, kidneys, immune systems, and suffering from circulatory problems, deformed faces, feet and tendons, intestinal blockages, and diabetes have been documented. Cloning seems more likely to cause problems when the cloned animals have been previously subjected to genetic engineering. Yet it is only through cloning that productive flocks and herds can

be quickly built from one or two founder transgenic/knockout stock. The treatment and ultimate fate of surrogate mother and egg-donor cattle, and other farmed animals used as mere instruments of commercial biotechnology, call for the most rigorous humane standards and their effective enforcement by the United States and other governments.

Conclusions

Is the incorporation of genetically engineered and cloned farmed animals into conventional, industrial agriculture ethically, economically and environmentally acceptable? Health and environmental experts, conservationists, and economists are calling for a reduction in livestock numbers globally, and for more sustainable, organic, and ecological farming practices, including more humane and free range animal production methods. They see no place for cloned livestock and agricultural bioengineering if there is to be a viable future for sustainable agriculture. We should all ask what farm animal cloning and genetic engineering have to do with feeding the poor and hungry and developing a sustainable and socially just agriculture locally and globally. The use of farm animals as medical models of human diseases, and as sources of new pharmaceutical and other medical products from livers to hearts for xenotransplantation into humans raises a host of scientific and ethical questions. It may not be a sustainable or effective path for medicine to take, profitability not withstanding. From a bioethical perspective, it places the human in the role of genetic parasite, which, from a cultural and evolutionary perspective, may not make for a better or desirable future.

Further Reading

Fox, M. W. 2001. *Bringing life to ethics: Global bioethics for a humane society.* Albany: State University of New York Press.

Fox, M. W. 2004. *Killer foods: What scientists do to make better is not always best.* Gulford, CT: The Lyons Press. See also www. doctormwfox.org.

Griffin, H. 1997. Briefing notes on Dolly. Roslin Institute Press Notice PN97–03. December 12, 1997. www.roslin.ac.uk/downloads/12–12–97-bn.pdf.

Loi, P, Clinton, M., Vackova, I. et al. 2006. Placental abnormalities associated with postnatal mortality in sheep somatic cell clones. *Theriogenology* 65(6): 1110–21.

National Research Council of the National Academy of Sciences. 2002. *Animal biotechnology: Science based concerns.* Washington, DC: The National Academies Press. www. nap.edu/books/0309084393/html/.

Niemann, H., Kues, W., and Carnwath, J. W. 2005. Transgenic farm animals: Present and future. *Rev.sci. tech. Off. Int. Epiz.* 24: 285–298.

Pew Initiative on Food and Biotechnology. 2004. Issues in the regulation of genetically engineered plants and animals. www.pewag biotech.org/research/regulation/Regulation. pdf.

Steinfeld, H. P., Gerber, P., Wassenaer, T., Castel, V., Rosales, M., and de Haan, C. 2006. *Livestocks' long shadow: Environmental issues and options.* Washington, DC: United Nation's Food and Agriculture Organization.

Wells, D. N. 2005. Animal cloning: problems and prospects. *Revue Scientifique et Technique (International Office of Epizootics)* 24(1):251–64.

Michael W. Fox

GENETIC ENGINEERING: GENETHICS

Genethics is the application of moral or social values to genetics. The field of genetics was born with the experiments of Gregor Mendel on generations of pea plants in the 1850s. The young field of genetics was promptly put on ice until the early years of the 20th century. Later geneticists expanded their experimental organisms to include plants and animals, both human and nonhuman.

Within the last decade, the techniques of genetics have advanced greatly, allowing us to identify genes for cancer, mental illness, obesity, and a host of other traits and diseases. Although we can map and identify the gene(s) for such characteristics, our ability to treat them lags far behind.

Genethics is typically applied to humans, particularly with relevance to and rejection of eugenics approaches which advocate selective breeding of humans. However, there is no reason we should not apply similar principles to other animals. Nonhuman animals are currently the experimental organisms of choice for research geneticists interested in human diseases and other traits. The reason is simple: the experimental work necessary to understand the genetic basis of a characteristic is often invasive and typically involves the rapid breeding of large numbers of offspring, procedures which cannot readily be applied to humans. For example, in research that focuses on the genetics of a behavior in mice which may be similar to alcoholism in humans, it is necessary to inject mice with a standard dose of alcohol so that researchers can assess its effect on them. Animals also have to be euthanized to allow for analyses of tissue or biochemistry that are lethal.

There are three types of genetic research that involve animals. The first is the use of animal models for human genetic diseases. These include diseases caused by abnormalities in single genes, such as cystic fibrosis, sickle cell anemia, and Huntington's disease, as well as polygenic (many gene) diseases such as cancer, heart disease, and alcoholism. Next come the genome projects, which have as their goal the identification of all the genes of a given organism. Currently genome projects have been completed on thousands of species including bacteria, viruses, plants, invertebrates, and many vertebrates such as cow, dog, opossum, mouse, and rat, in addition to the human genome. Finally there is transgenic research, also known as recombinant DNA technology, which moves genes from one organism into another. This area of research initially allowed the insertion of human genes into bacteria, primarily for the purpose of producing the protein specified by the human gene, for example, insulin. Now, many human genes are being moved into a variety of mammalian species both for production and to study the function of the human gene.

As genetic technology and statistical interpretations improve, more scientists are beginning to study humans in order to elucidate the genetic bases of human genetic conditions. As the potential to work directly on humans becomes more feasible, it is possible that we'll see a reduction in the use of animal subjects.

Further Reading

Guidelines for Ethical Conduct in the Care and Use of Animals: http://www.apa.org/science/anguide.html.

Luedke, D. 2000. Animals & Research. A 5-part special to the *Seattle Post-Intelligencer*: http://seattlepi.nwsource.com/opinion/anml4.shtml.

Lynn, R. 2001. *Eugenics: A reassessment*. Westport, CT: Greenwood Publishing Group.

Tannenbaum, J., and Rowan, A.N. 1985. Rethinking the Morality of Animal Research. *Hastings Center Report,* Volume 1 (October 1985), 32–43.

Beth Bennett

GLOBAL WARMING AND ANIMALS

Over the last 100 years, the average global surface temperature has increased approximately 0.8°C. This warming has been quite fast, and the rate of increase is continuing to escalate, significantly faster than when the globe warmed about 6°C from the last ice age (18,000 years ago) to our current warm interglacial period (12,000 years ago). The average rate of warming over this 6,000 year time period was about 0.01°C per decade. The rate of warming within the last 150 years is already significantly higher than the entirety of this prehistoric change.

With this rise of 0.8°C, wild animals are already exhibiting discernible changes. This is because all living things are affected by temperature in one way or another. Several types of changes have already been seen in the wild, including shifts in ranges boundaries (e.g., moving north in the Northern Hemisphere) and/or shifts in the density of individuals from one portion of their range to another (e.g., the center of the abundance pattern moving up in elevation), shifts in the timing (i.e., phenology) of various events primarily occurring in spring and/or autumn, changes in genetics, behavior, morphometrics (e.g., body size or egg size), or other biological parameters, and extirpation or extinction, the latter of which is the final irreversible change. Given what is known about the physiological requirements of species, these changes are consistent with those expected with increasing ambient temperatures.

Numerous studies have found that wild animals and plants on all continents are already exhibiting discernible changes in response to regional climate changes. A primary concern about wild species and their ecosystems is that they are not only having to adapt to rapidly warming temperatures, but they are also having to cope with other human-caused stresses: pollution, land-use change, invasive species, and others problems. The synergistic effects of these stresses combined with rapid warming are greatly influencing the resilience, that is, the ability to return to the same condition after a stress, of many species, communities, and ecosystems. Another major concern for the survival of species is explained in the *Summary for Policy Makers of Working Group I of the Fourth Assessment Report of the Intergovernmental Panel on Climate Change* (IPCC). Here we learn that if we do not change our reliance on fossil fuels, the global temperature could rise as much as 6.4°C and even beyond if we stay on the energy path we are currently traveling.

Changes in Ranges

As the globe warms, we find that species in North America are extending their ranges north and up in elevation, because habitats in these areas have now warmed sufficiently to allow temperature-restricted species to colonize. This dispersal of species forced by rapidly rising temperatures, however, is frequently slowed and often blocked by numerous other human-made stresses, such as land-use changes, invasive species, and pollution. Dispersing individuals must not only find suitable habitat through which to travel, but appropriate habitat in which to colonize. This is relatively easy for highly mobile species like butterflies, birds, and bats, but certainly scorpions, salamanders, shrews and the like will have trouble navigating across highways

and through farm fields or cities. Consequently, individuals that are moving have to navigate around, over, or across freeways, agricultural areas, industrial parks, and cities.

Species near the poleward side of continents, such as South Africa's fynbos, will have no habitats into which they can disperse as their habitat warms. The same is true for species living near the tops of mountains. Additionally, species living in these areas will be further stressed by species dispersing into their habitats from farther inland or farther down the mountain. Because of heat stress and the new species with which they must interact, many species currently on the poleward side of continents and near the tops of mountains are highly likely to go extinct unless humans manage to relocate them. Those species facing extinction unless aided by humans are called functionally extinct.

Throughout prehistoric and more recent times, species have been found to move independently from other species in their community or ecosystem; species move at different rates and in different directions, depending on their unique metabolic, physiological, and other requirements. Such independent movement results in a disruption of biotic interactions such as predator-prey relationships. For example, if the range of a predator shifts and the range of its prey does not, a population balance becomes disrupted —a perceived benefit if the prey is an endangered species. If, however, the prey is a food-crop pest, then humans could certainly see the increase in its population as detrimental.

Progressive acidification of oceans due to increasing atmospheric carbon dioxide is just now beginning to be understood, and the findings are surprisingly grim. The pH of the oceans has dropped around 0.3 over the last 100 years, with the steepest drop beginning around the mid-1970s. Carbonic acid, which is causing the lowering of pH, is not only hindering species from laying down needed calcareous structures, but this lower pH is eroding calcareous structures that have already been generated, such as the shells of clams and snails. Indeed, by the year 2100 ocean pH is very likely to be lower than during the last 20 million years.

Changes in Timing   Species on every continent are already shifting in the timing (i.e., phenology) of various events primarily occurring in spring, but also to some extent in the autumn. Frogs are breeding earlier, cherry blossoms bloom earlier, and leaves turn color later. Over the last 30 years, around 115 species (plants and animals together) from locations around the globe were found to be altering the timing of a spring event earlier by around five days per decade. Only 6 of the 115 species (~five percent) showed a later change in timing of their spring events.

Rapid phenological changes of species are of concern, because for over tens of thousands of years or more, animals have been adjusting to the timing of other species around them. For example, as the planet warms, farmers may have to change the timing of their planting and might even change the type of crop grown. Either of these changes could provide an insect with a food resource that was previously limited, thereby allowing the population size to grow. If the insect feeds on the nectar from the flowers of the crop, then the farmer could experience a benefit owing to the plants being

pollinated. If, however, the insect feeds on the tissue of the crop plant, then the increasing size of the insect population could be seen as a detriment that must be countered in some manner, for example, with pesticides. In wild communities, changes in timing could mean that a food source of a species is not available at the time it is needed. This in turn could cause the species stress, either in time and energy looking for food, or in competitive interactions with others over the little food available. Such stress may lead to lower fecundity rates which, if not rectified, could lead to extinction.

Changes in Genetics, Behavior, and Other Traits   The third type of change is of traits that are reported relatively infrequently: genetics, behavior, and other species' traits. An example of a behavioral change is the foraging habits of polar bears. Now, instead of hunting seals, they are by necessity increasingly foraging in garbage dumps. Some species that rely on seal kills, the Ross and Ivory gulls, may not be getting the food they need to sustain their population numbers.

Extirpation and Extinction   The escalating rise in average global temperatures over the past century has put numerous species in danger of extinction. Functionally extinct species, or species we can anticipate as likely to go extinct unless humans come to their aid, include those that cannot move to a different location by themselves as the temperature increases, due to either lack of available habitat or the inability to access it. For example, in Australia the Mallee emu-wren is quite sedentary (rarely moving farther than 5 or 6 km), with a small fragmented range that is frequently threatened by fires. This small bird cannot move until its habitat moves, which will likely be much slower than the speed the emu-wren will need, given the rate of temperature increase. Unless humans intervene and translocate individuals to suitable habitat farther south, this bird will most likely go extinct within the next 25–50 years. Unfortunately, only about 2,000 km² of suitable native habitat are available today, two- to three-year-old spinifex grass is needed to create suitable habitat farther south. After the birds are moved to a new habitat, preventing a fire cycle with a frequency of less than 10 to 15 years is necessary to ensure that both the habitat and the emu-wren survive.

Many factors are needed for a successful managed relocation: money, knowledge of how to move a species successfully, the ability to introduce individuals in a manner that allows establishment of a group but at the same time ensures that it does not become an invasive species and cause the extinction of other species, land, personnel, or negate political will. Also absent is the long-term commitment needed to monitor even a small percentage of the functionally extinct species we know of today. Consequently, many biologists believe we are standing at the brink of a mass extinction that would be caused by one species—us.

Roughly 20–30 percent of known species could likely be at increasingly high risk of extinction if global mean temperatures increase 2–3 C above pre-industrial temperatures (1.3–2.3°C above current temperatures). Given that there are around 1.7 million identified species on the globe, somewhere between 340,000 and 570,000 species could be committed to extinction primarily due to our carelessness. Extinctions are virtually certain to reduce our societal options,

such as adaptation responses, medicine, and others.

If we do not change our present trajectory of using carbon-intensive energy, then the global average temperature could go above 4°C, which could commit 40 to 50 percent of known species to extinction. In addition to endangering a large number of our ecosystem services (e.g., pollinating our crops), loss of any species is irreversible, and as such it is an unethically high price to pay. Indeed, many people pay higher insurance premiums for lesser catastrophes with much lower probabilities of happening.

In recent years, it has been pointed out that, especially in the United States, what each of us does adds up. Suggestions have included driving highly fuel-efficient cars; not using incandescent light bulbs anywhere; using more efficient roofing materials; using highly energy efficient windows, heaters, air conditioners and appliances; using materials that do not need to be shipped long distances; and making sure all materials are harvested sustainably.

Further Reading

Intergovernmental Panel on Climate Change (IPCC). 2001. *Climate change 2001: Impacts, adaptations, and vulnerability.* New York: Cambridge University Press.
The Royal Society. 2005. Ocean acidification due to increased atmospheric carbon dioxide. Policy document 12/05.

Terry L. Root

THE GREAT APE PROJECT

The Great Ape Project aims to grant basic moral and legal rights to nonhuman great apes—chimpanzees, bonobos, gorillas and orangutans. Since its establishment, many other organizations strive for the recognition of great ape rights as well. This has resulted in some remarkable changes. Several countries have imposed a ban on invasive biomedical research with great apes, and the United States, where most research with great apes occurs, has stopped killing so-called surplus great apes and instead now relocates them in sanctuaries.

The Great Ape Project was launched in London on June 14, 1993 by Peter Singer, philosopher at Princeton University, and Paola Cavalieri, philosopher and editor of the Italian journal *Etica & Animali*. On that day the book *The Great Ape Project: Equality beyond Humanity* was released, which contains contributions from more than thirty subscribers to "A Declaration on Great Apes." This declaration demands the extension of the moral community of equals to include all human and nonhuman great apes. Like us, nonhuman great apes are intelligent beings with a rich and varied social and emotional life. Therefore, it is argued, we should consider them our moral equals; we ought to respect their basic interests in the same way we respect similar human interests. The protection of these interests needs to be assured through the endorsement of three basic rights, namely the right to life, the protection of individual liberty, and the prohibition of torture. Among the early supporters of the Great Ape Project are zoologists/primatologists Marc Bekoff, Richard Dawkins, Roger and Deborah Fouts, Jane Goodall, Adriaan Kortlandt, Lyn Miles, Toshisada Nishida and Francine Patterson and philosophers Dale Jamieson, James Rachels, Tom Regan, Bernard Rollin, and Steve Sapontzis.

Why this focus on great apes? There appear to be three major reasons, namely our close relationship with nonhuman great apes, their rich mental lives, and the expectation that the cost to stop their exploitation is relatively limited and thus quite feasible. Though the Great Ape Project directs its attention to great apes, many of its contributors see this as a first step in the process of extending the community of equals. Indeed, many are prominent advocates for other animals as well.

The use of great apes for biomedical research is meeting increasing moral and legal resistance. Over the last decade, several countries have forbidden the use of nonhuman great apes for invasive biomedical research, namely Austria, Australia, Japan, the Netherlands, New Zealand, Spain, Sweden, and The United Kingdom. Among these countries, only Austria and the Netherlands used great apes for biomedical research, and these have since been moved to sanctuaries and zoos. At the time of writing, the European Union is considering imposing a ban on great ape experiments in all of its member states (Harrison, 2008).

The United States is virtually the only country which still uses great apes for biomedical research and testing. The majority of the approximately 1,200 chimpanzees still used for research are housed in six research facilities. In 2000, President Bill Clinton signed the CHIMP Act into law, which states that chimpanzees no longer needed for research should not be killed, but moved into sanctuaries, and that the government needs to assume the largest part of funding needed for their lifetime care. A 2007 amendment to the CHIMP Act prohibits using these chimpanzees for research ever again. In 2008, the Great Ape Protection Act was introduced to end biomedical research using the remaining chimpanzees in U.S. laboratories. Several animal advocates

A silverback mountain gorilla seen in the Virunga National Park, near the Ugandan border in eastern Congo. (AP Photo / Jerome Delay)

and organizations are working to end such research; among these in particular the efforts by the New England Anti-Vivisection Society through its Project R&R: Release and Restitution for Chimpanzees in US laboratories campaign is notable (www.releasechimps.org).

The special attention to great apes over the last fifteen years seems to have had an impact on the zoo community as well. Whereas many zoos favor the killing of surplus animals, an exception is to be made for great apes. In 2001, the book *Great Apes & Humans: The Ethics of Coexistence*, was published to respond to the Great Ape Project. In this book, Michael Hutchins and colleagues of the American Zoo and Aquarium Association comment:

As great ape zoo populations mature, the question arises of what to do with older, postreproductive individuals. Animal rights proponents argue that zoos have a responsibility to care for captive-bred animals from "the cradle to the grave." In the case of great apes, we agree. Despite arguments to the contrary (. . .) and the fact that it is legal, euthanasia of healthy great apes is not generally accepted in the professional zoo community as an option for controlling populations. (Hutchins et al., 2001, p. 352)

One is left wondering what the general zoo policy would have been without the growing influence of the movement for great ape rights.

A tremendous challenge for those who defend the interests of great apes is to deal with the enormous threats faced by the remaining great apes in the wild. There may be no viable populations remaining within the next two decades. Major threats are the logging of forests, hunting for meat—the bushmeat crisis—and diseases such as Ebola. The United Nations has launched the Great Apes Survival Project (GRASP) "to lift the threat of imminent extinction" faced by gorillas, chimpanzees, bonobos, and orangutans (see www.unep.org/grasp). Conservation organizations refer in particular to the importance of conserving species in their ecological role, and in their aesthetic, scientific, and economic value. Organizations such as the Great Ape Project add a special dimension by stating that each great ape is a valuable individual who needs to be protected because of his welfare interests as an individual. The Great Ape Project hopes for the passing of a declaration of great ape rights by the United Nations, similar to declarations for children, women, and the disabled.

Further Reading

Anonymous (Ed.) (s.d.) *Serving a life sentence for your viewing pleasure! The case for ending the use of great apes in film and television.* Washington DC: The Chimpanzee Collaboratory.

Anonymous (Ed.) (2003). *The evolving legal status of chimpanzees.* Reprinted from *Animal Law*, 9. Portland: Lewis & Clark Law School.

Anonymous (Ed.) (2003). *The Great Ape Project census: Recognition for the uncounted.* Portland: Great Ape Project (GAP) Books.

Cavalieri, P. (Ed.) (1996). *Etica & animali, 8* (Special issue devoted to The Great Ape Project).

Cavalieri, P. & Singer, P. (Eds.) (1993). *The Great Ape Project: Equality beyond Humanity.* London: Fourth Estate.

Harrison, P. (2008, November 5). Great Ape Debate leads to EU testing ban proposal. http://www.reuters.com/article/environmentNews/idUSTRE4A45TL20081105 (accessed December 26 2008)

Hutchins, M., Smith, B., Fulk, R., Perkins, L., Reinartz, G., & Wharton, D. (2001). Rights or welfare: A response to The Great Ape Project. In Beck, B. B., Stoinski, T. S., Hutchins, M., Maple, T. L., Norton, B., Rowan, A., et al. (Eds.), *Great apes & humans: The ethics of coexistence.* Washington and London: Smithsonian Institution Press.

Peterson, D. (2003). *Eating apes.* Berkeley, Los Angeles and London: University of California Press.

Peterson, D., & Goodall, J. (1993). *Visions of Caliban: On chimpanzees and people.* Boston and New York: Houghton Mifflin Company.

Singer, P. (2006, May 22). *The great ape debate unfolds in Europe.* http://search.japantimes.co.jp/cgi-bin/eo20060522a1.html (accessed July 15, 2006).

Wise, Steven M. (2000). *Rattling the cage: Toward legal rights for animals.* Cambridge, MA: Perseus Books.

Koen Margodt

GREAT APES AND LANGUAGE RESEARCH

Language research with nonhuman great apes (chimpanzees, bonobos, gorillas, and orangutans) allows for unique interaction between nonhuman animals and humans. In principle, it offers a distinctive window to the understanding of these animals' mental lives and welfare preferences; however, to some in the academic world, ape language research is considered to be highly controversial.

From the late 19th century until around the 1950s, several attempts were undertaken to teach nonhuman great apes to talk. These yielded very little success, and their failure has been attributed to anatomical differences in the vocal tracts of nonhuman great apes and humans. All of this changed in 1966, when Allen and Beatrice Gardner pioneered the teaching of American Sign Language (ASL) to the chimpanzee Washoe. When Washoe was four years old, the Gardners reported that she had reliably acquired at least 132 ASL signs. As they wanted to exclude the risk of inadvertent cueing, the Gardners tested Washoe and other ASL chimpanzees individually, requiring them to name objects shown on slides. Two uninformed observers recorded their signs. The chimpanzees usually provided more than 80 percent correct responses, and inter-observer agreement was around 90 percent (Gardner & Gardner, 1969; Gardner & Gardner, 1989). In the 1970s, Project Washoe was taken over by Roger and Debbie Fouts. Similar ASL projects were started with other great apes, such as the gorilla Koko by Francine Patterson, the chimpanzee Nim by Herbert Terrace, and the orangutan Chantek by Lyn Miles. Different communication methods were used as well. David and Ann Premack taught the chimpanzee Sarah to communicate by means of plastic symbols, and Sue Savage-Rumbaugh uses a computer console with arbitrarily designed geometric forms or lexigrams for her research with the bonobo Kanzi and other great apes.

In particular, toward the end of the 1970s, ape language research came under heavy fire. The single most significant blow was provided by psychologist Herbert Terrace of Columbia University. Terrace came to question his former research with the chimpanzee Nim after analyzing videotapes of Nim and his teachers. In an article published in 1979

in *Science,* Terrace and his colleagues wrote that the majority of Nim's utterances (87 percent) immediately followed a human's utterance or so-called adjacent utterances. Also, nearly 40 percent of these utterances were classified as partial imitations of what the human teacher had signed (Terrace et al., 1979). However, what remained an unfortunate blind spot in the article was the fact that the majority of Nim's utterances were either spontaneously initiated by Nim (13 percent) or composed of novel signs (40.6 percent), signs that differed from those used by the human teacher.

It is also important to take into account the highly controlled training conditions and Nim's increasingly problematic psychological state. Nim was taught sign language for five to six hours a day in a concrete classroom of barely six square meters. Terrace later "wondered how I and the other teachers could have spent so much time in these oppressive rooms." (Terrace 1979, 1987, p. 209). Though chimpanzees develop strong social bonds that may last a lifetime, Nim had some sixty teachers within only four years. Even his eight principal caregivers were present for only parts of these four years, and Terrace was too busy with many other occupations to be present enough for Nim's developmental well-being. Al four of Nim's main caregivers at the Delafield house left around August and September 1976. In particular, when Laura Petito left, Nim became depressed and inconsolable (Terrace 1979, 1987, p. 108). Terrace recognized that "undoubtedly the loss of Nim's immediate family at Delafield at a critical stage of his growth had a permanent adverse effect on his social, linguistic, and emotional development" (Terrace 1979, 1987, p. 139). Nevertheless, at least four of the ten videotapes used for the *Science* article were recorded between September 1976 and September 1977. As a consequence, their scientific reliability is highly questionable.

Research with other great apes has resulted in different findings than those of Terrace and his colleagues. The total of spontaneous and novel utterances for the bonobo Kanzi, the gorilla Koko, and the orangutan Chantek, range between 50 percent and more than 90 percent. Several of the language-research apes were reported to engage regularly in spontaneous self-signing, for example, during play; this behavior has been confirmed by independent observers. Jane Goodall describes a visit to the Temerlins, where she "watched as [the chimpanzee] Lucy, looking through her magazine, repeatedly signed to herself as she turned the

pages . . . She was utterly absorbed, paying absolutely no attention to either Jane [Temerlin] or me." (Goodall in Peterson & Goodall, 1993, p. 204). Roger and Debbie Fouts state that the chimpanzee Washoe spontaneously taught the use of ASL to her adopted chimpanzee son, Loulis. Not only did she demonstrate to him the correct signs, but on several occasions she also molded his hands into the proper signing configuration. For six years, the researchers made only seven signs in Loulis's environment (such as "who" and "where"). Loulis, nevertheless, mastered 55 signs by the end of the study period.

The well-known linguist Steven Pinker at the Massachusetts Institute of Technology has suggested that "the apes had not learned *any* true ASL signs." (Pinker, 1994, p. 337). His position is

Nim, a chimpanzee who was taught sign language, signals that he wants a drink during lunch in his Columbia University classroom. (AP Photo/Jerry Mosey)

based mainly upon the remarks of a deaf man who testified anonymously in Arden Neisser's *The Other Side of Silence* (1983). This man had worked with chimpanzees that were staying with the Gardners only a few years after Washoe had left with Roger Fouts. The witness accepted fewer of the signs made by the chimpanzees as true ASL signs. What he does not mention in his testimony is that some of the signs accepted by the Gardners are variations of the ASL signs used by deaf humans. The Gardners have always been explicit about this. For example, in a 1969 article for *Science,* they clearly describe exactly how some of Washoe's signs differ from default ASL signs. One of those signs—the sign for "more"—was rejected by the deaf man for not being an ASL sign.

It should be mentioned as well that deaf people had to fight a fierce emancipation battle before ASL became recognized as a full language. Several of these people clearly felt deeply humiliated by the ASL research with nonhuman apes. Neisser comments:

> The entire issue of chimpanzee sign language is a painful one for the deaf. There is simply nothing in it for them—nothing from which they might be able to take comfort or find dignity, but only the opposite. The image of an ape signing echoes the ancient and familiar charge that their language is only suited for the beasts. (Neisser, 1983, p. 16)

Unfortunately, critics like Pinker fail to mention this dimension.

In sharp opposition to the anonymous testimonial referred to by Pinker, it is remarkable that the pioneering ASL authority William Stokoe recognized the ability of nonhuman great apes to master ASL signs. This linguist, who taught at Gallaudet College, the first college for deaf people in the world and was the first author of *A Dictionary of American Sign Language* (1965), saw how, during a walk, Washoe formed ASL signs such as "cow" (the animals were far away in the fields, barely visible to Stokoe) and "flower" (before she ate it). Stokoe concluded his considerations on the ape language experiments by stating: "I find that the critics who attack the experiments have failed to provide any solid basis for denying what the animals have demonstrated" (Stokoe, 1983, p. 157).

Joel Wallman has written that a distinction needs to be made between making trained gestures to obtain a reward, and symbolic communication. The best criterion in favor of the latter, according to Wallman, is the ability to use displaced reference, that is, to communicate about things removed in time or space (Wallman, 1992). Multiple instances support the suggestion that nonhuman great apes can meet this criterion. The most convincing example is, perhaps, a systematic research project undertaken by Charles Menzel at the Language Research Center in Georgia. On various occasions, Menzel hid objects under sticks, beyond the reach of the adolescent chimpanzee Panzee. The next day, Panzee spontaneously tried to draw the attention of uninformed caregivers. She persistently made vocalizations, moved repeatedly in the direction of her outdoor enclosure, formed the sign "hide" (by covering her eyes with her hand), pointed in the direction of the hidden objects, and tried to communicate by selecting the appropriate lexigrams on her keyboard, such as the symbols for "stick", "hide," and "blueberries". She thus successfully initiated symbolic communication with uninformed humans about objects removed in time (she had to recall the object that had been hidden the day before) and space (these were beyond her sight and reach) (Menzel, 1999).

Some of the reports by ape language researchers suggest that nonhuman great apes may be remarkably creative in producing new signing combinations. A famous example is the combination "water bird," which was formed by Washoe upon seeing a swan. Critics have remarked that these were simply independent signs for separate objects, not a novel signing combination to describe the swan; however, in support of Washoe, it has been asserted that she consistently signed "water bird" for swans, whether they were in or out of the water (Lieberman, 1984). Also, such criticism may be less easily applied to combinations such as "white tiger" by the gorilla Koko, to indicate a zebra, "rock berry" by Washoe, for a Brazil nut, "cry hurt food" by the chimpanzee Lucy, for radishes, and "eye drink" by the orangutan Chantek, for contact lens solution.

What about the presence of syntax or grammar? Most language-trained apes seem to produce combinations of around three signs, though these may also consist of up to six or seven symbols. To meet the requirement of syntax, there must be indications of linguistic rules; in other words, the combinations of signs or lexigrams must reveal some order. Some indications indeed point in the direction of a rudimentary syntax. In Washoe's signing, for example, the subject precedes the action in almost 90 percent of her combinations. Washoe thus typically signs "you me go" or "you me out," but "out you me Dennis" is the exception. Roger Fouts writes that Washoe understands differences of meaning according to the position of the subject and object (Fouts & Mills, 1997). He illustrates this with the

examples "me tickle you" and "you tickle me." The chimpanzee Ai has learned to indicate on a computer console, through keys, the quantity, color, and kind of objects shown by Tetsuro Matsuzawa. Ai is familiar with lexigrams, Arabic numbers, and Japanese *kanji* characters. Although she was free to choose the order of the keys, she nearly always selected color/object/number and object/color/number among six possible alternatives (Matsuzawa, 1989).

Sue Savage-Rumbaugh emphasizes that we should look not only at the combinations one can produce, but also at the comprehension of such combinations. In a test with 660 different sentences, the bonobo Kanzi reacted properly to 72 percent of the requests (a two-and-a-half-year-old human child responded correctly to 66 percent of these sentences). He understood quite complex sentences, such as "You can have some cereal if you give Austin your monster mask to play with." When asked "Can you throw a potato to the turtle?" he did not make the mistake of throwing both items or throwing the turtle toward the potato. Some of his reactions were quite surprising, though; for example, when asked to put water on the carrots, he threw them outdoors in the rain (Savage-Rumbaugh & Lewin, 1994).

Whether we can say that nonhuman great apes can learn language depends, ultimately, upon how language is defined. Nonhuman great apes appear to be capable of using several hundred symbols in a meaningful way. There are also indications of rudimentary syntax. This suggests that what makes humans unique in connection with language may simply be a difference in degree of complexity. Marc Hauser, Noam Chomsky, and Tecumseh Fitch hypothesize that recursion is the only uniquely human component of the faculty of language. This capacity allows us to produce an, in principle, infinite number of combinations with a limited set of elements. For example, any possible longest sentence can still be made longer by adding "Mary thinks that . . ." (Hauser et al., 2002). Some commentators have suggested that the linguistic capacities of nonhuman great apes have resulted in redefining language in terms of what distinguishes humans from nonhuman apes, thus keeping language by definition beyond the reach of nonhuman apes. We may only wonder how important recursion will become in language definitions during the coming years.

Further Reading

Candland, D. K. 1993. *Feral children & clever animals: Reflections on human nature.* New York and Oxford: Oxford University Press.

Cavalieri, P., & Singer, P. (Eds.). 1993. *The great ape project: Equality beyond humanity.* London: Fourth Estate.

Fouts, R., & Mills, S. T. 1997. *Next of kin: My conversations with chimpanzees.* New York: Avon Books.

Gardner, R. A., & Gardner, B. T. 1969, August 15. Teaching sign language to a chimpanzee. *Science, 165,* 664–672.

Gardner, R. A., & Gardner, B. T. 1989. Cross-fostered chimpanzees: I. Testing vocabulary. In P. Heltne & L. Marquardt (Eds.), *Understanding chimpanzees*, 220–233. Cambridge, MA and London: Harvard University Press and The Chicago Academy of Sciences.

Hauser, M. D., Chomsky, N., & Fitch, W. T. 2002, November 22. The faculty of language: What is it, who has it, and how did it evolve? *Science, 298,* 1569–1579.

Lieberman, P. 1984. *The biology and evolution of language.* Cambridge, MA and London: Harvard University Press.

Matsuzawa, T. 1989. Spontaneous pattern construction in a chimpanzee. In P. Heltne & L. Marquardt (Eds.), *Understanding chimpanzees*, 252–65. Cambridge, MA and London: Harvard University Press and The Chicago Academy of Sciences.

Menzel, C. 1999. Unprompted recall and reporting of hidden objects by a chimpanzee (Pan troglodytes) after extended delays. *Journal of Comparative Psychology, 113, no. 4,* 426–34.

Neisser, A. 1983. *The other side of silence: Sign language and the deaf community in America.* New York: Alfred A. Knopf.

Patterson, F., & Linden, E. 1981. *The education of Koko.* New York: Holt, Rinehart and Winston.

Peterson, D., & Goodall, J. 1993 *Visions of Caliban: On chimpanzees and people.* Boston and New York: Houghton Mifflin Company.

Pinker, S. 1994. *The language instinct.* New York: Harper Perennial.

Savage-Rumbaugh, S., & Lewin, R. 1994. *Kanzi: The ape at the brink of the human mind.* London: Doubleday.

Stokoe, W. C. 1983. Apes who sign and critics who don't. In J. de Luce & H. Wilder (Eds.), *Language in primates: Perspectives and implications*, 147–58. New York: Springer-Verlag.

Temerlin, M. 1975. *Lucy: growing up human.* Palo Alto, California: Science and Behavior Books.

Terrace, H. 1979, 1987. *Nim: a chimpanzee who learned sign language.* New York: Columbia University Press.

Terrace, H. S., Petitto, L. A., Sanders, R. J., & Bever, T. G. 1979, November 23. Can an ape create a sentence? *Science, 206,* 891–902.

Wallman, J. 1992. *Aping language.* Cambridge: Cambridge University Press.

Koen Margodt

H

HORSE SLAUGHTER

Since 2001 there has been a concerted push to ban the slaughter of American horses for human consumption in Europe and Asia. The biggest equine welfare issue since passage of the Wild Free-Roaming Horses and Burros Act of 1971 (Public Law 92-195), the anti-horse slaughter effort has become a pivotal one in the animal protection world, with the result that horses are no longer being slaughtered on American soil. Yet in the absence of a comprehensive federal ban, tens of thousands of American horses are exported annually to Canada and Mexico for slaughter (National Agricultural Statistics Service, USDA). It is a trade that the horse slaughter industry and traditional agribusiness interests have fought to keep alive.

In fact, the entire debate has become extremely controversial, with two very distinct and entrenched sides telling two very different stories. Those advocating for a ban contend that horse slaughter is de facto animal cruelty, a predatory business that operates solely to turn a profit, while those wishing to maintain horse slaughter paint the practice as a humane disposal system, a necessary evil without which unwanted horses would suffer from neglect.

Multiple polls show that the majority of Americans consider horse slaughter to be inhumane and support an end to the foreign-driven trade (Public Opinion Strategies, 2006; McLaughlin & Associates, 2004; Voter/Consumer Research, 2003; Mason-Dixon Polling & Research, Inc., 2003). This comes as little surprise; Americans don't eat horses nor do they raise them for their meat. In reaction to public opinion, federal and state lawmakers have offered, and in some cases passed, legislation prohibiting horse slaughter. The courts have also been brought into the fray. Meanwhile, the slaughter continues.

Legal/Legislative Background

In 2007, the most recent year for which official numbers are available, 121,459 American horses were sent to slaughter, including more than 90,000 that were exported to Mexican, Canadian and Japanese abattoirs. While these numbers are far less than the nearly 350,000 horses slaughtered in America in 1989, there has been an upward surge in the number of horses enduring this fate annually since 42,312 were slaughtered on U.S. soil in 2002 (National Agricultural Statistics Service, USDA). The trend appears to correlate with the campaign to ban horse slaughter at the federal level, and would seem to be an effort by the foreign-owned horse slaughter industry to reap as much profit from the U.S. market as it can before a federal law is passed prohibiting the practice.

In this photo released by a protest group calling itself "Respect for Horses," demonstrators hold a banner above the severed head of a horse in Melbourne, Australia. The activists were protesting the use of horses in horse racing to coincide with Australia's biggest race, the Melbourne Cup. Thousands of horses who do not make the grade as race horses face slaughter. (AP Photo/Respect for Horses)

It can only be a matter of time before such a statute comes into existence. Under state law, the country's three remaining horse slaughter plants (all European-owned) were closed in 2007. In Texas, the BelTex and Dallas Crown plants were shut down when a 1949 law (Texas Agricultural Code, Chapter 149) prohibiting the sale of horsemeat was upheld as valid after a protracted legal battle (Empacadora de Carnes de Fresnillo sa de CV et al. v. T. Curry, District Attorney, Tarrant County, TX et al.). In Illinois, the Cavel International plant was shuttered after the state passed a law banning horse slaughter (Illinois Public Act 95-0002). While Cavel challenged the statute's legality, it was upheld by the courts (Cavel International, Inc. et al. v. Madigan). California voters also approved a ballot initiative in 1998 banning horse slaughter (California Prop. 6), though it was perhaps merely symbolic, given that there were no horse slaughterhouses operating in the state at the time. Still, the state witnessed a 34 percent drop in horse theft in the year following the law's enactment (California Livestock and Identification Bureau), presumably because fewer horses were stolen and shipped out of state for slaughter.

While American horses have long been exported for slaughter, even when U.S. plants were operational tens of thousands were exported annually, the plants' closures in conjunction with the lack of a comprehensive federal law have resulted in a sharp increase in the number of horses being exported to foreign slaughterhouses. Absent a federal statute, the potential also exists for horse slaughterhouses to open in states with less restrictive laws than those of Texas, Illinois, and California. In 2008, South Dakota's state senate considered but ultimately rejected legislation to facilitate the construction of a horse slaughterhouse (Senate Bill 170, South Dakota State Legislature, 2008 Legislative Session).

Multiple federal bills seeking to ban the slaughter of horses for human consumption and their export for the same purpose have been taken up by Congress since 2002. The first incarnation of the American Horse Slaughter Prevention Act (H.R. 3782) was referred to the House Agriculture Committee, where it eventually died without consideration (Thomas, Library of Congress). Revised and reintroduced in subsequent years (Animal Welfare Institute), the bill has gone to more receptive committees and even passed the full U.S. House of Representatives in 2006 by a landslide vote of 263 to 146 (U.S. House Roll Call No. 433, 109th Congress, Second Session), but has thus far failed to pass into law.

Taking a slightly different route in 2005, Congress passed and President George W. Bush signed into law as part of a larger agriculture spending bill (Public Law 109-97) a funding restriction that was designed to temporarily halt horse slaughter. The move was circumvented by the U.S. Department of Agriculture and, in the face of a legal challenge by humane groups (The Humane Society of the US et al. v. Cavel International et al.), the slaughter continued until the plants were closed under state law.

In the second half of the 110th Congress, the Prevention of Equine Cruelty Act (H.R. 6598) was introduced by John Conyers, Jr., chairman of the House Judiciary Committee. A streamlined version of its predecessors, the bill seeks to criminalize horse slaughter and related activities by amending Title 18 of the U.S. Code. The Committee approved the bill by a voice vote in September 2008, but once again, Congress failed to act on it before the session ended.

Horses at Risk

Because horses can live more than 30 years and are expensive to maintain, they are often sold multiple times in their lives, each time placing them at risk of ending up at the slaughter plant. While a handful of horses are purposely sold to slaughter by their owners and many others are stolen, most arrive at the slaughterhouse via livestock auction, where unsuspecting sellers enter their animals into the auction ring only to have the animal bought by a killer buyer, one of the middlemen who supply slaughterhouses. All types and breeds of horses are at risk of slaughter including racehorses, workhorses, wild horses, and family horses. Despite the fact that U.S. plants are no longer in operation, killer buyers continue to purchase and haul horses from livestock auctions around the country to the slaughterhouses that have now relocated to Mexico and Canada.

Humane Euthanasia and Carcass Disposal

The number of American horses going to slaughter represents just over one percent of the total U.S. equine population of 9.2 million (American Horse Council).

It is also dwarfed by the annual equine mortality rate in the United States, which is figured at approximately 5–10 percent or 460,000–920,000 horses (Veterinarians for Equine Welfare). Thus, the vast majority of horses that die every year are not slaughtered, but either do so of natural causes or are euthanized by a licensed veterinarian at their owner's request and expense. The procedure, which is painless, can be performed on location so that the horse may meet a peaceful death in familiar surroundings. The average cost for chemical euthanasia and carcass disposal is $225 (Veterinarians for Equine Welfare). Disposal options include rendering, composting, burial, or incineration. In rendering, the carcass of a humanely euthanized animal is processed into useful byproducts without any of the suffering endured in transport to and during slaughter.

These facts and figures are often cited by animal advocates who contend that ending horse slaughter would not result in a glut of unwanted horses, as horse slaughter proponents argue. Indeed, a Colorado State University study commissioned by the US Department of Agriculture revealed that more than 92 percent of horses going to slaughter are healthy horses, and thus, presumably, marketable (Grandin et al., 1999).

The Slaughter Process

Multiple studies and reports have shown horse slaughter to be quite brutal (Animal Welfare Institute, Veterinarians for Animal Welfare, The Humane Society of the United States, Humane Farming Association), with suffering beginning long before the horse reaches the kill box. Some veterinarians contest that rough handling, loud noises, a foreign environment,

overcrowding, and the smell of blood can cause the horses, who are unaccustomed to being handled as pure livestock, to endure great fear (Veterinarians for Equine Welfare). Government inspection reports obtained through the Freedom of Information Act show that rough and improper handling certainly does occur, and results in tremendous suffering (Animal Welfare Institute).

In the now-defunct U.S. plants, the standard operating procedure called for the use of the captive bolt gun. When it is administered correctly, application of the gun to the head so that the retracting captive bolt strikes directly into the brain, the horse is unconscious prior to being strung up and bled out. Yet undercover footage reveals that the technique is not always implemented correctly. Great pain and distress have ensued as a result, as described by Dr. Nicholas Dodman in reference to footage he reviewed from a Canadian horse slaughter plant:

> Because of the unsuitability of the slaughter setup, captive bolt operators were often trying to hit a moving target and in some cases were unable to locate the kill spot on the horses' forehead because the horse had turned around, slumped down, or moved backward in the kill box. I observed several horses being improperly "stunned." Mouthing, tonguing, and paddling of the feet were not uncommonly seen as horses were dragged away to be hung up and bled out. Some of these horses were likely still conscious as they were being bled. This experience is not significantly different than often occurred at horse slaughter plants operating in the U.S. (Dr. Nicholas Dodman, BVMS,

MRCVS in testimony before the U.S House of Representatives Judiciary Subcommittee on Crime, Terrorism and Homeland Security, July 31, 2008)

Concerns are even greater in Mexico. A 2007 investigation by *The San Antonio News-Express* revealed that the use of the *puntilla* knife on horses prior to slaughter is common practice in Mexican slaughter plants, such as a facility currently owned by BelTex, formerly operating in Texas. Footage shows horses being stabbed repeatedly in the neck with these knives prior to slaughter. Such a barbaric practice simply paralyzes the animal. The horse is still fully conscious at the start of the slaughter process, during which it is hung by a hind leg, the throat slit, and the body butchered.

Transport Issues

Concerns have been voiced by both sides about the harsh transport conditions endured by slaughter-bound horses. In fact, the 1996 Farm Bill (Public Law 104-127) instructed the Secretary of Agriculture to develop regulations governing the transportation of horses to slaughter. The resulting regulations (CFR Parts 70 and 88), however, were regarded as inadequate by the humane community, covering only the final leg of the journey to slaughter, and allowing horses to be hauled on journeys lasting more than 24 hours without food, water, or rest on double-decked cattle trailers. Not only are these double-deckers inhumane for transporting horses because of their low ceiling height, but they are also dangerous and have been involved in a number of tragic accidents. The regulations also allow the transport of horses that are partially blind, have broken legs, or are heavily pregnant. As a result, the regulations have been reopened and are currently under review (Docket APHIS-2006-0168), but it is unclear when a new rule will be issued. Regardless, it will not pertain to horses once they are shipped over our national borders.

The Future

One thing most observers of the protracted battle to end horse slaughter agree on is that Congress will pass a comprehensive law in the relatively near future that will effectively end the slaughter of American horses for human consumption. Where they tend to disagree is on what the long-term effects will be. Opponents of the legislation say horses will be abandoned *en masse.* Animal protection advocates contend that there are sufficient resources to deal with any unwanted horses through placement or veterinarian-administered euthanasia. What can be said with certainty is that the debate has engendered a very real discussion in the wider equestrian world about the need for responsible horse ownership and breeding, a discussion that can only be good for America's horses.

Further Reading
American Horse Council. http://www.horse council.org.
Animal Welfare Institute. http://www.awion-line.org.
Grandin, T., McGee, K., & Lanier, J. 1999. Survey of trucking practices and injury to slaughter horses. Department of Animal Sciences, Colorado State University.
How Americans feel about horse slaughter. Public Opinion Strategies, 2006.
Humane Farming Association. http://www.hfa.org.
Kentucky voter survey. Voter/Consumer Research, 2003.

Sandberg, Lisa. 2007. Horse slaughters taking place on the border. *San Antonio News Express*, September 30, 2007.

Texas statewide voter survey on horse slaughtering. 2003. Mason-Dixon Polling & Research.

The Humane Society of the United States. http://www.hsus.org.

USDA National Agricultural Statistics Service: http://www.nass.usda.gov/Statistics_by_Subject/index.asp.

VA voters support the stopping of slaughtering horses for human consumption. 2004. McLaughlin & Associates.

Veterinarians for Equine Welfare. 2008. Horse slaughter: Its ethical impact and subsequent response of the veterinary profession.

Chris Heyde and Liz Ross

HUMAN EFFECTS ON ANIMAL BEHAVIOR

Humans are a unique species, and a very curious and inquisitive group of mammals. We're here, there, and everywhere, and our intrusions, intentional or not, have significant impacts on animals, plants, water, the atmosphere, and inanimate landscapes. Thus, we need to consider how we influence the lives of animals, how we must protect them, and what important questions to ask. We are the most dominant species the Earth has even known. When humans influence the behavior of animals, the effects are referred to as anthropogenic. There are many ethical issues surrounding our effects on the lives of animals outside of laboratories and apart from research projects. Here we consider some of the issues that center on animal protection. Many of the topics discussed are also considered in other essays in this encyclopedia.

The relationship between humans and animate and inanimate nature is a complex, ambiguous, challenging, and frustrating affair. While we do many positive things for animals, we also make the lives of animals more difficult than they would be in our absence, and we make environmental messes that are difficult to fix. On the positive side, in October 2006 the German parliament unanimously voted to ban seal products from the country because of the way in which seals are clubbed to death during mass slaughters. Whiteface Mountain, located in the Adirondacks in upstate New York, changed the configuration and design of ski trails to eliminate the negative impact on an elusive bird called Bicknell's thrush that nests there. Bicknell's thrushes are not an endangered or even a threatened species, but rather a species of special concern.

Scientists are also increasingly concerned about how we affect deep-sea communities that frequently do not receive this sort of attention. Ecotourism also has many sides to it, and is getting more detailed attention, so that we come to better understand the positive and negative aspects of our intrusions into animals' lives and the ecosystems in which they live.

We also influence the behavior of the urban animals with whom we share our homes, and their presence also enriches our lives. We must remember that our land is their land. too.

When wild animals become accustomed to the presence of humans it is called habituation, and numerous animals have changed their daily routines because of our intrusions into their homes. Often predators and their prey become bolder, and this causes problems for everyone, humans and animals alike. Mountain lions, for example, have become very habituated to humans in many communities in the western United States, and this has caused people to launch campaigns to rid themselves of these magnificent animals. Yet attacks, while slightly on the rise, are still very rare. I once almost stepped on a male mountain lion while backing my car up and telling my neighbor that there was a lion in the area. On another occasion, thinking that a tan animal running towards my car was my neighbor's dog, I opened the car door only to see that it was a lion, not a dog, coming my way. Once, sitting in my living room reading, I saw a big black animal move slowly across my deck, seemingly without a care in the world. Then I heard some noise at my sliding glass door. I got up and went to the door, only to see a male black bear trying to open it. When he saw me, he stepped back, looked at me, and walked off my deck, went to my neighbor's house, and fell sleep under her hammock.

Because of the widely varying settings in which we interact with animals, we sometimes just do not know what to do when human interests compete with those of other beings, which happens almost every second of every day worldwide. Many people claim to love nature and to love other animals, and then, with little forethought, concern, or regret, go on to abuse them in egregious ways too numerous to count. Many of the animals we want to study, protect, and conserve experience deep emotions, and when we step into their worlds we can harm them mentally as well as physically. They are sentient beings with rich emotional lives. Just because psychological harm is not always apparent, this does not mean we do no harm when we interfere in animals' lives. It is important to keep in mind that, when we intrude on animals, we are influencing not only what they do but also how they feel.

Coexistence Is Difficult

Often we become at odds with the very animals with whom we choose to live when they become nuisances, dangerous to us or to our pets, or destroy our gardens and other landscapes. Thus, we have to make difficult decisions about whose interests and lives to favor, theirs or ours. A more aware public no longer believes that human interests always trump the interests of other animals; we have to factor in all of the variables to make the best choices on a case-by-case basis. For example, in some areas of Boulder, Colorado where I live, people choose to coexist with prairie dogs, whereas in other locales some people want to kill these family-oriented rodents because they are a nuisance to those building shopping malls, parking lots, soccer fields, and more homes. Killing prairie dogs, however, does not really solve the problem, and many believe we need to figure out the most humane solutions, so that people can pursue their interests and prairie dogs do not have to suffer because of our inability to limit growth.

Humans are generally motivated to care about other animals, because we assume that individuals are able to experience pain and suffering. Fortunately, very few people want to be responsible for adding pain and suffering to the world, especially intentionally. However, in our interactions with other animals, we often cause unintentional pain, suffering and death, usually for human ends. In addition, because humans interact with animals in an increasing number of settings as we expand our own horizons, it is becoming more common to debate whether or not to cull or kill members of a species because they may be involved in the transmission of disease to other animals or humans. For example, badgers in the United Kingdom play a role in the transmission of bovine tuberculosis that infects cattle. A move to cull badgers to control the spread of this disease was met with substantial public resistance; 96 percent of about 47,000 people polled throughout England said no to the planned cull, many favoring better farming practices. Years ago this sort of response was not very usual; people either ignored the problem or favored the wellbeing of humans or domestic livestock. This example, along with the treatment of prairie dogs, shows that as time passes more and more people are showing concern for how we interact with other animals.

Consider also the reintroduction of grcy wolvcs to Ycllowstone National Park, an area in which humans exterminated wolves about eight decades ago because of their predatory habits. The project is considered by many people to be successful, in that numerous wolves now roam the Yellowstone ecosystem. However, in the process some of the wolves who were moved from Canada and Alaska have died, and the newcomers have killed numerous coyotes in various parts of the park. Did we do harm when we removed wolves from one area to bring them to another locale? Are we robbing Peter to pay Paul? Should we favor ecosystems and species over individuals? These are some of the difficult questions with which conservation biologists are faced. Some people argue that individual wellbeing should come before the fate of a given species or the integrity of an ecosystem, whereas others believe that it is acceptable for a few individuals to die for the good of the species as a whole.

There also are other questions that need to be considered, because not everyone favors bringing wolves back to Yellowstone. Ranchers and farmers believe that wolves are responsible for significant losses of livestock due to predation, although available data do not support this claim.

Consider also the reintroduction of Mexican wolves in New Mexico, and how federal gunners are free to wipe out the Nantac pack, despite the fact that these wolves haven't stabilized or reached suitable numbers to increase the likelihood that they will survival. The federal predator control program has been responsible for reducing the population of wild Mexican wolves from 55 at the end of 2003 to 44 at the end of 2004, and 35 at the end of 2005. During May 2006, federal gunners killed 11 wolves, including six pups from one pack.

To sum up, the big questions with which we must be concerned include whether it is permissible to move individual wolves from areas where they a have thrived, and place them in areas where they might not have the same quality of life, for the perceived good of their species, and whether it is permissible to interfere in large ecosystems that have existed in the absence of the species to be reintroduced, and remove animals from an ecosystem in which they play an integral role.

Many animal behavioral scientists believe that the major guiding principle is that the lives of the animals whom humans are privileged to study should be respected, and when we are unsure about how our activities will influence them, we should err on the side of the animals, and not engage in these practices until we know the consequences of our acts. This precautionary principle will serve the animals and us well. Indeed, this approach could well mean that exotic animals so attractive to zoos and wildlife parks need to be studied for a long time before they are brought into captivity. For those who want to collect data on novel species to be compared to other perhaps more common animals, the reliability of the information may be called into question unless enough data are available that describe the normal behavior and species-typical variation in these activities.

We must continue to develop and improve general guidelines for research on free-living and captive animals. These guidelines must take into account all available information. Professional societies can play a substantial role in the generation and enforcement of guidelines, and many journals now require that contributors provide a statement acknowledging that the research conducted was performed in agreement with approved regulations. Guidelines should be forward-looking as well as regulatory. Much progress has already been made in the development of guidelines, and the challenge is to make them more binding, effective, and specific. Fortunately, many people worldwide are working to improve our relationships with other animals.

That many animals have subjective and inter-subjective communal lives, that is, they live in social groups and other animals are in their thoughts and feelings, and a personal point of view on the world that they share with other individuals, seems beyond question. In his development of an anthro-harmonic perspective on human-nonhuman relationships, Stephen Scharper, who studies the relationship between religion and environmental ethics, notes that "intersubjectivity is a fundamental reality of all human existence." Harmonic means of a integrated nature, which "acknowledges the importance of the human and makes the human fundamental but not exclusively focal." Working towards an anthro-harmonic understanding of human-nonhuman relationships in the future is a good road to travel.

What Should We Do?

Inquiries about how we interact with other animals raise a host of big ethical questions, such as why care about other animals? Who are we or who do we think we are in the grand scheme of things? How should we go about wielding our almost limitless power when we interact with other individuals, populations, species, and ecosystems? Are there any shoulds? Yes, there are; however, just because we can do something does not mean we should. Should be we concerned with the wellbeing of individuals, populations, species, or ecosystems? Can we reconcile a concern for individuals with a concern for higher and more complex levels of organization?

First and foremost in any deliberations about other animals must be deep concern and respect for their lives and the worlds within which they live, respect for who they are in their worlds, and not respect motivated by who we want them to be in our anthropocentric scheme of things. Can we really believe that we are the only species with feelings, beliefs, desires, goals, expectations, the ability to think about things, the ability to feel pain, or the capacity to suffer? Other animals have their own points of view, and it is important to appreciate, honor, and respect them when we interact with them. Ethics and scientific research are not incompatible.

The Best and Worst of Times for Animals

In many ways these are the best of times and the worst of times for many species of animals, the best, in that more and more people around the world are truly concerned about how we effect the lives of the animals with whom we share space, and the worst in that the global population of humans is increasing steadily at unprecedented rates, and there is less and less space for us to live without intruding into the lives of other animals.

Humans are a powerful force in nature, and obviously we can change a wide variety of behavioral patterns in many diverse species. Coexistence with other animals is essential. By stepping lightly into the lives of other animals, humans can enjoy the company of other animals without making them pay for our interest and curiosity. There is much to gain and little to lose if we move forward with grace, humility, respect, compassion, and love. Our curiosity about other animals need not harm them. The power we potentially wield to do anything we want to do to animals and nature as a whole is inextricably tied to responsibilities to be ethical humans beings. We can be no less.

Further Reading

Bekoff, M. 2006. *Animal passions and beastly virtues: Reflections on redecorating nature.* Philadelphia: Temple University Press.

Bekoff, M., ed. 2007. *Encyclopedia of human-animal relationships.* Westport, CT: Greenwood.

Bekoff, M. 2007. *The emotional lives of animals: A leading scientist explores animal joy, sorrow, and empathy and why they matter.* Novato, CA: New World Library.

Bekoff, M., and Jamieson, D. 1996. Ethics and the study of carnivores: Doing science while respecting animals. In J. L. Gittleman (ed.), *Carnivore behavior, ecology, and evolution,* Volume 2, 15–45. Ithaca, NY: Cornell University Press.

Caro, T. M., ed. 1998. *Behavioral ecology and conservation biology.* New York: Oxford University Press.

Cronin, W., ed. 1996. *Uncommon ground: Rethinking the human place in nature.* New York: W. W. Norton and Company.

Festa-Bianchet, M., and Apollonia, M., eds. 2003. *Animal behavior and wildlife conservation.* Washington, DC: Island Press.

Goodall, J., and Bekoff, M. 2002. *The ten trusts: What we must do to care for the animals we love.* San Francisco: HarperCollins.

Public says "no" to badger cull. http://news.bbc.co.uk/2/hi/science/nature/5172360.stm.

Scharper, S. 1997. *Redeeming the time.* New York: Continuum.

Siebert, C. 2006. Are we driving elephants crazy? *New York Times Magazine,* October 8. http://www.nytimes.com/2006/10/08/magazine/08elephant.html?ex=1160884800&en=b2676c7a2fa539e1&ei=5070&emc=eta1

Venting concerns: Exploring and protecting seep-sea communities. *Science News,* October 7, 2006; http://www.sciencenews.org/articles/20061007/bob7.asp.

Whiteface mountain and Bicknell's thrushes: http://select.nytimes.com/gst/abstract.html?res=F40B12FA385A0C778EDDA10894DE404482

Wilmers, C. C., and Post, E. 2006. "Predicting the influence of wolf-provided carrion on scavenger community dynamics under climate change scenarios." *Global Change Biology* 12: 403–409.

Marc Bekoff

HUMANE EDUCATION

Humane education is about kindness and respect. Most clearly identified with George Angell, the founder of the Massachusetts Society for the Prevention of Cruelty to Animals, it is based on the assumption that if children learn to care for and respect animals they will develop an empathetic or feeling personality that will guide them in their relations with people as well.

The general theme of being kind to animals was present in the very earliest publications printed for children. In the late 1700s and early 1800s, a number of stories and books for children talked about the mistreatment of animals. In *The Life, Adventures, and Vicissitudes of a Tabby Cat*, published in 1798, there is a description of a cat having its tail cut off with a pair of scissors by a terrible young man. Other stories told of stealing birds' eggs from nests, and the abuse of horses. The stories often had a strong moral theme that emphasized empathizing with the animals, and the evildoers came to a bad end because of their treatment of animals. This type of story would culminate with the publication of *Black Beauty* by Anna Sewell in 1877.

Early animal protection work did include elements of humane education. In the 1850s, M. DeSally published "Method of Teaching Kindness to Animals" in the *Bulletin Annuel de la Societe Protective des Animaux*. It was difficult for education to receive a high level of attention when an enormous amount of rescue and law enforcement work was required. George Angell, who had a background as a teacher, placed a major emphasis in the early work of the MSPCA on promoting humane education. He understood that to teach children kindness would be the best way to prevent cruelty to animals, and people.

When Angell began to formalize our understanding of humane education in the 1870s, he found fertile ground in the American educational system at the time. *McGuffey's Newly Revised Eclectic Reader,* published in 1843, included many stories about animals and nature. In that same era, the common school philosophy of Horace Mann maintained the important role that public education could play in providing students from many different backgrounds with a common sense of culture and morals. Most valuable at the time was the concept that schools could play a significant role in helping to solve major social problems.

In 1882 Angell began to organize Bands of Mercy in schools across the country. These clubs encouraged children to learn about animals and to do things to help animals. By 1883, when Angell addressed a meeting of the National Education Association, there were already 600 Bands of Mercy with 70,000 members in schools throughout the country. Angell founded the American Humane Education Society (AHES) in 1889, ". . . to carry Humane Education in all possible ways, into American schools and homes." One method was sponsoring the publication of literature with a humane message. It was Angell who brought the classic *Black Beauty* to American children. AHES also promoted Bands of Mercy across the country. By 1923 there were over 140,000 Bands of Mercy with a membership of over four million children! Twenty states, recognizing the importance of humane education for society in general, passed laws requiring its practice in the schools by 1922. Edwin Kirby Whitehead published the first humane education textbook in 1909, *Dumb Animals and How to Treat Them*, and Flora Helm followed with a *Manual of Moral and Humane Education.*

At the same time, the humane movement suffered the pains of evolution in a changing society. Many of the earliest humane societies, including the ASPCA and MSPCA, had been inspired by the need to protect the many horses used for transportation and work in America's cities and towns. As carriage and cart horses disappeared from streets and roads, the humane movement came to grips with new roles and challenges.

Stamp from 1964 encourages the humane treatment of animals. (Dreamstime.com)

In the 1960s, America shook off the effects of the Great Depression and Two World Wars. People once again began to question their relationships with one another and the environment. New educational philosophies emerged. Earth Day and the developing environmental movement gave rise to environmental education, and humane educators were poised to move forward with new opportunities.

New efforts have included curriculum development, teacher training, and teaching materials for classroom use. Most humane societies offer humane education programs, recognizing that the only certain way to prevent cruelty to animals is help children learn the meaning of kindness.

Further Reading

Angell, George T. 1884. *Autobiographical sketches and personal recollections.* Boston: Franklin Press: Rand, Avery & Co.

Bank, Julie, and Zawistowski, Stephen. 1994. The evolution of humane education. ASPCA *Animal Watch,* Fall.

Good, H. G. 1956. *A history of American education.* New York: The Macmillan Co.

Spring, Joel. 1985. *The American school 1642–1985.* New York: Longman.

Steele, Zulma. 1942. *Angel in a top hat.* New York: Harper & Brothers Publishers.

Wells, Ellen B., and Grimshaw, Anne. 1989. *The annotated black beauty* by Anna Sewell. London: J. A. Allen & Company Limited.

Stephen L. Zawistowski

HUMANE EDUCATION, ANIMAL WELFARE, AND CONSERVATION

Conservation education is beginning to be recognized as one of the critical components of preserving life on earth (Orr, 2004). The emerging field of conservation psychology is the study of human behavior and the achievement of positive and enduring humane conservation goals. Recent work by Susan Clayton and Gene Myers (2009), in their textbook *Conservation Psychology: Understanding and Promoting Human Care for Nature,* is evidence of a growing interest and need for new methods of understanding psychology and behavior in terms of conservation.

In China, disturbing levels of animal abuse and neglect (Song, 2004) as well as the staggering loss of native wildlife there (Elvin, 2004) prompted the need for an intervention program to address these issues. Out of a five-year collaboration between an American conservation educator (S. Bexell) and her Chinese colleagues, a program was developed to help young people form emotional bonds with animals. Through learning about the behavior, minds, and emotions of animals, this group hoped children would develop humane attitudes and behavior toward animals, potentially leading to the development of a wildlife conservation ethic and more compassionate personal attitudes toward conservation. Because they wanted the program to have broad applicability, it was developed so conservation and humane education practitioners could apply it in multiple cultures. Animal abuse and neglect, as well as wildlife losses, are human problems and certainly not just Chinese phenomena. They are global tragedies needing urgent attention.

The program consisted of a camp experience developed for children ages 8-12 to encourage the acquisition of correct knowledge about animals, care about animals, a propensity for environmental stewardship, and compassionate behavior toward animals. The program was designed to take children along a "continuum of care." To facilitate this process, students first met small animals (rabbits, guinea pigs, hamsters, parakeets, and tortoises) as *individuals* (and not merely members of a species) and were allowed to recognize them as individuals with personalities and feelings similar to humans. They also met exotic captive animals (including giant pandas, red pandas, zebras, golden monkeys, giraffes, and lemurs) as individuals. The desired outcome was that students would begin to care about these animals as individuals. Stemming from this the conservation educators hoped that students would then begin to care about the environment that their new animal friends depend on. Finally, the educators hoped students would care enough about animals and their living space (for example home environments, captive situations, and natural habitats) to change their own behavior to care for and to protect animals (whether captive or wild) and their environs based on the knowledge and skills learned during camp. It was also hoped that students would develop a new and heightened empathy and compassion for animals to make it more likely that they would take better care of and protect individual animals.

For the program to be successful the conservation educators determined several essential curricular components for the camp experience: (1) extended personal interactions with animals; (2) "multiple points of contact" that provided the opportunity for children to interact and/ or study the same individual animals over time to facilitate the human-animal bond through mutual trust and respect; (3) hands-on animal care by participants; (4) observation and interpretation of animal behavior; (5) encouragement of empathy with animals through teaching about animal minds (emotions and pain) and behavior by respected adults;

(6) conversations with conservation and animal care experts; (7) specific skills and knowledge about appropriate pets and animal care; and (8) provision of knowledge and skills to enable effective communication to others about animals, their welfare, and conservation.

Why a camp? The educators believed the format of the program was critical for their goals to be achieved. They designed it as a camp experience to provide extended contact time between the students and the animals, and extended time with positive role models, as well as peers. The duration of the camp, five days and four nights, also provided more time for students to acquire a depth of knowledge and skills. Lastly, they believed that time in nature (camping in tents and exploring nature were curricular components) was also important for developing a humane conservation ethic (e.g. Louv, 2005).

To test the hypothesis that exposure to animals would enhance development of humane attitudes and a positive conservation ethic (Myers, 2007; Myers and Saunders, 2003), this program was evaluated to determine its effectiveness (Bexell, 2006). Through evaluation, Bexell and colleagues found statistically significant self-reported increases of knowledge, level of care, and propensity for animal and environmental stewardship. They also found the students showed (1) significant increases in actual knowledge and, in agreement with qualitative data collected, an increase in the breadth and depth of accurate knowledge of animals; (2) care for animals; and (3) ways in which they could and wanted to take action for the welfare and conservation of animals. These findings support the efficacy of a camp program where personal experiences with animals spark interest

in learning and promoting human-animal bonds that support caring behavior and the willingness to take conservation action. The findings also support the hypothesis that empathy with animals can be a precursor to wildlife and environmental stewardship (Myers, Saunders and Bexell, 2009).

The foundation of the camp curriculum is based in large part on human universals of compassion, morality, and solid scientific knowledge about animals and natural systems. Many scientists believe that humane and conservation education programs need to be designed to help children overcome socially and culturally imposed distancing from animals, and it's hoped that the research and curriculum foundations of the program designed in China will help to shape the future of conservation and humane education not only in China but also globally. The camp described provides another cultural lens into the field of conservation psychology and shows an element of hope for future generations.

See also China: Animal Rights and Animal Welfare; China: Moon Bears and the Bear Bile Industry

Further Reading

Bexell, S. M. 2006. Effect of a wildlife conservation camp experience in China on student knowledge of animals, care, propensity for environmental stewardship, and compassionate behavior toward animals. Unpublished Doctoral Dissertation, Georgia State University.

Clayton, S., and Myers, Jr. O. E. 2009. *Conservation psychology: Understanding and promoting human care for nature.* West Sussex, UK: Wiley-Blackwell.

Elvin, M. 2004. *The retreat of the elephants: An environmental history of China.* New Haven: Yale University Press.

Louv, R. 2005. *Last child in the woods: Saving our children from nature-deficit disorder.* Chapel Hill, NC: Algonquin Books.

Myers, Jr. O. E. 2007. *The significance of children and animals: Social development and our connections to other species*, 2nd ed. West Lafayette, IN: Purdue University Press.

Myers, Jr. O. E. & Saunders, C. D. 2003. Animals as links to developing caring relationships with the natural world. In P. H. Kahn Jr. & S. R. Kellert (Eds.), *Children and nature: Psychological, sociocultural and evolutionary investigations.* (pp. 153–178). Cambridge, MA: MIT Press.

Myers, O. E., Saunders, C. D., & Bexell, S. M. 2009. Fostering empathy with wildlife: Factors affecting free-choice learning for conservation concern and behavior. In J. H. Falk, J. E. Heimlich and S. Foutz (Eds.) *Free-choice learning and the environment.* Lanham, MD: AltaMira Press. Pp. 39–55.

Orr, D. W. 2004. *Earth in mind: On education, environment, and the human prospect.* Washington, D.C.: Island Press.

Song, W. 2004. Traditional Chinese culture and animals. *Animal legal and historical center* (online). www.animallaw.info/nonus/articles/arcnweiculturalatt2005.htm.

Sarah M. Bexell, Olga S. Jarrett, Xu Ping, and Feng Rui Xi

HUMANE EDUCATION MOVEMENT

Humane education explores all the challenges facing our planet, from human oppression and animal exploitation, to materialism and ecological degradation. It explores how we might live with compassion and respect for everyone, not just our friends and neighbors, but all people; not just our dogs and cats, but all animals; not just our own homes, but the Earth itself, our ultimate home. Humane education inspires people to act with kindness and integrity, and provides an antidote to the despair many feel in the face of entrenched and pervasive global problems, and persistent cruelty and abuse towards both people and animals. Humane educators cultivate an appreciation for the ways in which even the smallest decisions we make in our daily lives can have far-reaching consequences. By giving students the insight they need to make truly informed, compassionate, and responsible choices, humane education paves the way for them to live according to abiding values that can lend meaning to their own lives while improving the world at the same time. Additionally, and perhaps most important, humane education encourages students to become engaged citizens and problem-solvers for a better world.

The term humane education originated in the late 19th century, as founders of SPCAs and child protection organizations (often the same people) realized the importance of teaching children the principles of kindness and respect for others, both human and nonhuman. For many decades in the late 20th century, humane education became synonymous with elementary-level school programs that primarily taught children about kindness toward and care of companion animals. As the crisis of dog and cat overpopulation grew, humane education began to focus on the importance of spaying and neutering. With the emergence of dog fighting as a popular sport among some communities, humane education programs often discussed the cruelty inherent in dog fighting as well as offering bite prevention presentations.

In the 1990s, several humane education programs emerged that expanded the then-limited perception of humane education, returning to its roots. These programs focused on the definition of the word humane (meaning having what are considered the best qualities of human beings), and applied this definition to our relationships with everyone: animals, people, and the earth.

Humane education now encompasses animal protection education, environmental and sustainability education, media literacy, character education, and social justice education. It is the only educational movement that currently does so. Drawing connections between all forms of oppression and exploitation, humane education empowers and inspires students to be changemakers who not only have the skills to connect the dots between various problems and forms of abuse, but also to find solutions that work for everyone.

Quality humane education accomplishes its goals through the use of four elements. They are:

1. Providing accurate information so that students understand the consequences of their decisions as consumers and citizens

2. Fostering the 3 Cs: curiosity, creativity, and critical thinking, so that students can evaluate information and solve problems on their own

3. Instilling the 3 Rs: reverence, respect, and responsibility, so that students will act with kindness and integrity.

4. Offering positive choices that benefit oneself, other people, the earth, and animals, and the tools for problem-solving, so that students are able to help bring about a better world

Humane education achieves these goals through interactive and engaging teaching techniques that model compassion, respect, and openness.

Providing Accurate Information

In order to make the kindest and wisest choices, we need knowledge. For example, unless we know about the problem of dog and cat overpopulation, the abuse of farmed animals in factory farms, the plight of women and children working in sweatshops, the dangers of certain products and chemicals to the environment, or the escalating travesty of worldwide slavery, to name a few, we cannot make informed, conscious, and humane choices that help solve these growing problems. With knowledge, however, individuals, businesses, and governments are able to make choices that do not cause suffering and destruction, but instead create a more peaceful, humane world. Humane educators help their students by offering them accurate information so that they can make wise and compassionate decisions both personally and as emerging members of a democracy.

Fostering the 3 Cs: Curiosity, Creativity, and Critical Thinking

Humane educators do more than expose students to hidden truths. They teach the critical thinking skills necessary to evaluate information, as well as foster curiosity and creativity so that students pursue lifelong learning and imaginative, yet practical, solutions to difficult problems. When one visits a school where humane education is in progress, one may find students analyzing popular advertising or reading pamphlets from opposing groups, trying to separate fact from opinion. Students may be working together to develop creative answers to challenges often portrayed in either/or terms, crafting persuasive essays on various issues, tracing the effects on animals, people, and the environment of certain products and behaviors, or coming up with ideas for everything from proposed legislation to meaningful disclosure on product labels. Humane educators inspire their students to think about, consider, and creatively and positively respond to norms and attitudes that are often accepted without question, from what is served in the cafeteria, to how and where the school's sports uniforms are produced, to the use and disposal of paper in the school, to dissection in biology classes, and much more.

Instilling the 3 Rs: Reverence, Respect, and Responsibility

Without the 3 Rs of reverence, respect, and responsibility, the acquisition of knowledge and improved critical and creative thinking by themselves will generally fail to inspire a person to take the necessary steps toward solving problems and making kinder and more positive choices in their lives and communities.

Reverence is an emotion akin to awe. What people revere, they tend to honor and protect. If young people have reverence for life, for other humans, for animals, and for the beautiful planet Earth, they are more likely to find the will to make choices that diminish harm to others and create more peace. Respect is an attitude people bring to the world; it is reverence manifested in interactions. Responsibility is respect turned into action. When young people are filled with reverence, and when they feel respect for others, taking responsibility for their actions and choices is an inevitable next step.

How do humane educators cultivate the 3 Rs? Through age-appropriate activities, reflections, field trips, opportunities to meet people and animals who've been exploited or abused, stories, pictures, and films, humane educators awaken the hearts and souls of their students and ignite their love for this earth, its people, and its animals. They spark students' innate empathy, so that respect follows easily and the motivation to take responsibility, in age-appropriate ways, is the likely result.

Providing Positive Choices

Humane educators do not tell students what to think or what to do, which would be the opposite of teaching critical and creative thinking, but they do make sure that students know that they have choices that can improve or diminish the world, end suffering or contribute to it, solve problems or perpetuate them.

This fourth element of humane education is the one that makes the rest meaningful. If students are exposed to the problems in the world and the suffering and destruction that abound, but are given no tools or choices to make a difference, they may become cynical and apathetic, exactly the opposite outcome from what humane education tries to achieve. When, instead, humane educators introduce students to innovative ideas and inspiring successes, and provide examples of ways in which individuals, communities, corporations, and governments can make a lasting positive contribution, they pave the way for young people to become visionary entrepreneurs, leaders, change agents, and engaged citizens in both small and large ways.

When these four elements come into play, young people not only become aware of the challenges facing animals, people, and our planet, but also learn to trust that they can make a difference, and they become more enthusiastic and committed citizens. Their education becomes deeply meaningful, and their lives may take on a purpose greater than simply good grades or a future lucrative career. For those students who see the future as bleak, humane educators offer hope, meaning, and solidarity, empowering such students to create a better future for themselves as well as for others, drawing links between the oppression of other species and oppressive systems in our society that affect those who are disenfranchised.

Humane education has the capacity to change the world by educating a new generation to be caring, compassionate, and responsible. As humane education is integrated into curricula, and as humane educators are hired by schools in the same numbers as math or language arts teachers, students will gain the knowledge, opportunity, and will to live with more respect for others, be they other humans, other animals, or the ecosystems that support us all.

Further Reading

Bekoff, Marc. 2000. *Strolling with our kin.* Jenkintown, PA: AAVS.

Bigelow, Bill, and Peterson, Bob. 2002. *Rethinking globalization: Teaching for justice in an unjust world.* Milwaukee, WI: Rethinking Schools Press.

Cornell, Joseph. 1979. *Sharing nature with children.* Nevada City, CA: Dawn Publications.

Lickona, Thomas. 1991. *Educating for character: How our schools can teach respect and responsibility.* New York: Bantam.

Orr, David. 1994. *Earth in mind.* Washington, DC: Island Press.

Seed, John, Macy, Joanna, Fleming, Pat, and Naess, Arne. 1988. *Thinking like a mountain: Toward a council of all beings.* Gabriola Island, BC, Canada: New Society Publishers.*

Selby, David. 1995. *EARTHKIND: A teachers handbook on humane education.* Staffordshire, UK: Trentham Books.

Stoddard, Lynn. 2003. *Educating for human greatness.* Brandon, VT: Holistic Education Press.

Van Matre, Steve. 1990. *Earth education: A new beginning.* Greenville, WV: Institute for Earth Education.

Weil, Zoe. 1990. *Animals in society: Facts and perspectives on our treatment of animals.* Jenkintown, PA:AAVS.

Weil, Zoe. 1994. *So, you love animals: An action-packed, fun-filled book to help kids help animals.* Jenkintown, PA: Animalearn.

Weil, Zoe. 2003. *Above all, be kind: Raising a humane child in challenging times.* Gabriola Island, BC, Canada. New Society Publishers.

Weil, Zoe. 2004. *The power and promise of humane education.* Gabriola Island, BC, Canada. New Society Publishers.

Weil, Zoe. 2009. *Most good, least harm: A simple principle for a better world and a meaningful life.* Hillsboro, OR: Beyond Words/ Atria.

Weil, Zoe, Claude, and Medea. 2007. *The hellburn dogs.* Herndon, VA: Lantern Books.

Zoe Weil

HUMANE EDUCATION MOVEMENT IN SCHOOLS

Humane education is a pedagogical concept that centers on inculcating the ethic of kindness to animals through formal or informal instruction of children, although it is sometimes used to describe efforts to reach people of all ages. Its modern origins trace back to John Locke's environmentalist theory of mind, as outlined in his *Essay Concerning Human Understanding* (1690) and *Some Thoughts Concerning Education* (1693). The concept that virtuous character could be formed through the ideas, impressions, and experiences of youth soon prompted the emergence of an entire publishing industry for children's literature. The kindness-to-animals ethic was one of the most common themes in such works, and had special resonance in the 19th century as a means for inculcating standards of bourgeois gentility such as empathy and moral sensitivity. By the time animal protection societies formed in England, Europe, and North America, humane education was already an established instrument of youth socialization.

During the post-Civil War period, the formation of character came to be seen by American moral reformers, including humane advocates, as a driving dynamic for social change. The promotion of humane education as a solution to numerous social ills drew animal protection into closer alignment with other reform movements of the era, especially child protection and temperance. These movements in particular all shared a deep concern about the implications of cruelty and violence for individuals, the family, and the social order.

The early decades of the 20th century saw the passage of compulsory humane education requirements in a number of states, the production of humane anthologies and textbooks, and the emergence of the professional humane educator, usually an employee of a local society for the prevention of cruelty to animals. However, as local societies became bogged down with the overwhelming challenges of municipal animal control during the middle decades of the 20th century, humane education became less of a priority. Given its limited resources and declining influence, the movement's efforts to institutionalize humane education within teacher-training institutions and school systems largely failed.

Nevertheless, the kindness-to-animals ethic continued to resonate as a theme in children's literature and other cultural forms, and how ever restricted its influence, humane education helped to reinforce the notion that wanton acts of individual cruelty against animals were the sign of a maladapted and sick personality, while a kind disposition toward animals became more recognized as an attribute of the well-adjusted individual.

In recent decades, the locus of human education continues to be the animal care and control community. Many organizations and agencies offer education programs at the municipal or county level, sometimes involving partnerships with schools or other youth-oriented institutions. For many reasons, however, elementary and secondary schools and colleges have yet to integrate humane education into their curricula. Companion animal issues predominate over other concerns in the content of humane education programs for reasons having to do with agency mission, institutional sensitivities, the perception of humane education as a special interest, and the view that certain issues are not age-appropriate for young people.

Although many animal advocates are quick to cast humane education as crucial to the advancement of their movement's objectives, it remains an underemphasized and underfunded component within animal protection. It is not a major programmatic focus of any of the larger national organizations, and at the local level must compete for priority with other needs, including the most basic ones associated with operating a shelter, finding homes for animals, and keeping humane agents in the field.

Contemporary humane education suffers from a further disadvantage in the lack of definitive empirical proof to demonstrate its effectiveness. There

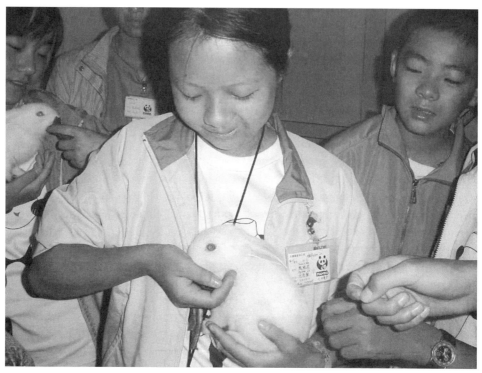

A student in the humane education summer camp feeding appropriate food, in this case lettuce, to San Maio ("Three Whiskers"), a domesticated rabbit. (Sarah M. Bexell)

A young girl in the humane education summer camp offering security, respect, and love for Tiao Tiao ("Jumpy"), a domesticated guinea pig. (Sarah M. Bexell)

is relatively little evidence to show that humane education programs actually increase children's knowledge about or improve their attitudes and behavior toward animals, and none to show that such gains carry into adulthood. Intuition, anecdotal evidence, and a few formal studies suggest its promise, but there is an urgent, ongoing need for formal evaluation and assessment of humane education with respect to both content and methodology.

As at other times in the past, the current emphasis on character education, in the form of core or consensus values that transcend political, cultural, and religious differences, promises to increase opportunities for the expansion of humane education teaching. The advent of service learning mandates and the growth of social networking sites may also provide new opportunities for youth engagement. Still, in the absence of a stronger programmatic and financial commitment from the animal protection movement, and better efforts to establish humane education within institutions of higher learning, such progress cannot be assumed.

Bernard Unti

HUMANE EDUCATION: THE HUMANE UNIVERSITY

Woodrow Wilson, the only U.S. President who had a Ph.D. and taught at the college level, noted that "it is easier to move a cemetery than it is to change a curriculum at a university." Of course, the curriculum does change, slowly and cautiously. The changes not only reflect new knowledge, but new definitions of what is important to know. One area remarkably ignored is our relationship to animals. Animals have shared the homes of people in all cultures ever since those people lived in villages more than 15,000 years ago, and today more than 60 percent of American households have a nonhuman animal sharing their dwelling.

An increasing proportion of people believe that companion, laboratory, and farm animals should receive the best possible health care, including the latest advances in science and technology. One approach is to develop a focused course of study for students involved in a variety of fields of inquiry, addressing not only animal welfare, but also issues related to the conservation of endangered animals and their environments. Such a curriculum has been developed at Purdue University. Like any curriculum, it reflects the strengths of the faculty and the concerns of the present student body.

In 1982, Purdue University developed the Center for Applied Ethology and Human-Animal Interaction at its School of Veterinary Medicine, to promote interdisciplinary activities in the university and serve as a focal point for the exchange of ideas, and the development of new information related to human-animal interactions, and disseminate information in an unbiased manner to students, scientists, consumers, and agricultural groups. In 1997, the center's name was changed to the Center for the Human-Animal Bond, to better reflect our relationship with companion animals, and perhaps all animals.

The primary objectives of the program are to educate undergraduate students about the social, ethical, biological, behavioral, and economic aspects of animal care and use, provide students with a scientific and philosophic care and use, and train students to resolve conflicts concerning the humane use of animals, and to become leaders in policy development and implementation.

Today, more than half of all the veterinary schools in North American have centers dedicated to research and education about the human-animal bond. Perhaps it is time for this area of study to be part of higher education in general.

There is ever-growing concern and interest for our environment, the well-being of animals, and the quality of our interactions with animals; this course of study provides the knowledge and skills to communicate and act on these issues. It also stimulates research to improve human and animal well-being:

www.vet.purdue.edu/chab/; www.the press.purdue.edu/Newdirectionsinthe human.html

Further Reading

Beck, A. M, & Katcher, A. H. 1996. *Between pets and people: The importance of animal companionship* Rev. ed. West Lafayette, IN: Purdue University Press.

Beck, A. M, & Katcher, A. H. 2003. Future directions in human-animal bond research. *American Behavioral Scientist,* 47(1): 79–93.

Beck, A.M., & Martin, F. 2008. Current human-animal bond course offerings in veterinary schools. *Journal of Veterinary Education* 35(4): 483–486.

Glickman, N. W., Glickman, L. T., Torrence, M. E., & Beck, A.M. 1991. Animal welfare and societal concerns: an interdisciplinary curriculum. *Journal Veterinary Medical Education* 18(2):60–63.

Pritchard, W. R. (ed.). 1988. *Future directions for veterinary medicine.* Durham, NC: Pew National Veterinary Education Program, Institute of Policy Sciences and Public Affairs.

Alan M. Beck

HUMANE SOCIETY OF THE UNITED STATES

See Humane Education Movement

HUNTING, HISTORY OF IDEAS

Although prehistoric people needed to hunt to survive, hunting has had little economic significance throughout most of the history of Western civilization. Its importance in Western thought derives chiefly from its symbolic meaning. That meaning has much to do with how we define hunting and distinguish it from butchery. Hunting is not simply a matter of killing animals. To count as quarry, the hunter's victim must be a wild animal. For the hunter, this means that it must be hostile: unfriendly to human beings, intolerant of their presence, and not submissive to their authority. The hunt is thus by definition an armed confrontation between the human domain and the wilderness, between culture and nature. The meanings that hunting has taken on in the history of Western thought reflect the varying values ascribed to culture and nature in this artificial confrontation.

Throughout Western history, the hunter has been seen as an ambiguous figure, sometimes a fighter against the wilderness and sometimes a half-animal participant in it. The meaning of hunting accordingly varies with the meanings ascribed to the wilderness. For the Greeks and Romans, forests were generally threatening and frightening places. In early Christian thought, the wilderness was a sort of natural symbol of hell, and the wild animals living there in rebellion against man's dominion were seen as typifying demons and sinners in rebellion against God. But this image was undermined by the counterimage of the hermit saint in the wilderness, attended by friendly wild animals that the saint's holiness had restored to the docility of Eden.

Other medieval changes in the symbolic meaning of wild places and creatures reflect changes in the social status of hunting. From the 10th century on, Europe's forests dwindled as improved techniques of agriculture fostered a surge in human population growth. Hunting gradually became the exclusive privilege of the aristocracy, who put the remaining forest patches off limits as hunting preserves and ruthlessly punished any peasants caught taking game. Deer, the symbolic inhabitants of the wilderness, became the main objects of the aristocratic hunt, and took on an air of nobility in both folk ballads and high culture.

It was not until the early 1500s that the chase began to be viewed as cruel and to be invoked as a symbol of injustice and tyranny. Erasmus condemned the hunt in 1511 as a bestial amusement. In 1516, Thomas More denounced it in *Utopia* as "the lowest and vilest form of butchery . . . [which] seeks nothing but pleasure from a poor little beast's slaughter and dismemberment." Similar revulsion toward hunting is evident in the essays of Montaigne and in the plays of Shakespeare. Anti-hunting sentiment also crops up in 16th-century hunting manuals, which from 1561 on contain rhymed complaints by the game animals denouncing the senseless cruelty of Man the Hunter.

The rise of anti-hunting sentiments in the 1500s reflected rising doubts about the importance of the boundary between people and animals. In 1580, Montaigne denied the existence of that boundary and concluded that "it is [only] by foolish pride and stubbornness that we set ourselves before the other animals and

sequester ourselves from their condition and society." The erosion of the animal-human boundary in Western thought was accelerated by the scientific revolution of the 1600s and the associated mechanization of the Western world. Animal suffering came to be more widely regarded as a serious evil, and hunting was increasingly attacked as immoral.

The romantic movement of the late 1700s brought about a radical transformation in Western images of wilderness. In romantic thought, nature ceased to be a system of laws and norms and became a place, a holy solitude in which one could escape man's polluting presence and commune with the Infinite. Romantic art and literature picture the hunter sometimes as a poet with a gun participating in the harmony of nature, for example, James Fenimore Cooper's Natty Bumppo, but more often as a despoiler of nature and animal innocence, for example, Samuel Taylor Coleridge's Ancient Mariner.

Western hunting has always been a characteristically male activity, often regarded as valuable training for the military elite and praised as a prototype of the just war. In the context of 19th-century European imperialism, this tradition gave birth to a third stereotype of the huntsman, the colonial White Hunter who dons a pith helmet and leads an army of servile natives on safari to assert his dominion over the conquered territory's land, animals, and people. At the height of Europe's empires in the late 1800s and early 1900s, a love of hunting commonly went hand in hand with imperialist politics, and anti-imperialism was often associated with anti-hunting sentiment. This link between hunting and the political right has persisted into our own time.

During the 20th century, the romantic idea of the sanctity of nature and the

Nietzschean and Freudian picture of man as a sick animal have interacted to yield a vision of the wilderness as a place of timeless order and sanity, in opposition to the polluted and unstable domain of civilization and technology. However, hunters tend to regard the hunt as a healing participation in the natural order, what the hunting philosopher José Ortega y Gassett described as "a vacation from the human condition," whereas opponents of hunting see it as an armed assault on the harmony of nature.

Both attitudes are grounded in the romantic image of nature as a place with no people in it. If we reject that concept of nature and adopt instead a more scientific and pre-romantic conception of human beings and their works as part of nature, the distinction between wild and domestic animals evaporates. Hunting thereby loses its rationale and appears to us, as it did to More, as nothing but a species of butchery practiced for amusement. However, doing away with the opposition between the human and natural domains poses problems as well for the philosophy of animal rights.

The rights view generally assumes that the moral order and nature are separate realms and that what wild animals do to each other is a matter of moral indifference. But if the boundaries between people and animals and between culture and nature are imaginary, it is not clear why we should have a duty to prevent a wolf from eating a baby but not from eating a rabbit.

See also Wildlife Abuse

Further Reading

Anderson, J. K. 1985. *Hunting in the ancient world*. Berkeley: University of California Press.

Cartmill, M., 1993. *A view to a death in the morning: Hunting and nature through history*. Cambridge, MA: Harvard University Press.

MacKenzie, J. M. 1988. *The empire of nature: Hunting, conservation, and British imperialism*. Manchester: Manchester University Press.

Ortega y Gassett, J. 1972. *Meditations on hunting*. New York: Scribner's.

Thiebaux, M. 1974. *The stag of love: The chase in medieval literature*. Ithaca, NY: Cornell University Press.

Matt Cartmill